Contents

Confronting AIDS

A World Bank Policy Research Report

Confronting AIDS

Public Priorities in a Global Epidemic

Published for the World Bank
OXFORD UNIVERSITY PRESS

Oxford University Press

OXFORD   NEW YORK   TORONTO
DELHI   BOMBAY   CALCUTTA   MADRAS   KARACHI
KUALA LUMPUR   SINGAPORE   HONG KONG   TOKYO
NAIROBI   DAR ES SALAAM   CAPE TOWN
MELBOURNE   AUCKLAND

and associated companies in

BERLIN   IBADAN

© 1997 The International Bank for Reconstruction
and Development / THE WORLD BANK
1818 H Street, N.W.
Washington, D.C. 20433

Published by Oxford University Press, Inc.
200 Madison Avenue, New York, N.Y. 10016

Manufactured in the United States of America
First printing October 1997

Cover credits: Mother and child, Curt Carnemark/World Bank; condom box, Population Services
International (PSI); Anti-AIDS Club signboard, Warren Parker/PSI; clinic worker and patient,
UNAIDS/Yoshi Shimizu.

The boundaries, colors, denominations, and other information shown on the maps in this vol-
ume do not imply on the part of the World Bank Group any judgment on the legal status of
any territory or the endorsement or acceptance of such boundaries.

Library of Congress Cataloging-in-Publication Data

Confronting AIDS : public priorities in a global epidemic.
*   p.   cm. — (A World Bank policy research report)*
*   Includes bibliographical references.*
*   ISBN 0-19-521117-0*
*   1. AIDS (Disease)—Prevention—Government policy.   I. World Bank.*
II. Series.
RA644.A25C6339   1997
362.1'969792—DC21   97-17434
*   CIP*

⊗ *Text printed on paper that conforms to the American National Standard*
for Permanence of Paper for Printed Library Materials, Z39.48-1984

Text Figures

Box Figures

Text Tables

Foreword

AIDS HAS ALREADY TAKEN A TERRIBLE HUMAN TOLL, NOT only among those who have died but among their families and communities. Short of an affordable cure, this toll is certain to rise. Ninety percent of HIV infections are in developing countries, where resources to confront the epidemic are most scarce. But the course of the epidemic is not carved in stone.

This book argues that the global epidemic of HIV/AIDS can be overcome. National governments have unique responsibilities in preventing the further spread of HIV and in mitigating the impact of AIDS. But governments alone cannot overcome the epidemic, nor have they always risen to the task. Nongovernmental organizations and other groups in civil society, including people living with HIV, have played and must continue to play a critical role in shaping government action and in bringing prevention and care to people that governments cannot easily reach. The international community can also do much to support developing countries and regions in financing programs to ensure prevention and improved equity in access to care. It can also support the production and dissemination of information worldwide, and invest in research on prevention approaches, vaccines, and low-cost, effective prophylaxis and treatment that can be used in developing countries.

This report is itself an example of the potential benefits of international cooperation in response to the epidemic. The preparation of this volume by World Bank researchers has benefited greatly from the technical inputs, advice, and financial support provided by the Joint United Nations Programme on HIV/AIDS (UNAIDS) and the European Commission. This research report makes a valuable contribution to the international debate on the role of government in addressing the AIDS epidemic in developing countries. The report's recommendations are those of the authors and do not necessarily reflect the positions of our respective institutions.

The world can overcome HIV. Given the necessary information, means, and a supportive community, individuals can and do alter their behavior to reduce the risk that they will contract and spread HIV. But there are certain actions that only governments can take, and political commitment has often been lacking. The costs of inaction are potentially enormous. Policymakers who demonstrate commitment, by working in creative ways with people most severely affected by HIV/AIDS, have a unique opportunity to contain a global epidemic and save millions of lives.

James D. Wolfensohn
President
The World Bank

João de Deus Pinheiro
Member of the
European Commission

Peter Piot
Executive Director
Joint United Nations
Programme on AIDS

Introduction

DEVELOPING COUNTRIES SIMPLY CANNOT IGNORE THE HIV/AIDS epidemic. According to UNAIDS, about 1.5 million people died from AIDS in 1996. Each day about 8,500 people, including 1,000 children, become newly infected. About 90 percent of these infections occur in developing countries, where the disease is likely to exacerbate poverty and inequality. But HIV/AIDS is not the only problem demanding government attention. In the poorest countries especially, confronting AIDS can consume scarce resources that could be used for other pressing needs. How can developing country governments and the international community identify the public priorities in confronting this global epidemic?

This book provides information and analysis to help policymakers, development specialists, public health experts, and others who shape the public response to HIV/AIDS to design an effective strategy for confronting the epidemic. It draws upon three bodies of knowledge: the epidemiology of HIV; public health insights into disease control; and especially public economics, which focuses on assessing tradeoffs in the allocation of scarce public resources.

The report offers persuasive evidence that, for the 2.3 billion people living in parts of the world where the epidemic is still nascent, an early, active government response encouraging safer behavior among those most likely to contract and spread the virus has the potential to avert untold suffering and save millions of lives. Even where the virus has spread widely in the general population, prevention among those most likely to contract and spread it is still likely to be the most cost-effective way to reduce infection rates.

Of course, national governments are not alone in their fight against the epidemic. Bilateral and multilateral donors have provided leadership and major funding for national AIDS prevention programs, especially in the poorer developing countries, as well as for basic research for a vaccine and a cure. Local and international nongovernmental organizations have often assisted and sometimes led the fight against the epidemic. Governments can greatly improve the effectiveness of their response by working collaboratively with these actors.

But only the government has the means and mandate to provide what economists call *public goods*. In the case of HIV/AIDS, these include information about the distribution of infection and behaviors that spread it and knowledge of the costs and effectiveness of prevention and mitigation programs. Similarly, governments have a unique responsibility to reduce the negative *externalities* from risky behavior, by encouraging safer behavior among those most likely to pass the virus to others.

Although sound, these policies can be politically difficult. Indeed, because the spread of HIV involves private behaviors that many people deplore—-multiple sexual partners and injecting drug use—-governments that attempt to reduce the spread of HIV by these activities may appear to their constituents to condone immoral acts. Governments must make clear that the best way to protect everyone from HIV is to help people who engage in the riskiest behavior to avoid infection.

Because resources are scarce, one must think through how best to allocate them. The consequences of these decisions for particular individuals can be enormous. And there are painful dilemmas. In countries where HIV has spread widely, the epidemic will greatly increase the demand for health care and the need for poverty assistance. Governments of poor countries face the challenge of responding to the new needs of the AIDS-affected poor while not neglecting the needs of the poor who suffer from other illnesses and other causes of poverty. Drawing on the experience of countries that have faced these dilemmas, the report suggests responses that are both humane and affordable.

Confronting AIDS: Public Priorities in a Global Epidemic is the sixth in a series of Policy Research Reports designed to bring findings of World Bank research on a key development issue to a wide audience. It is a product of the staff of the World Bank; the judgments made in the report do not necessarily reflect the views of the Board of Directors or the governments they represent.

Joseph E. Stiglitz
Senior Vice President
  and Chief Economist
Development Economics
The World Bank

The Report Team

THE PRINCIPAL AUTHORS OF THE REPORT WERE MARTHA Ainsworth and Mead Over. Nina Brooks and Samantha Forusz wrote many of the boxes, compiled the statistical appendix, and provided research assistance. Kathleen Mantila provided additional research assistance. Deon Filmer produced the material for box 3.3 and other results based on Demographic and Health Survey data. Tim Brown and Werasit Sittitrai contributed box 3.11. Eduard Bos generated the projections for box 6.1. Julia Dayton and Michael Merson contributed appendix A. Lawrence MacDonald edited the report. The report was produced under the direction of Lyn Squire and Joseph Stiglitz.

The editorial-production team was led by Deirdre T. Ruffino, with additional help from Paola Brezny, Joyce Gates, Audrey Heiligman, Brenda Mejia, and Anthony Pordes. Jeffrey Lecksell produced the maps in chapters 1 and 2. Grace O. Evans provided support in production of the manuscript, with assistance from Thomas Hastings and Jim Shafer. Secretarial support was also provided by Joanne Fleming.

Acknowledgments

THIS REPORT BENEFITED FROM CLOSE COLLABORATION, technical review, and financial support from the European Commission HIV/AIDS Programme (EC) and by the Joint United Nations Programme on HIV/AIDS (UNAIDS). We would like to express particular appreciation and gratitude to Drs. Lieve Fransen of the EC and Stefano Bertozzi of UNAIDS for their exceptional personal contributions in sponsoring workshops and review meetings, commissioning background studies, and reviewing the draft report.

We are indebted to the European Commission HIV/AIDS Programme for sponsoring most of the background papers and an authors' workshop in Limelette, Belgium, in June 1996. Comments from the workshop participants were very helpful: Tony Barnett, David Bloom, Marijke Bontinck, Jean-Claude Deheneffe, Dominique Dellicour, Deon Filmer, Michel Garenne, Paul Gertler, Dik Habbema, King Holmes, Roberto Iunes, José Antonio Izazola, Wattana Janjareon, Emmanuel Jimenez, Tony Klouda, Tiékoura Koné, Sukontha Kongsin, Michael Kremer, Ajay Mahal, Allechi M'bet, Rekha Menon, Anne Mills, Martina Morris, Phare Mujinja, Amadou Noumbissi, I. O. Orubuloye, Nicholas Prescott, Pamela Rao, Innocent Semali, Zmarak Shalizi, Donald Shepard, Lyn Squire, John Stover, Paula Tibandebage, Inge Van Den Bussche, Peter Way, Marc Wheeler, Alan Whiteside, and Debrework Zewdie. A complete list of the background papers is provided at the end of this report.

The important technical contribution of the following UNAIDS experts is gratefully acknowledged: Bai Bagasao, Michel Caraël, Renu Chahil-Graf, Suzanne Cherney, Mark Connolly, Sally Cowal, Isabelle de Vincenzi, José Esparza, Purnima Mane, Peggy McEvoy, Rob Moodie, Joseph Perriëns, Peter Piot, Joseph Saba, Bernhard Schwartländer, Werasit Sittitrai, and Country Programme Advisers who responded to the survey highlighted in chapter 3. We are also grateful to UNAIDS for sponsoring a review meeting on the first full draft of the report in

Geneva, and to the following developing country policymakers who provided extensive comments at that time: Akan Akanov (Kazakhstan), Papa Fall (Senegal), Mary Muduuli (Uganda); and Jaime Sepúlveda (Mexico).

In addition to the individuals mentioned above, many others inside and outside the World Bank provided valuable contributions or comments: Peter Aggleton, Sevgi Aral, Natalie Béchu, Seth Berkeley, Dorothy Blake, John Bongaarts, Kenneth Bridbord, Denis Broun, Tim Brown, Richard Bumgarner, Tony Burton, Anne Buvé, Julia Dayton, David de Ferranti, Jacqueline Dubow, Richard Feachem, Steven Forsythe, Mark Gersovitz, Ronald Gray, Jacques du Guerny, Salim Habayeb, Jeffrey Hammer, David Heymann, Philip Harvey, Richard Hayes, Estelle James, Dean Jamison, Prabhat Jha, Christine Jones, Arata Kochi, Kees Kostermans, Maureen Lewis, Samuel Lieberman, Bernard Liese, Georges Malempré, Jacques Martin, Raymond Martin, Clyde McCoy, Tom Merrick, Michael Merson, David Metzger, Norman Miller, Susan Mlango, Stephen Moses, Philip Musgrove, Jeffrey O'Malley, Junko Otani, Cheryl Overs, David Paltiel, Lant Pritchett, Hnin Hnin Pyne, Bill Rao, Wendy Roseberry, Lewis Schrager, Thomas Selden, Guy Stallworthy, Karen Stanecki, Daniel Tarantola, Kitty Theurmer, Anne Tinker, Dominique van de Walle, Carina Van Vliet, Maria Wawer, Roger Yeager, and Fernando Zacarías. The opinions and conclusions expressed in this report are nevertheless those of its authors and do not necessarily reflect positions of the World Bank, its member governments, or other collaborating or sponsoring institutions.

The financial assistance of the governments of Australia and Switzerland is gratefully acknowledged.

Definitions

Data Notes

HISTORICAL DATA IN THIS BOOK MAY DIFFER FROM THOSE in other World Bank publications if more reliable data have become available, if a different base year has been used for constant price data, or if countries have been classified differently.

The former Zaire is referred to as the Democratic Republic of the Congo (Congo DR), and Hong Kong (China, as of July 1, 1997), is sometimes referred to as Hong Kong.

- All *dollar ($)* amounts are current U.S. dollars unless otherwise specified.
- *Billion* is a thousand million

Abbreviations and Acronyms

AIDS	Acquired immune deficiency syndrome
AIDSCAP	AIDS Control and Prevention Project
CSM	Condom social marketing (*see* Glossary, below: Social marketing of condoms)
DALY	Disability-adjusted life year
DHS	Demographic and Health Survey
DOTS	Directly observed treatment short course (for tuberculosis)
EC	European Commission
FSU	Former Soviet Union
GAPC	Global AIDS Policy Coalition
GDP	Gross domestic product (*see* Glossary, below)
GNP	Gross national product (*see* Glossary, below)
GPA	Global Programme on AIDS

HIV	Human immunodeficiency virus
IDU	Injecting drug user
IEC	Information, education, and communication
MSM	Men who have sex with men (*see* Glossary, below)
NEP	Needle exchange program
NGO	Nongovernmental organization (*see* Glossary, below)
OECD	Organization for Economic Cooperation and Development
ODA	Overseas Development Administration (U.K.)
OI	Opportunistic infection
PAHO	Pan American Health Organization
PCP	*Pneumocystis carinii pneumonia*
PSI	Population Services International (private firm)
SOMARC	Social Marketing for Change (private firm)
STD	Sexually transmitted disease
TB	Tuberculosis
UNAIDS	Joint United Nations Programme on HIV/AIDS
UNDP	United Nations Development Programme
UNESCO	United Nations Educational, Scientific, and Cultural Organization
UNFPA	United Nations Population Fund
UNICEF	United Nations Children's Fund
USAID	U. S. Agency for International Development
WHO	World Health Organization
WHO/GPA	World Health Organization/Global Programme on AIDS

Glossary

Adverse selection: the selection into an insurance pool of people likely to have higher claims than others.

Assortative sexual mixing: the extent to which people with similar numbers of sexual partners pair with each other.

Asymptomatic: infected by a disease agent but exhibiting no medical symptoms; subclinical.

Commercial sex: the selling of sexual services for compensation; prostitution.

Concurrent partnership: partnerships that overlap in time.

Disassortative sexual mixing: the extent to which people with many sexual partners pair with people with few partners.

Discordant couple: a couple in which one partner is infected with HIV and the other is not.

Endemic: usually prevalent; persistent at relatively constant levels.

Epidemic: a sudden unusual increase in cases that exceeds the number expected on the basis of experience.

Epidemic, concentrated: an HIV epidemic in a country in which 5 percent or more of individuals in groups with high-risk behavior, but less than 5 percent of women attending urban antenatal clinics, are infected.

Epidemic, generalized: an HIV epidemic in a country in which 5 percent or more of women attending urban antenatal clinics are infected; infection rates among individuals in groups with high-risk behavior are also likely to exceed 5 percent in countries with a generalized HIV epidemic.

Epidemic, nascent: an HIV epidemic in a country in which less than 5 percent of individuals in groups with high-risk behavior are infected.

Epidemiology: the study of the distribution and determinants of disease and injury in human populations.

Externality: an unpriced side effect on a third party of a transaction between two parties.

Gross domestic product: a crude measure of national economic well-being: aggregate expenditure by the residents of a country or final goods and services for consumption, investment, and government services.

Gross national product: an alternative to gross domestic product for measuring national economic well-being. Adds to the gross domestic product income obtained by nationals from labor or property outside the country and deducts the income of foreign nationals residing in the country.

High-risk behavior: unprotected sexual intercourse (i.e., without a condom) with many partners, or sharing of unsterilized injecting equipment.

HIV-positive: having antibodies to HIV.

Incidence of HIV: the number of new cases of HIV in a given time period, often expressed as a percentage for a given number of the susceptible population.

Low-risk individuals: individuals practicing behavior that puts themselves and their partners at low risk of HIV infection; depending on the extent to which they mix with high-risk individuals, however, they may nevertheless be at high risk of becoming infected.

Men who have sex with men: homosexual, bisexual, and heterosexual men who have sex with other men.

Merit good: a good (or service) whose consumption by the poor is valued by society as a whole.

Moral hazard: the increase in the average loss incurred by people who are insured compared with those who are not; term used by health insurance companies to refer to the increased demand for health care exhibited by the insured.

Nongovernmental organization: for-profit firms and private nonprofit organizations.

Opportunistic illness: an illness that affects people with weak immune systems.

Pandemic: an epidemic occurring simultaneously in many countries.

Prevalence of HIV: the number of people with HIV at a point in time, often expressed as a percentage of the total population.

Public good: a good or service having the following two attributes: (1) consumption by one person does not diminish the amount available to others, and (2) excluding people from consuming the good is impossible or costly.

Reproductive rate: the average number of susceptible people infected by an infected person over his or her lifetime.

Seroprevalence: the prevalence of an infection as detected in blood serum.

Sex worker: someone who offers sexual services for money.

Social marketing of condoms: programs designed to raise condom use by improving the social acceptability of condoms, making them more widely available through nontraditional outlets and offering them for sale at subsidized prices.

Susceptible: vulnerable to becoming infected.

Symptomatic: exhibiting sufficient symptoms to require medical treatment.

Summary

TWO DECADES AFTER THE APPEARANCE OF THE human immunodeficiency virus (HIV), an estimated 30 million people have contracted the virus, and 6 million have died of acquired immune deficiency syndrome (AIDS). About 90 percent of infections occur in developing countries, where the disease has already reduced life expectancy, in some cases by more than a decade. HIV is already widespread in many countries in Sub-Saharan Africa and may be on the verge of exploding in other regions. Because most people who develop AIDS are adults in the prime of life, the disease exacts a heavy toll on surviving family members, especially children, and may exacerbate poverty and inequality. Clearly, the human toll of the epidemic is great. But low-income countries face a multitude of pressing human needs. How should developing country governments and the international community respond?

In answering this question, *Confronting AIDS: Public Priorities in a Global Epidemic* draws on three bodies of knowledge: the epidemiology of HIV; public health insights into disease control; and especially public economics, which focuses on assessing tradeoffs in the allocation of scarce public resources. In relying primarily on public economics, we do not intend to deny the validity of other points of view. Much has been written on the epidemic from the perspectives of public health, medical science, and human rights. This Policy Research Report differs by addressing the epidemic in a way that fits more closely the perspective of decisionmakers outside the health sector who shape national efforts to combat the disease. To this audience, the report argues that AIDS is a large and growing problem and that governments can and should confront the epidemic by preventing new infections and by mitigating the impact of infections that occur. It finds that some policies will be much more effective than others in reducing the spread of HIV and mitigating its impact, and it provides a framework that helps to distinguish among activities that can be undertaken by households and the private sector, including nongovernmental organizations (NGOs), those that should be initiated by developing country governments, and those that should be most strongly supported by the international development community.

Although there are clear arguments in favor of government intervention to slow the spread of HIV, social norms and politics make AIDS policy uniquely challenging. This is especially true during the early stages of the epidemic, when the advantages of government intervention are greatest but the potential severity of the problem is not yet apparent. The report argues that governments have a mandate to endorse and subsidize risk-reducing preventive interventions, especially among those most likely to contract and spread HIV, while protecting them from stigmatization.

This report is a strategic document. It has been written to inform and motivate political leaders, policymakers, and development specialists to support the public health community, concerned civil society, and people living with HIV in confronting the AIDS epidemic. Some readers will already know a great deal about public policy and HIV/AIDS; others may be considering the disease from a policy perspective for the first time. It is just as relevant for countries in the earliest stage of the epidemic as it is for those that have suffered the ravages of the disease for more than a decade. Although the report offers examples of programs from many countries, some of which have worked remarkably well, it is not intended as a how-to guide for designing and implementing specific programs. There are many other sources of such information, and summarizing them is beyond the scope of the report. Rather, the report offers an analytical framework for deciding which government interventions should have high priority for addressing the HIV/AIDS epidemic in developing countries and, based on that framework, advocates a broad strategy that can be adapted by countries according to their resources and the stage of their epidemic.

Chapter 1 AIDS: A Challenge to Government

THIS CHAPTER PROVIDES BASIC INFORMATION ABOUT THE nature of HIV/AIDS, the extent of the epidemic, and its current and likely future impact on such measures of well-being as life expectancy, health, and economic growth. Because AIDS strikes adults in their economic prime and, despite recent medical advances, is almost always fatal, the disease reduces average life expectancy (sometimes dramatically), increases the demand for medical care, and is likely to exacerbate poverty and inequality. The relationship between economic development policies and HIV is complex: cross-country data and other evidence indicate that the AIDS epidemic is likely to both affect and be affected by economic development.

Nevertheless, policymakers have often been reluctant to intervene. Faced with competing demands for scarce public resources, and aware that HIV/AIDS is spread primarily through private sexual and drug-injecting behavior, governments may conclude that the disease is not a public priority. Drawing on well-accepted principles of the role of government, which have been the subject of the discipline of public economics, the chapter explains why governments must be actively involved in the fight against AIDS.

Starting from the view that government has a mandate to advance economic well-being and to promote a fair distribution of society's output, the chapter applies public economics to argue that government cannot leave the battle against HIV/AIDS to the private sector. First, in countries that wish to subsidize most of the cost of health care, AIDS will generate enormous government health care expenditures; this alone is sufficient justification for early, effective prevention. Second, whenever a transaction between two parties imposes negative effects, or *externalities,* on a third party, as is the case when a sexual encounter between two people increases the risk of HIV infection to their other partners, public economics argues for government intervention. Third, the provision of information about the state of the epidemic or about the effectiveness of alternative remedies meets the economist's definition of a *public good*; that is, something that benefits society but that private entrepreneurs have insufficient incentive to produce on their own. Public economics argues that governments can often enhance the welfare of society by ensuring the adequate provision of such services. Fourth, fairness and compassion for the poor warrant a government role in both preventing and mitigating the epidemic. Finally, governments influence social norms and promulgate legislation that affect the rights of both the HIV-infected and the uninfected. Measures that protect the powerless from prejudice, bigotry, and exploitation will simultaneously help to protect everyone from the AIDS epidemic.

Chapter 2 Strategic Lessons from the Epidemiology of HIV

I N SOME COUNTRIES, HIV HAS INFECTED ONLY A TINY PERCENT-
age of the population and its effects are all but invisible; in others
the virus has spread so widely that few families have been spared
the tragedy of AIDS illness and death. What accounts for these differences? In reviewing how HIV spreads in populations and the behavioral and biological factors behind the epidemic, this chapter identifies
important principles for an effective response, based on the epidemiology of HIV. These provide the foundation for considering government
priorities for preventing the spread of HIV (chapter 3).

In order for HIV to sustain itself in a population, an infected person
must, on average, transmit the virus to at least one other person over his
or her lifetime. Both biological and behavioral factors affect the rate of
spread of HIV through the population. The key biological factors include the long asymptomatic period of HIV, the risk of infection per
contact by different modes of transmission, and cofactors, such as infection with other sexually transmitted diseases (STDs). However, HIV
transmission can be slowed dramatically by changes in behavior: reducing the number of sexual and drug-injecting partners, using condoms
during sexual intercourse, and using sterilized injecting equipment. Until
there is a vaccine or cure affordable to developing countries, the most effective way to arrest the epidemic will be by enabling individuals to reduce the risky behavior that may lead to their infection and the spread of
HIV. The specific measures that can be taken to reduce risky behavior at
both the individual and societal levels are discussed in chapter 3.

The epidemiology of HIV/AIDS suggests two important objectives
for public programs to slow and stop the spread of HIV:

Act as soon as possible. Nearly half of the world's population lives in
areas where HIV is rare, even among people whose behavior might put
them at high risk of infection. By investing in prevention when few people are infected with HIV, *before* AIDS becomes a significant health
issue, governments can contain the epidemic at relatively low cost. Even
in countries where the virus has already spread widely, effective prevention *now* can save the lives of many people who would otherwise have
become infected.

Prevent infection among those most likely to contract and spread HIV.
Not everyone in the population who contracts HIV is equally likely to
spread it to others. People with the highest number of partners and the

4

lowest levels of protective behavior (such as use of condoms and of sterile injecting equipment) are the most likely to contract and inadvertently to spread HIV. Each case of HIV infection directly prevented among people who practice these high-risk behaviors will indirectly prevent many secondary infections in the rest of the population—a kind of "multiplier" effect. Others in the population who practice lower-risk behavior by having few partners, consistently using condoms, or using sterilized injection equipment are unlikely to spread HIV, even if they contract HIV themselves. The likelihood that an individual will contract and spread HIV is determined by the level of the individual's risk behavior. Behavioral studies show that observable individual characteristics, such as occupation, age, or sexual orientation, can partially predict risk behavior and therefore can be useful in guiding prevention efforts. However, those with the riskiest behavior vary from country to country and over time. For example, sex workers have large numbers of sexual partners and, if they do not use condoms, are among those who are highly likely to contract and inadvertently spread the virus. However, in places where condom use in commercial sex has become the norm, others may be more likely to contract and spread HIV.

The chapter concludes with an overview of the level and distribution of HIV in developing countries, by region. In countries with "nascent" epidemics, HIV prevalence is very low, even among people whose behavior would put them at high risk of contracting it. In countries with "concentrated" epidemics, HIV has risen to high levels among those practicing the riskiest behaviors and is set to spread more widely in the rest of the population. In countries with "generalized" epidemics, HIV prevalence is high even among those whose behavior is unlikely to spread HIV to others. The stage of the epidemic has important implications for government priorities in preventing the spread of HIV; these are discussed in chapter 3.

Chapter 3 Efficient and Equitable Strategies for Preventing HIV

CAN PUBLIC POLICY AFFECT THE VERY PRIVATE BEHAVIORS that spread HIV? If so, what course of action should governments pursue as a priority to have the largest impact? This chapter addresses these two key issues.

Despite the private nature of the behaviors that spread HIV, governments *do* have options for influencing decisions among those most likely to contract and spread the virus. Public policy can directly influence individual high-risk behavior, either by lowering the "costs" of safer behavior (for example, by subsidizing information of various types, condoms, and access to clean injecting equipment) or by raising the "costs" of behavior that can spread HIV (for example, by attempting to restrict prostitution or the use of injected drugs). The chapter highlights examples of successful programs of the first type. Although the second approach is sometimes politically appealing, enforcement actions can exacerbate the epidemic by making it harder to reach those most likely to contract and spread the virus and encourage them to adopt safer behavior.

An important complementary approach is to promote behavioral change indirectly through policies that remove social and economic constraints to adopting safer behavior. One set of activities involves promoting social norms conducive to safer behavior, including improving the social acceptability of condoms. A second set aims to improve the status of women, whose lower social and economic status reduces their ability to insist upon sexual fidelity and to negotiate safe sex. These policies include those to expand female education and employment opportunities; to guarantee basic inheritance, property, and child custody rights; and to outlaw and severely punish slavery, rape, wife abuse, and child prostitution. Finally, policies that reduce poverty will ease the economic constraints faced by the poor in paying for essential HIV prevention services, such as STD treatment and condoms. Many of these actions address fundamental development objectives and have numerous other benefits besides slowing the spread of HIV. Their benefits are sometimes difficult to quantify, but they are highly complementary to policies that directly affect the costs and benefits of risky behavior.

What prevention strategy should governments pursue to have the maximum impact with limited resources? In keeping with the principles of public economics, governments should either ensure financing for or implement directly those interventions that are essential to stopping the spread of HIV but that private individuals or firms would not have sufficient incentive to pay for on their own. As noted in chapter 1, three major areas in which this is likely to be the case are reduction of the negative externalities of risky behavior, provision or regulation of public goods, and protection of the poor from HIV infection. Programs that address these issues will improve the efficiency and equity of government prevention efforts. In addition, following the principles of epidemiology discussed in chapter 2, program effectiveness will be improved if governments act as soon as possible and if they succeed in preventing infection among those most likely to contract and spread HIV. Thus both public economics and epidemiological principles argue strongly for giving priority to measures that prevent infection among those most likely to contract and spread HIV. The effect of specific program components may be direct or indirect and their impact immediate or long term, but their effectiveness in slowing the epidemic will depend on the extent to which they contribute to this goal. These recommendations are not meant to limit the scope of government involvement if there are ample resources and public will to undertake even more. Rather, the intention is to point out the minimum set of activities that all governments should engage in to improve the efficiency and equity of prevention programs, and a rational order in which to expand activities if resources permit.

Governments have many tools for implementing this strategy, such as direct provision of services, subsidies, taxes, and regulatory powers. Meeting any one objective will often require a combination of complementary interventions. To maximize the impact of scarce resources, public prevention programs should avert as many secondary HIV infections as possible per public dollar spent. Furthermore, priority should be given to interventions that augment (not substitute for) private sector services. HIV prevention programs often have considerable benefits for society beyond those of preventing the epidemic; these benefits and the synergies between interventions and policies should be taken into consideration in evaluating costs and benefits. Some interventions, such as reproductive health and HIV/AIDS education in schools, offer widespread social benefits in addition to benefits for HIV prevention, are inexpensive, and are therefore often a sound investment. Programmatic targeting criteria are imperfect, and reaching people at high risk of contracting and spreading HIV can be difficult. The cost-effectiveness of government programs for HIV prevention often can be improved by working with NGOs and those severely affected by the epidemic in the design and implementation of programs.

This broad prevention strategy based on epidemiology and public economics offers guidance for countries at all stages of the epidemic. For example, both epidemiology and the need to reduce negative externalities of high-risk behavior argue for heavily subsidizing safer behavior among those most likely to contract and spread HIV. This action alone may be sufficient to dramatically slow the spread of a nascent epidemic. In countries with concentrated and generalized epidemics, preventing HIV among those with the highest chances of contracting and spreading the virus is still essential to slowing the epidemic and is likely to be highly cost-effective. However, in addition, behavioral change among others who practice risky behavior will be necessary to reverse the course of the epidemic. As the epidemic spreads, the cost-effectiveness of prevention among those who practice moderately risky behavior increases. With respect to the equity of HIV prevention programs, in areas where HIV has not yet spread widely, governments can protect the poor best by taking appropriate early action to prevent an epidemic. In countries with generalized epidemics, governments can ensure that the poor have access to the knowledge, skills, and means to prevent HIV.

While the chapter identifies some basic principles underlying an efficient and equitable national strategy for preventing the spread of HIV, it remains for individual countries to identify the specific combination of programs, policies, and interventions to pursue this strategy in a cost-effective way. Program choices are necessarily country-specific because the costs and effectiveness of interventions are likely to vary widely across settings, depending on factors such as the stage of the epidemic, underlying patterns of sexual and drug-injecting behavior, social and economic constraints on safe behavior, local costs, and implementational capacity. The characteristics and accessibility of those most likely to contract and spread HIV are also highly country-specific.

To what extent are governments already pursuing the strategy suggested by this chapter? Many developing countries have launched HIV prevention programs, representing a constellation of interventions, but little is known about the extent to which they collectively have reached those at highest risk of contracting and spreading HIV and enabled them to adopt safer behavior. A review of the limited evidence found the following.

First, basic data on the patterns of HIV infection and sexual behavior, essential for making sensible decisions about allocating resources among alternative preventive interventions, are deplorably scarce. Many governments, particularly those with nascent or undocumented epidemics, need to expand their collection and analysis of data about HIV infection levels in various groups and about the nature and extent of behavior patterns that could spread the virus. This information is essential for establishing an operational definition of those most likely to contract and spread HIV. In countries with concentrated or generalized epidemics, governments need to ensure that costs and effects of interventions are more closely tracked to improve the cost-effectiveness of prevention.

Second, despite the best efforts to date, programs to change the behavior of those most likely to contract and spread HIV reach too few of them. Few national programs appear to have systematically assessed the coverage of government and NGO prevention programs—that is, the proportion of people most likely to contract and spread HIV who are reached by prevention interventions. Occupational groups such as the military and police, whose members in many places have more sexual partners on average than the rest of the population, are relatively easy and inexpensive for government to reach. Yet programs to provide members of these groups with condoms and prevention information are often lacking or inadequate.

Finally, the effectiveness of government programs in ensuring access to prevention for the poor has rarely been evaluated. For example, the social marketing of condoms (promoting the sale of subsidized condoms) has been very effective in increasing condom use. However, the extent to which these programs are benefiting the poor, are raising condom use among those with the highest rates of partner change, and are supplementing rather than crowding out private condom supply has not been established.

Taken together, chapters 2 and 3 argue that the effectiveness of government HIV prevention programs depends critically on the extent to which they reduce the risk behavior of those most likely to contract and spread HIV. Chapter 3 concludes that the greatest impediments to improved effectiveness of government HIV prevention programs are the lack of political will: first, to collect the data on HIV prevalence, risk behavior, and cost-effectiveness necessary to mount effective programs and, second, to work constructively with those most likely to contract and spread HIV.

Chapter 4 Coping with the Impact of AIDS

WHILE SOME COUNTRIES STILL HAVE THE OPPORTUNITY to avert a full-scale AIDS epidemic, others already find themselves facing the consequences of widespread HIV infection. What can be done that is effective and affordable to help people with AIDS in developing countries? What will be the consequences of AIDS morbidity and mortality for health systems and poverty? And what can society and governments do to mitigate those impacts? These are the three issues addressed in chapter 4.

The first and most basic impact of HIV/AIDS is on those who contract the disease. The chapter discusses how medication to relieve symptoms and treat opportunistic infections can ease suffering and prolong the productive lives of people with HIV, sometimes at low cost. But as the immune system collapses, available treatments become increasingly expensive and their efficacy less certain. Antiretroviral therapy, which has achieved dramatic improvements in the health of some individuals in high-income countries, is currently unaffordable and too demanding of clinical services to offer realistic hope in the near term for the millions of poor people infected in developing countries. An analysis of alternative treatment and care options concludes that community-initiated care provided at home, while often shifting costs from the national taxpayer to the local community, also greatly reduces the cost of care and thereby offers hope of affordably improving the quality of the last years of life of people with AIDS.

Second, the epidemic will increase demand for medical care and reduce its supply at a given quality and price. As the number of people with HIV/AIDS mounts, access to medical care will become more difficult and more expensive for everyone, including people not infected with HIV, and total health expenditure will rise. Governments will likely be pressured to increase their share of health care spending and to provide special subsidies for the treatment of HIV/AIDS. Unfortunately, because of the scarcity of resources and the inability or unwillingness of governments to increase public health spending enough to offset these pressures, either of these policies may exacerbate the impact of the epidemic on the health sector and may make it more difficult for the majority who are not infected with HIV to obtain care. However, there are things that governments can do. Governments should ensure that HIV-infected patients benefit from the same access to care as other patients with comparable illnesses and a similar ability to pay. Sometimes, because of discrimination, people with HIV are denied treatment or face barriers to care that others do not encounter. In other situations, people with HIV receive subsidized access to advanced therapies while people sick with other severe and difficult-to-treat diseases lack comparable access to therapies of similar cost. Although patients with HIV-related illnesses need and should receive a different mix of services than those with, say, cancer, diabetes, or kidney disease, they should pay the same percentage of their health care costs out of their own pockets as would patients with other diseases. Other measures that governments can and should undertake include providing information about the efficacy of alternative treatments for opportunistic illnesses and AIDS, subsidizing the treatment of STDs and infectious opportunistic illnesses, subsidizing the start-up of blood safety and AIDS care programs, and ensuring access to health care for the poorest, regardless of their HIV infection status.

The third major impact of the epidemic is on households and, in the aggregate, on the extent and depth of national poverty. Households and extended families cope as best they can with the loss of prime-age adults to AIDS. They reallocate their resources, for example, by withdrawing children from school to help at home, working longer hours, adjusting household membership, or selling household assets, and they draw on their friends and relatives for cash and in-kind assistance. Poorer households, having fewer assets to draw on, have more difficulty coping. Their children may be permanently disadvantaged by worsening malnutrition or withdrawal from school. However, in responding, governments and NGOs should not forget that low-income countries have many poor households that have not experienced an AIDS death but are nonetheless so poor that their children suffer similar disadvantages. At the same time, some households will have enough resources to cope with an adult death without government or NGO assistance. The government's equity objective will thus typically be served more effectively by targeting assistance based on both direct poverty indicators and the presence of AIDS in the household, rather than on either indicator alone. The chapter closes with specific recommendations to ensure that available resources reach the households that most need help by coordinating targeted poverty reduction efforts with programs to mitigate the impact of the epidemic.

Chapter 5 Working Together To Confront HIV/AIDS

NATIONAL GOVERNMENTS BEAR THE RESPONSIBILITY FOR protecting their citizens from the spread of the HIV epidemic and for mitigating its worst effects once it has spread. But they are not alone in the effort. Bilateral and multilateral donors have provided both leadership and major funding for national AIDS prevention programs, especially in low-income developing countries. Local and international NGOs have stepped forward to help, and sometimes to prod reluctant governments into action. The challenge for national governments is to define their role in the struggle against the epidemic in collaboration with these other actors.

This chapter turns from specific national policies to the strategic roles played by various actors in the policy arena. First, it examines the roles that national governments and donors have played in financing AIDS policies within developing countries, arguing that the governments of many low-income countries should confront the epidemic more forcefully, both directly and in collaboration with NGOs. Many types of NGOs are potential and actual contributors to this effort, including for-profit and nonprofit firms, broad-based private charities, and "affinity groups" of those affected by HIV/AIDS. Second, the chapter argues that, despite their substantial contributions to combating the epidemic, bilateral donors and multilateral organizations have invested too little in international public goods, including knowledge about prevention approaches and treatment methods and research on a vaccine that will work in developing countries. Furthermore, both bilateral and multilateral donors have a responsibility to coordinate their activities more effectively at the country level. Finally, the chapter discusses how public opinion and politics shape AIDS policy and how developing country governments can listen to and work with a variety of partners to minimize and overcome the obstacles to sound policies for fighting AIDS.

Chapter 6 Lessons from the Past, Opportunities for the Future

THE FINAL CHAPTER SUMMARIZES THE MAIN POLICY RECOMmendations of the report and discusses opportunities for countries to change the course of the epidemic at various stages.

For on-line information about the economics of HIV/AIDS, visit http://www.worldbank.org/aids-econ/.

CHAPTER 1

AIDS: A Challenge for Government

MORE THAN A DECADE AFTER THE HUMAN immunodeficiency virus (HIV) was first identified as the cause of acquired immune deficiency syndrome (AIDS), the disease has been reported in nearly all developing and industrial countries.[1] UNAIDS, the United Nations joint program dedicated to combating the AIDS epidemic, estimates that at the end of 1996 about 23 million people worldwide were infected with HIV and more than 6 million had already died of AIDS. More than 90 percent of all adult HIV infections are in developing countries (figure 1.1) About 800,000 children in the developing world are living with HIV; at least 43 percent of all infected adults in developing countries are women (AIDSCAP and others, 1996).

In many developing countries the HIV/AIDS epidemic is spreading rapidly. In major cities of Argentina, Brazil, Cambodia, India, and Thailand, more than 2 percent of pregnant women now carry HIV. These levels are similar to those found ten years ago in such African countries as Zambia and Malawi, where more than one in four pregnant women are now infected. In two African cities, Francistown, Botswana, and Harare, Zimbabwe, 40 percent of women attending antenatal clinics are infected. Figure 1.2 presents UNAIDS estimates of the number of new adult infections by region and over time. While new infections are thought to be leveling off in Sub-Saharan Africa as a whole, in some countries military conflict and civil unrest may be spreading the epidemic. Meanwhile, the disease is spreading rapidly in Asia. Extrapolation of the trends in figure 1.2 leads some observers to think that Asia may already have surpassed Africa in the number of new infections per year. In Latin American and the Caribbean countries the number of new infec-

Figure 1.1 Estimated Number of Adults with HIV/AIDS, by Region, 1997

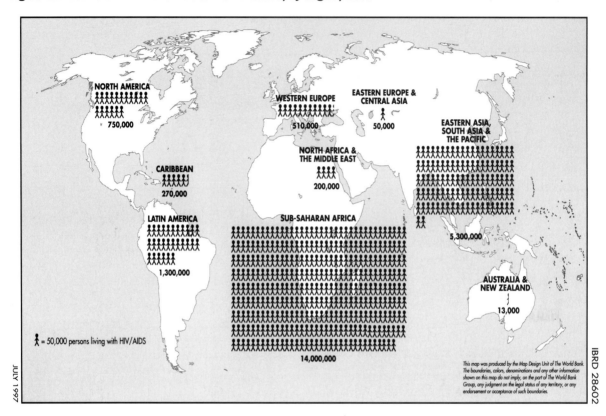

Source: UNAIDS data 1997.

tions has been steady at about 200,000 per year for several years, while the countries of eastern Europe and central Asia are experiencing the initial stages of rapid spread (not shown). Only in North America and western Europe has the number of new infections declined from its peak in 1986, but even here the future of the epidemic is unclear as it invades lower-income populations whose education and access to health care more closely resemble those of the developing world.

AIDS is clearly taking an immense and growing human toll. The disease is catastrophic for the millions of people who become infected, get sick, and, in stark contrast to the recent hopeful news of treatment breakthroughs, die. It is also a tragedy for their families, who, in addition to suffering profound emotional loss, may be impoverished as a result of the disease. Because AIDS kills mostly prime-age adults, it increases the num-

Figure 1.2 New Adult HIV Infections, by Region, 1977–95

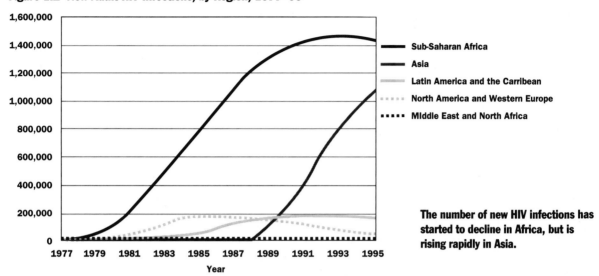

The number of new HIV infections has started to decline in Africa, but is rising rapidly in Asia.

Source: UNAIDS 1997.

ber of children who lose one or both parents; some of these orphans suffer permanent consequences, due to poor nutrition or withdrawal from school. Numbers cannot begin to capture the suffering caused by the disease. Each infection is a personal tragedy; box 1.1 describes the experience of one of the nearly 30 million people who have contracted HIV.

AIDS is not alone in causing human suffering, however. In low-income countries in particular, many urgent problems compete for scarce skills and resources. In the year 2000, malnutrition and childhood diseases that can be prevented or treated much more easily than AIDS are expected to kill 1.8 million children in the developing world; tuberculosis (TB) is expected to kill more than 2 million people; and malaria, about 740,000. Worldwide, annual deaths from smoking are expected to increase from 3 million in 1990 to 8.4 million in 2020, and nearly all of this annual increase is expected to occur in developing nations (Murray and Lopez 1996).[2] And disease is only one of many problems facing governments in improving the welfare of their citizens. About a billion people lack access to clean water, and about 40 percent of women and one-quarter of men in the developing world are illiterate. Throughout the world, inadequate transportation and communications hinder the efforts of billions of people to improve their lives.

Box 1.1 Pauline: One Woman's Story

PAULINE, THE YOUNGEST OF SEVEN CHILDREN in a farm family in Ghana, was in her early twenties when an older cousin promised her work as a waitress in Abidjan, Côte d'Ivoire, and offered to lend her the bus fare. Because she was unable to make enough money to support herself trading fish near her home, she readily accepted.

"When I got there I found there was no job as a waitress. My cousin said I must work as a prostitute to pay back the bus fare. I lived in a house with several other women doing the same work.

"I did the business in the bars. There were so many other women there I couldn't count them. Some days I had about four or five men—the number depends on your beauty. But I didn't like the work, so I did it until I had enough for my food and rent and then I'd stop for a few days.

"If you didn't pay your rent the landlord would seize your belongings and throw you out. I only wanted to save up enough income to come home.

"After three months I had enough money and came home. That was two years ago. After a year a boil grew under my arm."

Pauline's sister took her to a herbalist who sold her a potion. When that failed to relieve her worsening symptoms she went to a private doctor and was admitted to a hospital, where she remained for three months. The doctors did not tell her that she had AIDS, only that she "must not go with men."

"I never used a condom while I was in Abidjan. Men never asked for them. I never even heard of AIDS until I returned home and met friends who had it," she said.

At the time of the interview, in 1991, Pauline was very thin, had septic ulcers on her chest and shoulders, and suffered from constant itching on her arms and legs. She is one of the six million people who have so far died of AIDS.

Source: Hampton 1991.

Given these many other pressing problems, how much time, effort, and resources should governments devote to fighting AIDS? Views differ widely. Some people consider AIDS to be a late-twentieth-century version of the Black Death, which devastated Europe in the fourteenth century. According to this view, governments should do everything in their power to slow the epidemic. Others think that governments should do little or nothing, either because they think AIDS will not be a major problem in their country or because they think that governments are powerless to change the private behavior that spreads the virus. Most people would probably agree that governments should attempt to do *something*. But even among those who share this opinion, there are many views about which actions should be regarded as public priorities.

This book is addressed to policymakers, development specialists, public health personnel, and others in a position to influence the public response to HIV/AIDS. It provides an analytical framework for considering how society generally and government in particular should confront the epidemic. In doing so, it draws upon three bodies of knowledge: the epidemiology of HIV; public health insights into disease control; and especially public economics, which focuses on tradeoffs in the allocation of scarce public resources.

The report argues that AIDS is a large and growing problem and that governments can and should actively confront the epidemic. It finds that some policies will be much more effective than others, and it distinguishes between activities that can be undertaken by households and the private sector, including nongovernmental organizations (NGOs); those that should be initiated by developing country governments; and those that should be most strongly supported by donor governments and the international development community.

This chapter provides basic information that the remainder of the book will draw upon to analyze government roles in preventing and mitigating the HIV/AIDS epidemic. Following a short summary of the biology of HIV and how it spreads, we discuss the impact of AIDS on life expectancy and health and compare this with other health threats. We then set AIDS in the context of development, showing that the epidemic both affects and is affected by many aspects of economic growth. Drawing on this material, we analyze the various rationales for government involvement in confronting AIDS; this analysis will provide important guidance in identifying public priorities in the global HIV/AIDS epidemic addressed in later chapters. The chapter concludes with a discussion of why social norms and politics can make AIDS policy particularly difficult for governments.

What Is AIDS and How Is It Spread?

HIV IS A FATAL, SEXUALLY TRANSMITTED DISEASE (STD). After an initial week or two of flulike symptoms, the disease has no visible effects on the infected person during an asymptomatic period, which can be as short as two years or as long as 20. Although the average time without symptoms is about ten years in industrial countries, limited data suggest that it might be as short as five years among the poorest people in the poorest countries (Mulder 1996). Then, in all but a very small proportion of cases, the disease destroys the immune system. This leaves the infected person vulnerable to other infectious diseases, which are typically fatal within six to 24 months (box 1.2). As discussed in more detail in chapter 4, recent medical breakthroughs in treating HIV infection in high-income countries, although encouraging, are still very far from offering a technically feasible or affordable cure for the developing world.

Like other STDs, HIV is difficult to transmit except by sex or other direct contact with the bodily fluids of an infected person. The major modes of transmission are sexual intercourse, reuse of contaminated syringes by injecting drug users, infection via birth or nursing from mother to child, reuse of needles in medical settings, and transfusions of contaminated blood or blood products. HIV cannot be transmitted by a sneeze, a handshake, or other casual contact.[3]

About three-quarters of HIV transmission worldwide is through sex; of these sexual transmissions, about three-quarters involve heterosexual intercourse and one-quarter involve sexual relations between men. In developing countries, sex accounts for an even greater proportion of cases. In Sub-Saharan Africa, Asia, and the Caribbean, sexual transmission is overwhelmingly between men and women; less than 1 percent involve homosexual acts. In Latin America and Eastern Europe, however, sex between men still accounted for most sexual transmission as recently as the early 1990s (Mann, Tarantola, and Netter 1992).

The next most important means of transmission after sexual intercourse is the sharing of unsterilized needles among injecting drug users. Transmission through injecting drug use has been the primary mode of transmission in China and Southeast Asia, except in Thailand, where heterosexual transmission has outpaced transmission by needle sharing. Injecting drug use is also thought to account for about one-quarter to one-third of transmissions in Brazil and Argentina. HIV can spread through a population of injecting drug users extremely rapidly, in some locales infecting the majority within a few months.

The importance of mother-to-child transmission varies widely across countries. The major mode of infection among infants can occur in the uterus through contact with the mother's blood at birth or later through breastfeeding (see box 4.6). Since mother-to-child transmission can occur only if the mother is herself infected, it is most common in widespread heterosexual epidemics, such as in Sub-Saharan Africa. By one estimate, 15 to 20 percent of all HIV infections in Africa occur in infants infected by their mothers. Worldwide mother-to-child transmission accounts for about 5 to 10 percent of infections (Quinn, Ruff, and Halsey 1994).

Box 1.2 The Natural History of HIV/AIDS

ACQUIRED IMMUNE DEFICIENCY SYNDROME, OR AIDS, is the end stage of HIV infection. HIV destroys the immune system, and when the immune system becomes unable to protect the body against common, otherwise unthreatening diseases, AIDS may be diagnosed. HIV infection is caused by two strains of the human immunodeficiency virus, HIV-1 and HIV-2. Within HIV-1 are at least nine slightly different subtypes, each predominating in different parts of the world, although researchers have found increased dispersion in recent years. HIV-2, which is less infectious and progresses more slowly, is found primarily in West Africa, although it, too, is spreading to other regions. HIV-1 is the most common form of HIV, and is hereafter referred to simply as HIV.

Once introduced into the human body, HIV attacks mainly a subset of immune system cells, which bear a molecule called CD4. Specifically, the virus binds to two types of CD4-bearing cells: CD4+ T-cells and, to a lesser extent, macrophages. These cells perform various tasks critical to the normal functioning of the immune system. Macrophages engulf foreign invaders and prime the immune system to recognize these invaders in the future, and CD4+ T-cells organize the overall immune response by secreting chemicals to help other immune cells work properly. The mechanism—or mechanisms—by which HIV actually kills CD4+ T-cells is not well understood, but scientists do know that the immune system is able to check the onslaught of HIV to some extent, at least in the early stages of infection.

Like other viral infections, HIV infection can be characterized as a battle between the immune system and the invading virus. What makes HIV unusual is that it is a relatively even match for the human immune system, resulting in a long struggle of, on average, eight to ten years, during which HIV advances slowly but inexorably. HIV finally "wins" the battle when the infected person develops serious opportunistic illnesses: AIDS.

The battle between HIV and the immune system is fought in three general stages. The first, known as primary (or acute) HIV infection, begins at the time of infection and lasts until the body's initial immune response gains some measure of control over viral replication, usually within a few weeks of infection. During this period the CD4+ T-cell count drops dramatically, and between 30 and 70 percent of people experience flulike symptoms. These usually disappear within three weeks, as the CD4+ T-cell count rebounds.

The disease then enters its second stage, which is generally asymptomatic and accounts for about 80 percent of the time from infection to death. Only at the beginning of the second stage do antibodies to HIV become detectable in the bloodstream. Since most HIV tests work by detecting these antibodies, it is usually not possible prior to this stage to determine if a person is infected.

Most HIV-infected people remain clinically healthy during this stage, while the immune system wages an invisible but intense struggle against the virus. Each day, HIV destroys huge numbers of CD4+ T-cells. The bone marrow compensates by speeding up production of new cells, but the rate of replacement cannot quite keep up with the rate of loss. The CD4+ T-cell count, which is about 800 to 1,000 per cubic millimeter of blood in a noninfected individual, gradually declines by about 50 to 70 cells each year. When the total CD4+ T-cell count diminishes to around 200 per cubic millimeter of blood, the rate of decline accelerates and the individual becomes susceptible to opportunistic infections and other illnesses. This marks the beginning of the final stage of HIV infection—clinical AIDS.

Some of the illnesses that strike people with AIDS are communicable, such as TB; others, like HIV-related cancers, are not. Some are common infections that become unusually severe in people with AIDS, like sinusitis or pneumonia, while others are nor-

(Box continues on the following page.)

Box 1.2 *(continued)*

mally rare diseases that would not have taken hold at all had the person been HIV-negative. Some AIDS-associated infections can be treated with conventional antibiotics, particularly at the early stages of clinical AIDS. As the immune system continues to deteriorate, however, treatment becomes increasingly difficult and the number and variety of illnesses increases, leading to death. Box table 1.2 lists the main AIDS-associated illnesses diagnosed in developing countries.

The length of survival after infection depends on many factors, including the strain and subtype of the virus, the general state of the person's health, and access to medical treatment for opportunistic illnesses. Most research on this question has focused on the industrial world. Prior to the use of triple-drug therapies (see table 4.2), the median time from HIV-1 infection to death in industrial countries was around twelve years: the first two stages comprising eight to ten years and the final-stage, clinical AIDS comprising about fourteen to 25 months (Kitahata and others 1996).

Much less is known about the survival rates of HIV-infected people in developing countries, but both the time from infection to AIDS and the time from AIDS to death are believed to be much shorter,

Box Table 1.2 Opportunistic Illnesses Often Diagnosed in HIV-Infected People, Developing Countries

Name	Notes
Tuberculosis	Because latent TB is common among HIV-negative people in developing countries, it is the most common opportunistic infection there, occurring in 40 to 60 percent of the HIV-infected. As in people without HIV, TB usually occurs as a lung infection, although the likelihood of TB infecting other parts of the body is higher in the HIV-infected.
Pneumococcal disease	This bacterial infection is the most common cause of pneumonia in people without HIV and also causes bacteremia, sinusitis, and meningitis among the HIV-infected.
Pneumocystic carinii pneumonia	Although almost unknown among people with normal immune systems, this small parasite is the most common cause of pneumonia among HIV-infected people outside Africa.
Toxoplasmosis	Previously known as a cause of an occasional birth defect when it infects pregnant women, in people with AIDS it is a common cause of encephalitis, or infection of the interior of the brain, which causes seizures, coma, and death.
Candidiasis	Commonly known as oral or esophageal thrush, this fungus infection occurs in almost every person with HIV and makes swallowing painful.
Cryptococcosis	Although almost unknown in people without AIDS, this fungus infection occurs in about 5 percent of AIDS patients worldwide, usually as meningitis, an inflammation of the surface of the brain, which causes severe headache, fever, coma, and death.
AIDS-associated cancers	Common among upper-income people in developing countries (who have access to treatment for more common opportunistic illnesses).

Note: Co-infection with TB and one or more other opportunistic infections may be common in the developing world. Other important opportunistic infections, such as cytomegalovirus (CMV) and *mycobacterium avium* complex (MAC), do occur in developing countries, but are rarely diagnosed because of lack of resources.

Source: Morrow, Colebunders, and Chin 1989; *background paper,* Perriëns 1996.

Box 1.2 *(continued)*

with a total survival time from infection to death of perhaps around seven years. Aside from the generally poorer health and nutritional status of many in the developing world, lack of treatment for opportunistic infections that appear early in the course of AIDS is one factor in the shorter survival times. For example, people with HIV in developing countries are more likely than their counterparts in rich countries to succumb to TB, which is more prevalent and less likely to be treated in poor countries. In addition, TB has been associated with the faster evolution of HIV disease (De Cock 1993).

In all countries, AIDS is overwhelmingly fatal, but more than fifteen years after its emergence, it still has not been proven that HIV infection is *always* fatal. Rather it appears that survival after HIV infection follows a bell curve. Just as a few people progress to AIDS and die very quickly, at the other end of the curve are a few who have been infected with HIV for more than a dozen years but are still healthy. Medical researchers are very interested in these long-term nonprogressors as they are called, because they may shed some light on the characteristics of the immune system that could be boosted, for example, by a vaccine, to protect the average person against HIV infection. The prospects—and importance—of work on a vaccine are discussed in chapter 5.

HIV may also be spread through medical injections. In some of the poorest countries, injections are the preferred delivery system for a variety of medications, and the same syringe may be used on many people in one day without sterilization between injections. However, even in these countries, medical injections with dirty needles are thought to account for less than 5 percent of all HIV infections.

Transmission through blood transfusions, once a cause for concern in many countries, has been nearly eliminated in many high- and middle-income countries by routine screening of blood for transfusions. In developing countries, transmission through the blood supply has yet to be eliminated, especially where HIV prevalence rates are high among blood donors and where screening blood for HIV has not yet become routine. In Africa, young children may be given transfusions for treatment of malaria-related anemia, putting them at risk of acquiring HIV. But while transmission through transfused blood and other blood products greatly increases the risk of medical care and can rapidly spread HIV among specific populations—for example, among hemophiliacs in industrial countries in the 1980s—HIV transmission through transfusions has never accounted for more than about 10 percent of total HIV infections, even in developing countries.

The Impact of AIDS on Life Expectancy and Health

THE MOST OBVIOUS IMPACT OF AIDS IS ON LIFE EXPECTANCY and health. Measuring and predicting these impacts are difficult, not only because of the lack of quality data, but also because the relative size of an impact depends on many factors besides the spread of AIDS, including success in fighting other health problems. Available evidence discussed below suggests that in the most severely affected countries AIDS threatens to reverse a century of progress in the fight against infectious diseases. Elsewhere, it is likely to account for an increased share of the infectious disease burden. Even so, AIDS is only one of many health problems confronting people in developing countries. Indeed, the poorer the country, the more likely it is that other problems—including easily treated problems like malnutrition and diarrhea—account for a large share of the burden of disease.

Life Expectancy

Life expectancy is a basic measure of human welfare and of the impact of AIDS. From 1900 to 1990, dramatic progress in the fight against infectious disease raised life expectancy from 40 to 64 years in developing countries, narrowing the gap between these countries and industrial countries from 25 to 13 years. AIDS has slowed and in some countries reversed this trend. For example, life expectancy in Burkina Faso, a mere 46 years, is 11 years shorter than it would have been in the absence of AIDS (figure 1.3). Life expectancy in several other hard-hit countries also has been pushed back to levels of more than a decade ago. The impact of AIDS on life expectancy in Thailand is less, because its infection rate is less than that of the other countries in the figure.

Disability-Adjusted Life Years (DALYs)

AIDS accounted for about 1 percent of all deaths worldwide in 1990; this proportion is likely to rise to 2 percent of all deaths in 2020 (Murray and Lopez 1996). However, the proportion of total deaths caused by a disease is an imperfect representation of its burden on society, because it ignores illness and does not distinguish among the deaths of people of different ages. Murray and Lopez (1996) have estimated the cost of diseases in terms of disability-adjusted life years, or DALYs. Introduced by

Figure 1.3 The Current Impact of AIDS on Life Expectancy, Six Selected Countries, 1996

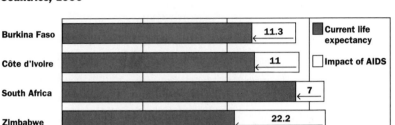

Life expectancy (years)

AIDS has already sharply reduced life expectancy in some countries.

Source: U.S. Bureau of the Census, 1996, 1997.

the *World Development Report 1993* (World Bank 1993c), a DALY includes the disability as well as the mortality effects of disease and uses age weights to discount the importance of infant and elderly deaths. In 1990, poor health resulted in the loss of about 265 DALYs per thousand persons per year in developing countries, almost twice the 124 DALYs per thousand per year lost in industrial countries. Since HIV/AIDS deaths entail substantial disability before death and disproportionally strike prime-age adults, HIV/AIDS has a larger impact on health measured as DALYs than when measured as a share of total deaths. However, the difference is not large: Murray and Lopez (1996) project that HIV/AIDS would account for almost 3 percent of all DALYs lost in developing countries in the year 2020, up from 0.8 percent in 1990 (table 1.1).[4] One reason that HIV/AIDS does not account for a larger percentage of lost DALYs is that other causes of death in developing countries also entail substantial disability and premature death. Further, some of the increased impact of HIV/AIDS is offset by the decreasing share of prime-age adults in the population associated with the demographic transition.

HIV/AIDS as a Share of Infectious Diseases

The contribution of HIV/AIDS to the disease burden looms larger when we focus attention on infectious disease. Such a focus is particu-

Table 1.1 Annual Burden of Infectious Disease and HIV, as Measured by Deaths and Lost DALYs, the Developing World, 1990 and 2020

	1990		2020	
	Deaths	*Lost DALYs*	*Deaths*	*Lost DALYs*
Annual burden of disease	(percentage of total)		(percentage of total)	
Infectious disease (as percentage of total burden)	30.7	24.5	14.3	13.7
HIV (as percentage of total burden)	0.6	0.8	2.0	2.6
HIV (as percentage of infectious burden)	2.0	3.2	13.6	19.3
HIV plus a portion of TB (as percentage of infectious burden)[a]	2.8	3.8	20.3	25.3
Total burden per 1,000 people	9.7	265.2	8.6	186.2
Infectious burden per 1,000 people	3.0	64.9	1.2	25.5
HIV burden per 1,000 people	0.1	2.1	0.2	4.5

a. The fourth row of the table is computed by adding 5 percent of the 1990 TB burden and 25 percent of the 2020 TB burden to the numbers for HIV. These percentages are the authors' estimates of the portion of HIV-negative TB deaths that would not have occurred had HIV-positive people not contributed to the spread of TB.

Source: Baseline scenario from Murray and Lopez (1996).

larly relevant to our overarching purpose—identifying the appropriate roles for developing country governments in the fight against AIDS—because economic theory, public health teaching, and long-standing practice all affirm that governments should play a significant role in preventing the spread of infectious disease.

By 2020 infectious diseases, which currently account for about 30 percent of deaths and one-quarter of lost DALYs in developing countries, will have declined to about 14 percent of both measures.[5] But the contribution of HIV/AIDS to the infectious disease burden in developing countries is projected to increase sharply, from about 2 percent of deaths and 3 percent of lost DALYs to about 14 percent of deaths and nearly one-fifth of lost DALYs. Moreover, because HIV is an increasingly important factor in the spread of TB, it is estimated that about one out of four TB deaths *among HIV-negative people* in 2020 would not have occurred in the absence of the HIV epidemic.[6] Adding a quarter of TB deaths among HIV-negative people to the deaths directly attributable to HIV/AIDS suggests that HIV/AIDS will be responsible for about one-fifth of all infectious disease deaths in low-income countries in the year 2020.[7] In addition, HIV is likely to be responsible for a portion of deaths from several other infectious diseases (figure 1.4).[8]

Figure 1.4 Breakdown of Deaths from Infectious Diseases, the Developing World, by Disease Category, 1990 and 2020

(percent)

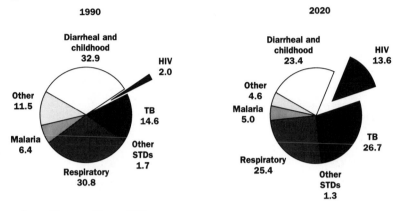

As the epidemic progresses, HIV will account for a greatly increased share of total deaths from infectious diseases in developing countries.

Source: Murray and Lopez 1996.

HIV/AIDS as a Major Killer of Prime-Age Adults

Because HIV/AIDS is sexually transmitted, AIDS usually strikes prime-age adults—often people who are raising children and are at or near the peak of their income potential. In the absence of AIDS, prime-age adults tend to be less vulnerable to sickness and death than children, adolescents, or old people. Accordingly, AIDS casts an even larger shadow on the health of prime-age adults and the welfare of their dependents. In 1990, HIV was already third after TB and non-TB respiratory infections as a cause of adult death in the developing world; by 2020, HIV will be second only to TB as a killer of prime-age adults in developing countries (figure 1.5). Adding one-quarter of TB deaths among HIV-negative prime-age adults makes HIV/AIDS the largest single infectious killer of prime-age adults in the developing world in 2020, responsible for half of all deaths from infectious disease among this important group.

The HIV/AIDS share of the adult infectious disease burden varies widely across developing regions. In Africa, where other infectious diseases decline less quickly than in other regions and HIV/AIDS infection rates are assumed to be leveling off in many areas, HIV/AIDS will account for about one-third of these deaths (figure 1.6). Because Latin America and the Caribbean countries are projected to make the most

Figure 1.5 Causes of Death from Infectious Diseases among People Ages 15 to 59, the Developing World, 1990 and 2020

(percent)

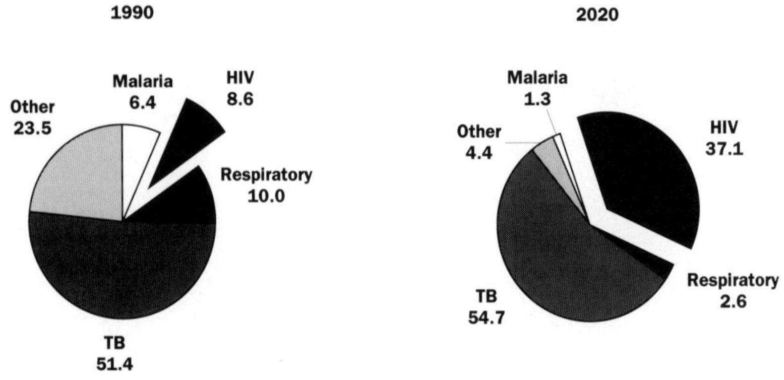

Between directly causing AIDS deaths and indirectly facilitating the spread of TB, HIV will be responsible for up to half of all adult deaths from infectious disease in the year 2020.

1990

Malaria
6.4

HIV
8.6

Other
23.5

Respiratory
10.0

TB
51.4

2020

Malaria
1.3

HIV
37.1

Other
4.4

Respiratory
2.6

TB
54.7

Source: Murray and Lopez 1996.

Figure 1.6 HIV/AIDS as a Percentage of the Infectious Disease Burden of Adults, the Developing World, 2020

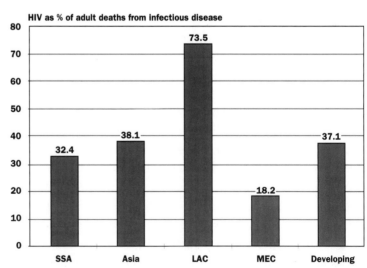

HIV as % of adult deaths from infectious disease

SSA	Asia	LAC	MEC	Developing
32.4	38.1	73.5	18.2	37.1

AIDS will account for a larger share of the infectious disease burden in regions where other infectious diseases are less of a problem.

SSA Sub-Saharan Africa
LAC Latin America and the Caribbean Countries (total)
MENA Middle East and North Africa
Source: Murray and Lopez 1996.

progress in reducing other infectious diseases, and HIV infection is predicted to continue to rise, HIV will be responsible for almost three-quarters of the infectious disease burden there.[9]

AIDS and Development

ALTHOUGH THE HEALTH IMPACTS OF THE DISEASE ALONE are ample cause for concern, there are additional reasons why the development community in general and policymakers in particular should be concerned about the HIV/AIDS epidemic. First, widespread poverty and unequal distribution of income that typify underdevelopment appear to stimulate the spread of HIV. Second, the accelerated labor migration, rapid urbanization, and cultural modernization that often accompany growth also facilitate the spread of HIV. Third, at the household level AIDS deaths exacerbate the poverty and social inequality that are conducive to a larger epidemic, thus creating a vicious circle. Policymakers who understand these links have the opportunity to break this cycle—through policies that are suggested below and analyzed in detail in the rest of the book.

Poverty and Gender Inequality Spread AIDS

While the determinants of an individual's sexual activity are subtle and complex, it is reasonable to expect that at the aggregate level social conditions would influence the frequency of risky sexual behavior and hence the size of the epidemic. One hypothesis is that poverty and gender inequality make a society more vulnerable to HIV because a woman who is poor, either absolutely or relative to men, will find it harder to insist that her sex partner abstain from sex with other partners or use a condom or take other steps to protect herself from becoming infected with HIV.[10] Poverty may also make a man more prone to having multiple casual partners, by preventing him from attracting a wife or by causing him to leave home in search of work. The idea that poverty and gender inequality exacerbate AIDS is supported by an exploratory analysis of national-level aggregate data on HIV infection rates.

Eight epidemiological, social, and economic variables can explain about two-thirds of the variation in cross-country HIV infection rates. Figure 1.7

Figure 1.7 Relationship of Four Societal Variables with Urban Adult HIV Infections, 72 Developing Countries, circa 1995

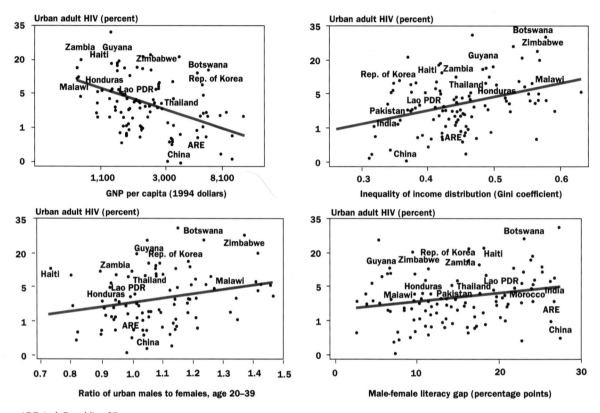

ARE Arab Republic of Egypt

Note: The vertical axis measuring HIV infection has been transformed to a logarithmic scale. Points on a given scatter plot represent the data for 72 individual countries after removing the effects of the other seven variables included in the regression analysis. Income inequality is measured by the Gini coefficient. The methodology and the detailed statistical results are presented in Over (*background paper,* 1997).

Source: Authors' estimates.

Poverty, inequality of income across households, and low status for women all contribute to the spread of HIV.

shows the associations of four of these variables with the percentage of urban adults infected with HIV.[11] The top two panels of the figure demonstrate that, holding constant the other variables, both low income and unequal distribution of income are strongly associated with high HIV infection rates. For the average developing country a $2,000 increase in per capita income is associated with a reduction of about 4 percentage points in the HIV infection rate of urban adults. Reducing the index of inequality from 0.5 to 0.4, the difference in inequality between, for example, Honduras and Malawi is associated with a reduction in the infection rate by about 3 percentage points.[12] These findings suggest that rapid and fairly distributed economic growth will do much to slow the AIDS epidemic.[13]

When examining the influence of gender inequality on HIV infection, one must hold constant as much as possible other cultural influences, such as Islam, which may be correlated with gender inequality across countries. The bottom two panels of figure 1.7 show that, after controlling for the percentage of the population that is Muslim (as well as per capita gross national product (GNP), income inequality, and four other societal characteristics), two measures related to gender inequality are associated with higher HIV infection rates. The first of these, the ratio of males to females in urban centers, varies remarkably across countries: some countries have fewer male urban residents than female, and others have 40 percent more males. Other things being equal, one might suppose that commercial sex would be more common in cities where men greatly outnumber women, and that HIV levels would therefore be higher. The evidence of the lower left panel of figure 1.7 is that cities in which men ages 20 to 39 greatly outnumber women do in fact have significantly higher HIV infection rates. For the average country, increasing the job opportunities for young women so that the ratio of males to females in urban areas falls, for example, from 1.3 to 0.9, would decrease the HIV infection rate by about 4 percentage points.

The second measure related to gender inequality included in the analysis is the gap between adult male and female literacy rates. Again, there is great variation across countries, the literacy rate among men being as much as 25 percentage points higher than among women in some countries. When women are much less literate than men, they may be less able to bargain effectively with men and thus be at greater risk in sexual encounters. Furthermore, illiterate women will have difficulty finding jobs and thus may depend more on sexual relationships for economic survival, again reducing their bargaining power. The lower right panel of figure 1.7 supports these ideas, suggesting that the average country that reduces the literacy gap between genders by 20 percentage points can expect urban HIV infection to be about 4 percentage points lower.

The Dynamics of a Growing Economy May Facilitate the Spread of AIDS

From the evidence of figure 1.7, a country that improves per capita income and reduces inequality, for example, by implementing investment policies that generate jobs and raise economic growth, will reduce its risk of suffering an AIDS epidemic or help to minimize an epidemic already under way. If, in addition, the country acts to close the literacy and urban employment gaps between men and women, HIV would have even more difficulty spreading. Unfortunately, some of the very processes that might achieve these goals can also stimulate the spread of AIDS. And other policies that sometimes accompany growth, without necessarily contributing to it, can likewise worsen the epidemic.

An open economy is recognized as a key requirement of rapid growth. Openness primarily refers to the ease with which entrepreneurs can move goods and capital across national borders. Yet a higher degree of openness to trade and financial flows is typically also accompanied by a greater degree of openness to movements of people, including immigration. Moreover, some studies have suggested that immigration itself contributes to economic growth. This would hardly be surprising, since immigrants are often among the hardest-working and most enterprising people in any country. However, a cross-country regression analysis suggests that countries with larger immigrant populations tend to have larger AIDS epidemics: other things being equal, a country in which 5 percent of the population is foreign-born can expect to have an infection rate about 2 percentage points higher than a country with no foreign-born.

Does this mean that governments should restrict immigration to avoid an AIDS epidemic? No, it does not. Indeed, if immigration is beneficial to economic growth, reducing immigration may slow it, a result that, in addition to many other negative effects, may stimulate the spread of AIDS. Attempting to screen out HIV-positive immigrants is unlikely to be very effective, since immigrants are more likely to become infected after they have arrived in a new country, when they are disconnected from the social networks of their former homeland, than before leaving home. Worse, attempts at screening potential immigrants may exacerbate the epidemic: if people infected with the virus evade screening and arrive illegally, identifying and reaching them with programs to prevent them from infecting others is extremely difficult.

Sometimes a specific project promises significant economic benefits but carries with it the danger of worsening the epidemic. An example of such a project—and of an effective response by the governments concerned—is the Chad-Cameroon oil pipeline project described in box 1.3. The challenge to governments, donors, and multilateral institutions is to recognize the potential AIDS hazard inherent in such a project and incorporate into the project design elements that eliminate or at least minimize and mitigate these problems. Economic development projects that

Box 1.3 AIDS and the Chad-Cameroon Oil Pipeline Project

THE CHAD/CAMEROON OIL PIPELINE IS THE FIRST large-scale infrastructure project supported by the World Bank to assess the potential for an adverse impact on STDs, including HIV/AIDS, and to incorporate prevention efforts in the project design.

The 30-year, $3.5 billion project, due to begin construction in 1998, involves development of oil fields in southern Chad and construction of an 1,100 km pipeline to port facilities on the Atlantic coast of Cameroon. A cooperative effort between the World Bank, the governments of Chad and Cameroon, and a consortium of private oil companies, the project promises the two countries substantial economic benefits.

But the project also involves a potential risk of exacerbating the HIV/AIDS epidemic. During the peak construction period, from 1998 to 2001, the project will draw an estimated 2,000 construction workers from Chad and Cameroon, and employ an additional 400 to 600 truckers who will travel the length of the pipeline. Most of the workers will be single and unaccompanied males. Those working in Chad will commute from their villages of origin, while those working along the pipeline in Cameroon will live in temporary barracks. Some areas along the proposed pipeline already have extremely high levels of HIV: a 1995 report from an area adjacent to the Chad/Central African Republic border, and directly on the proposed pipeline route, indicated that more than half of the sex workers and one in four truckers were already infected with the virus.

Alerted to these problems by an environmental assessment performed as part of the project prepara-

tion, the World Bank, the consortium, and the two governments involved have identified a package of measures to avoid exacerbating the HIV/AIDS epidemic in the project area. Since preliminary estimates suggest that effective interventions can be implemented for less than $1 million a year, the substantial returns on the project are more than sufficient to justify project execution despite these costs. Using baseline data and experience gained elsewhere in Africa, the consortium is developing a layered intervention strategy that includes:

- monitoring of the STD and HIV status of the workforce
- vigorous marketing of subsidized condoms
- information, education, communication (IEC)
- treatment of classic STDs
- interventions to modify high-risk behavior
- coordination with existing government and NGO programs, particularly those directed toward sex workers.

To supplement the work of the consortium in this area, the World Bank is preparing two technical assistance projects that will help the governments of Chad and Cameroon monitor and assess the health impact of the project. Implementing these programs will involve significant challenges, including the difficulties of reaching the highly mobile truckers and the sex workers they frequent.

Source: Caldwell and Caldwell 1993, pp. 817–48; Carswell and Howells 1989, pp. 759–61; Dames and Moore 1996; and Mwizarubi and others 1992.

do not generate sufficient net economic returns after covering the cost of mitigating negative impacts, including the spread of AIDS, should be rejected as undesirable—even if the gross returns are quite large.[14]

Sometimes a low-income society beginning rapid growth may face an increased risk of AIDS as a result of a broad shift from conservative social norms to more liberal attitudes; these attitudes often include greater individual freedom, especially for women. Lacking an objective measure of social conservatism, the regression used the percentage of the population that is Muslim as an imperfect proxy for a country's degree of social conservatism. Controlling for all other variables described above, a high degree of social conservatism is associated to a statistically significant degree with lower rates of HIV infection. This does not necessarily imply that governments should attempt to instill or maintain social conservatism simply to minimize HIV; such broad social values may in any event be very difficult for governments to shape. However, the evidence does suggest the usefulness of an explicit government education policy that would help young people entering a rapidly modernizing society to recognize and avoid risky sexual encounters.

A final factor in the regression analysis that is not associated with development but can be readily affected by government policy is the level of militarization. In developing countries, military forces are often based near urban centers and consist predominantly of young, unmarried men. Using a variable that measures the number of men in the nation's armed forces as a percentage of its urban population, the regression analysis shows that, even after controlling for the ratio of male to female urban residents, countries with more soldiers will have higher infection rates. For the average country, reducing the size of the military from 30 to 12 percent of the urban population will reduce seroprevalence among urban adults by about 4 percentage points. An alternative discussed in chapter 3, which may be more feasible (and is reasonable regardless of the size of the military) is a vigorous HIV prevention program covering everyone in the military.

AIDS Has Little Net Macroeconomic Impact

Because HIV/AIDS is spreading rapidly and is nearly always fatal, some observers have concluded that it will significantly reduce population growth and economic growth; a few have suggested that populations in badly hit countries will decline in absolute terms and that there will be an associated collapse in economic output (Anderson and others 1991, Rowley, Anderson, and Ng 1990). However, the available evidence suggests that the impact of AIDS on these variables, although varying across countries, will generally be small relative to other factors. Moreover, at a very crude level, declines in population growth due to HIV/AIDS will tend to offset declines in economic growth, so that the net impact on gross domestic product (GDP) growth per capita will generally be small.

HIV/AIDS is expected to reduce population growth rates in many countries, but in no country is an absolute decline in population expected. The latest projections suggest that declines in population *growth rates* from HIV/AIDS mortality will range from about 0.1 percentage point in Thailand to 2.3 percentage points in Botswana, and that the median country's population growth rate will decline by about 1 percentage point (U.S. Bureau of the Census 1997).[15] Over time, such a reduction in growth would result in a significantly smaller population than would have existed in the absence of AIDS. In Zambia, for example, the population is predicted to be 7 percent smaller in 2005 than it would have been without AIDS. In two countries with very severe AIDS epidemics, Botswana and Zimbabwe, the projections suggest that by 2010 the population will cease growing.

The impact of AIDS on economic growth is a much more complex issue than the impact on population growth. The inadequacy of GDP per capita as a measure of human welfare is striking when changes in per capita GDP are used to measure the impact of AIDS. Other things held constant, the death of people with higher incomes will reduce average income—even though the welfare of those who remain alive has not changed. Conversely, the death of those with lower incomes raises average income, without necessarily improving the lot of any surviving individual and despite the suffering and economic losses of the families of those who died. Further, increased spending on health care and funerals is included in GDP calculations. As a result, per capita GDP may increase, even though overall well-being has not improved and the incomes of survivors have been reduced.

With these caveats in mind, it is nonetheless possible to estimate the size of the epidemic's impact on individual incomes. This impact will depend on characteristics of the country, including the severity of the epidemic, the efficiency of the labor market, the proportion of the treatment cost of AIDS financed from savings, the distribution of HIV infection by productivity of worker, the time lost from work by the person with AIDS and others as a result of his or her sickness, and the effectiveness of household and community formal and informal insurance mechanisms.

Because AIDS kills prime-age adults, many of whom are at the peak of their economic productivity, the shock of AIDS to the labor market is one mechanism through which AIDS might affect growth. However, in economies with substantial unemployment, firms should find it easy to replace sick or deceased workers, particularly if they are not key personnel. Other things being equal, the impact of the AIDS epidemic will be small until the economy begins to grow and is constrained by labor supply rather than by insufficient demand. Box 1.4 provides evidence from a sample of 992 firms that the departures of lower-skilled workers due to AIDS may have had little effect on firm profits in five Sub-Saharan African economies.

Another factor that is likely to have a significant effect on the size of the macroeconomic impact of the epidemic is the percentage of AIDS treatment costs financed from savings. Since spending on AIDS treatment is likely to reduce the capital available for more productive investment, the higher the proportion of care financed from savings, the larger the reduction in growth resulting from the epidemic. If one takes these factors into consideration, a rough estimate would be that a generalized epidemic, as defined in chapter 2, would reduce per capita GDP by as much as half a percentage point per year.[16]

The importance of an impact of this size will vary depending on the country's underlying growth rate. In some very poor Sub-Saharan African countries, growth rates in GDP per capita that are already negative may deteriorate even further as a result of the AIDS epidemic. But some countries with severe AIDS epidemics, including Botswana, Thailand, and Uganda, have been growing rapidly. With per capita growth rates in excess of 5 percent per year, a reduction of 0.5 percent in per capita growth will not be crippling. For these countries, as for many other countries where the epidemic may peak at lower infection levels, the more serious consequences will be the impacts on public health spending and on poverty.

Poverty, Inequality, and Orphanhood

Although the macroeconomic effects of AIDS are likely to be small in most countries, severely affected countries will experience quite large impacts on their health sector and on the poor. The effect on the health sector will be to increase the price and reduce the availability of health care for everyone, which will tend to hurt the poor most. Furthermore, among the households that suffer an AIDS death, lower-income households will be less able than others to cope with the medical expenses and other impacts, including loss of income.

Box 1.4 Looking for the Impact of HIV/AIDS on a Sample of African Firms

IN COUNTRIES WITH GENERALIZED HIV EPIDEMICS, the mortality rate of prime-age working adults will eventually rise by a multiple of 2 to 10, depending on the baseline mortality rate in the country and the extent of HIV infection (table 4.3). Such increases should raise the firm's labor costs by requiring it to replace workers more frequently, to spend more on sickness and death benefits, and, perhaps, to implement AIDS education programs designed to prevent the workers from becoming infected. Whether these changes will have a measurable effect on firm profits depends on whether they are large relative to the other components of labor costs and whether labor costs themselves are large as a portion of total firm costs.

Although several studies have shown that AIDS increases the death rate of workers in specific firms, none has compared these death rates with the rates of worker attrition from other causes or has estimated the impact of the deaths on firm profits (Giraud 1992, Smith and Whiteside 1995, Baggaley and others 1994, Jones 1997). To analyze the impact of AIDS deaths within the context of overall firm performance, a background paper for this study analyzed data on worker attrition due to sickness and death that were collected as part of a survey administered to 992 firms from four segments of the manufacturing sector of five African countries (*background paper,* Biggs and Shah 1996).[1]

Box table 1.4 presents the data on HIV infection rate in the urban population of each of the five countries and the percentage of workers leaving their employment in 1994 due to sickness and death (right-most column). Clearly there is a strong correlation between these two variables at the country level. Zambia, with the highest measured infection rate, also has the highest rate of turnover due to sickness or death. Ghana is at the other extreme on both variables.

The impact of the higher levels of sickness and death would be large if the resulting attrition rates are large in relation to overall firm attrition or if it takes a long time to replace the workers. However, it appears that neither is usually the case. The average attrition rate from all causes is from 8 to 30 times larger than that from sickness and death. The time to replace deceased workers varies from an average of two weeks for unskilled workers to only three weeks for skilled workers, not enough to significantly raise costs. The only hint in this labor force

(Box continues on the following page.)

Box Table 1.4 Worker Attrition in Ghana, Kenya, Tanzania, Zambia, and Zimbabwe, Total and by Sickness or Death, 1994

Country	Urban HIV prevalence	Total in Sample		Percentage of workers leaving firm	
		Firms	Workers	Due to all causes	Due to sickness or death
Zambia	24.7	194	14,582	20.8	2.5
Zimbabwe	20.5	199	59,210	9.1	1.2
Kenya	17.1	214	17,126	7.7	0.9
Tanzania	16.1	197	14,611	19.3	0.6
Ghana	2.2	188	9,607	11.6	0.3
Total		992	115,136	11.9	1.15

Source: Seroprevalence data refer to low-risk, sexually active adults as reported in U.S. Bureau of the Census (database, 1997). Other data from RPED Panel Survey described in Biggs and Shah (*background paper,* 1996) A preliminary version of this table appears in National Research Council (1996, p. 237).

Box 1.4 *(continued)*

data that AIDS might be costly to firms is that they took 8 times longer on average to find a replacement for a deceased professional than they did for a skilled worker. However, even 24 weeks does not seem long to search for a skilled professional.

The ultimate question is whether labor force sickness and death visibly reduce firm profits. In an output-constrained firm, hiring more workers increases output. However, a firm that is suffering from a drop in demand for its product can raise profits (or decrease losses) by releasing workers. If some of the firms in the sample are experiencing an increase in demand and others a decrease, it would be impossible to disentangle these two effects, and the estimated impact of AIDS deaths would be meaningless.

One solution to this problem would be to assume that the departure of workers due to sickness and death is beyond the control of the firm, whereas the departure of other, presumably healthy, workers is partly decided by the firm. Under these assumptions the departure of a worker due to sickness and death is estimated by instrumental variable methods

to reduce the firm's value-added per worker by a statistically significant, but small, amount (*background paper,* Biggs and Shah 1996).

Of course, these results are far from definitive. First, they are specific to Africa and to economies that were doing rather poorly; for any sick worker many replacements were available. Second, even a random sample of 992 firms is a small sample in which to study adult mortality, especially if the events of most interest are the deaths of the most highly skilled workers—the professionals—of whom there are few in any firm. Nevertheless, until more definitive studies are performed, the evidence suggests that the impact of AIDS sickness and deaths is not a major determinant of the economic performance of the average firm in developing countries.

---

[1] The sectors were food processing, metal working, wood working, and textiles and garments. The firms were selected by a random process designed to assure they would be representative of the sectors from which they were drawn. The questionnaire was designed by economists, statisticians, and management specialists in order to learn about the causes of firm success in Africa. Shortly before it went to the field, questions were added on worker attrition.

We argue in chapter 4 that, because low-income households are more adversely affected by an AIDS death than other households, a severe epidemic will tend to worsen poverty and increase inequality.

One important way in which AIDS is likely to exacerbate poverty and inequality—and indeed, one of the most tragic effects of the epidemic—is the increase in the number of children who lose one or both parents. To be sure, AIDS is not the only cause of orphanhood: in some countries other causes of prime-age adult death may orphan many more children than AIDS. Nonetheless, as AIDS mortality rises, growing numbers of children will be orphaned by the disease; the resulting impact on orphanhood rates in three hard-hit countries is shown in box 1.5.

The impact of an adult death on surviving children is discussed at length in chapter 4. Here it is sufficient to note that, even if we set aside

Box 1.5 Orphans and AIDS

THE IMPACT OF A SEVERE AIDS EPIDEMIC ON maternal orphan rates can be seen in census data over the past 20 years from three East African countries (box figure 1.5). In the absence of AIDS, gradual improvements in maternal health over the past two decades would have reduced maternal orphan rates. Instead, we see that in Kenya the maternal orphan rate has remained nearly constant. In Tanzania the maternal orphan rate fell between the 1970s and late 1980s but then climbed rapidly to almost 3 percent in the 1990s. Finally, maternal orphan rates in Uganda show steady increases since 1969, a trend that can probably be attributed to a combination of AIDS and civil war. Because AIDS tends to be geographically clustered, maternal orphan rates are even higher in areas hard hit by the epidemic. Across fifteen villages in Rakai District of Uganda, for example, the maternal orphan rate in 1990 was 6.6 percent, double that for the rest of the country (Konde-Lule and others 1997).

Losing a parent can have profound consequences for any child, and these are likely to be worse in poor households. Governments and NGOs trying to mitigate the impact of AIDS should be careful to consider overall needs and avoid creating programs that favor AIDS orphans over other orphans who may be equally or even more needy.

Yet a consideration of the impact of the epidemic must also recognize that AIDS orphans often face uniquely severe problems. Very young orphans whose mothers are infected or die of AIDS have higher mortality rates than other orphans because roughly one-third of them are themselves infected

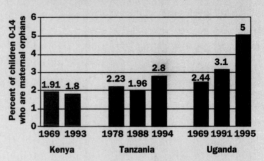

Box Figure 1.5 Trends in Maternal Orphan Rates, Three Hard-hit East African Countries, Various Years

Note: Maternal orphan rates in this figure include children who are two-parent orphans.

Sources: Kenya (1969), Tanzania (1988), and Uganda (1969) are based on census data as reported in Ainsworth and Over (1994a,b). Kenya (1993), Tanzania (1994), and Uganda (1995) are from DHS data. Tanzania (1978) and Uganda (1991) are from census data as reported by Hunter and Williamson (forthcoming).

with HIV at or around the time of birth. Also, AIDS orphans are more likely to become two-parent orphans because HIV is transmitted sexually. For example, in a population-based survey of rural areas of Masaka District, Uganda, 10 percent of all children under 15 had lost one or both parents (Kamali and others 1992). Fifteen percent of the surviving parents of single-parent orphans were infected with HIV, three times the parental infection rate among nonorphans. Finally, AIDS orphans may suffer social stigma from having lost their parents to a sexually transmitted disease.

the unmeasurable grief and psychological pain experienced by children who lose a parent, the measurable declines in nutritional status and reductions in schooling can cause profound and lasting damage to a child. These effects, which are likely to be greatest in the poorest families, can greatly reduce an individual's ability to aquire the skills and knowledge needed to escape poverty.

The Government Role in Confronting AIDS

GIVEN THE IMMENSE IMPACT OF HIV/AIDS ON LIFE EXPEC-
tancy and health, and the possibility that AIDS will exacerbate
problems of poverty and inequality, the need for governments
to confront the epidemic appears straightforward. Indeed, for many
people the human suffering caused by the epidemic is reason enough
for governments to be involved. However, there are other important
reasons for government involvement, some of which are not so readily
evident. Analysis of these rationales for government involvement is a
necessary foundation for considering *how* governments should confront
HIV/AIDS.

The Impact of AIDS on Public Health Spending

One economic rationale for government involvement in HIV prevention is very straightforward: prevention is much cheaper than treatment
and avoids the sickness and death that are the final outcome of the disease. This argument is particularly important in the many low-income
countries where governments remain committed to publicly financed
curative health care. In such countries, the high cost of AIDS treatment
starkly reveals the scarcity of resources.

Figure 1.8 illustrates the difficult tradeoff governments face. In the
figure, each country is represented by a point indicating, on the vertical
axis, the estimated total cost (public and private) per year of AIDS treatment and, on the horizontal axis, the national GNP per capita. Not surprisingly, the amount spent on treatment rises sharply with GNP per
capita. The upper of the two regression lines fits those points well and
suggests that in the average country the annual treatment cost of AIDS
is about 2.7 times GNP per capita. The second line in the figure (estimated from other data) demonstrates that for less than this amount the
average developing country could finance a year of primary education
for ten students. And this is just one of the many alternative productive
uses to which these financial resources could be put.

As the number of AIDS cases and treatment costs mount, it becomes
painfully evident that AIDS treatment is consuming public resources
that could have been used for other human needs. Yet governments may
find it very difficult to restrict financing for AIDS treatment without at
the same time reassessing their commitment to publicly financed health

Figure 1.8 Annual Treatment Cost for an AIDS Patient Correlated with GNP per Capita

(dollars)

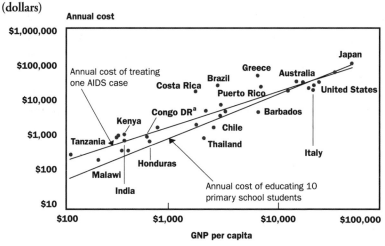

Note: The trendline for AIDS is: Annual cost = 2.7 × (GNP per capita)$^{0.95}$.

a. Formerly Zaire.

Source: Annual AIDS treatment costs are from Mann and Tarantola (1996) and Ainsworth and Over (1994a,b). Annual cost of educating ten primary school students are the authors' calculations based on data from 34 countries in Lockheed and others (1991).

Spending on AIDS treatment increases with GNP; on average, treating an AIDS patient for one year costs about the same as educating ten primary school students for one year.

care. Indeed, in many countries there are political pressures to subsidize AIDS treatment at a *higher* level than other health care services, and these pressures are likely to grow as the number of people with HIV increases. For all these reasons, a government that wishes to continue subsidizing health care should initiate aggressive prevention efforts as early in the epidemic as possible. Even governments that are attempting to reduce subsidies for curative care would do well to invest in HIV prevention, since political pressures for subsidized care may be very difficult to resist.

The Public Economics Rationale for Governments to Fight HIV/AIDS

Let us assume, for the sake of argument, that a government does not subsidize curative health care and that it is able to resist all pressures to do so. Given that HIV/AIDS is primarily transmitted by sex, would there still be a rationale for government intervention to reduce its spread?

The answer from a public economics perspective is yes. To understand why, it is useful to first consider the public economics rationale for government intervention against other contagious diseases, such as TB. If all markets worked perfectly, governments would not need to be involved in the fight against these diseases. Instead, each person at risk of infection would pay an appropriate share of the cost of reducing his or her risk. In reality, of course, there is no mechanism other than government through which individuals can pay this amount. Since a person infected with TB is likely to consider only the benefits to himself when deciding whether to pay for treatment, without government intervention people infected with TB would be cured less often than everyone else would prefer. Economists call the benefits of treatment that are not captured by the person paying for the treatment *external benefits* and the negative impacts on others of not getting treatment *external costs*. These *externalities*, if large, are a strong economic justification for government intervention.

A related problem can be best understood in the case of a vector-borne disease, such as malaria. Even if people know that draining a stagnant pool where the anopheles mosquito breeds will greatly reduce their chances of catching malaria, people may not voluntarily pay the costs of drainage, since everyone benefits regardless of who pays. Thus, each person may hope to benefit from the actions taken by others. Elimination of stagnant water is an example of what economists call a *public good*. Because individuals hope to gain from what others have paid for, a public good may not be produced at all unless government taxes everyone in order to finance its production.

In giving advice on how governments should spend scarce public resources, economists look for evidence of large externalities or public goods. Where these exist, markets are said to have failed, and public intervention to address the market failure is warranted. In the case of TB, malaria, and other diseases that may strike people regardless of their individual behavior, the public economics arguments for government intervention due to market failure are clear.

At first consideration, it may appear that externalities and public goods are not a significant concern in the case of STDs, including HIV. Since most STD transmission occurs as a result of a voluntary act between two people, each can weigh his or her risk and proceed only if the benefits of doing so outweigh the risks. If both parties agree to have unprotected sex even though they may catch an STD, why should the government intrude in these private decisions? The problem, of course, is that the decision of these two people has consequences for many others, endangering marital sex and procreation as well as more casual sexual relationships. Ideally the couple should take the interests of others into account when they decide whether or not to engage in unprotected intercourse. However, even if they agree to use a condom or otherwise reduce their risk of infection, they cannot demonstrate to potential future partners that they have behaved prudently. In public economics terms, there are external benefits associated with refraining from risky sex. Since an individual cannot capture these benefits, he or she will be less prudent than would otherwise be the case.[17] The result: higher STD infection rates and a greater risk of infection for everyone who is sexually active, even those who are monogamous, since most people cannot be sure that their partner is also monogamous. In such a situation, government intervention is justified if it can increase the incentives for the most sexually active individuals to practice safer sex (or for injecting drug users to adopt safer injecting behavior) to the point that their decisions more closely reflect consideration for the social consequences of risky behavior.

The arguments above for government intervention to prevent the spread of STDs apply even more strongly to HIV/AIDS. In addition to being sexually transmitted, HIV/AIDS has two characteristics that worsen the market failures associated with the disease—and suggest that governments should be particularly concerned about HIV prevention. The most obvious is that AIDS cannot be cured and is almost always fatal. Since adult deaths impose costs on other family members and the rest of society, as argued above and shown in chapter 4, they are an argument for goverment intervention. In addition, we have seen that HIV makes people vulnerable to other infectious diseases, including TB. Since individuals have little control over their exposure to TB, and since people who have HIV and TB can spread TB even to people who are HIV-negative, the link between HIV and TB further strengthens the arguments for a government role in controlling HIV.

While this link suggests that HIV/AIDS should receive special attention, the close epidemiological relationship between HIV and other STDs, discusssed in chapter 2, means that in practice any effective strategy against HIV will almost certainly involve stepped-up prevention efforts against other STDs, and vice versa. Since the monitoring problem described above applies equally to all STDs, governments would have a role in STD control even in the absence of HIV. Because HIV multiplies manyfold the external costs associated with a case of gonorrhea or genital ulcer disease, the presence of HIV strengthens the argument for government intervention to control the spread of all STDs.

The Government Role in Generating Information

The above argument for government intervention assumes that people already know of the risks of HIV, or have the means to find out what they need to know. However, this is often not the case. Thus there is yet another compelling rationale for a government role in confronting the epidemic: the provision of information to enable individuals to decide whether or not to change their behavior to reduce the chance of infection. In some countries, HIV/AIDS has now been present for two decades, long enough for most people to know that it adds a mortal risk to sexual liaisons; yet surveys show that an alarmingly large proportion of people in some countries still do not know how to protect themselves. In other societies, the disease is still new, an invisible danger spreading through an unsuspecting citizenry that, because of HIV's two-to-twenty-year asymptomatic period, has yet to be jolted by a sudden upsurge in AIDS deaths. In either society, only the government has the incentives and the capacity to generate the information that permits people to take the first steps toward self-protection.

Information about the state of the epidemic and how to avoid infection is a true public good. As in the case of malaria eradication, each individual who benefits from new information subtracts nothing from its value to others. Although it is possible to restrict access to information, for example, by selling it in magazines available only to subscribers, valuable information has a tendency to spread beyond those to whom it is sold. Therefore, private firms have less incentive to produce and sell information and will produce less of it than is socially desirable. This is particularly true of information from epidemiological monitoring of infection rates in various groups in society. In contrast to information generated by military surveillance, which is often particularly valuable when kept secret, the value of public health surveillance lies in announcing the results, so that people are aware of the disease in their midst and can take steps to protect themselves by reducing risky behavior.

The public role in the production of new information reaches beyond monitoring the epidemic to include various types of research that enable a more effective response. In all countries, governments will require country-specific information about how to identify and reach people at highest risk of becoming infected and spreading HIV to others. Research that improves the effectiveness of interventions throughout the country has important public good qualities, and therefore deserves government support. Some information relevant to prevention efforts, including bio-medical insights into the virus, is an international public good. Chapter 5 argues that the production of such information, especially research on a vaccine suitable for developing countries, deserves broad support from the international community.

AIDS and Human Rights

HIV/AIDS has created new human rights concerns and cast harsh new light on long-standing problems. Thus, the universally recognized obligation of governments to protect people from harm at the hands of others is a compelling reason for governments to play an important role in society's response to HIV.

Because people can be infected with HIV and transmit it to others for years before they become ill, the disease defines and creates a new minor-ity group in society. Government responses to the difficult task of bal-ancing the interests of the infected against the interests of others have var-ied widely. In Cuba, for example, HIV-infected people are confined to protect others from the risk of infection (Leiner 1994). At the other ex-treme, courts in the United States have upheld the right of an individual not to reveal his HIV infection, even to the extent of prohibiting govern-ment authorities from informing a woman of her husband's HIV infec-tion after his death (Burr 1997). Some prevention strategies manage to avoid tradeoffs between the rights of the infected and the rights of the uninfected, yielding benefits to both; we present evidence of the success of such approaches in chapter 3. More difficult are choices that arise in allocating public health care expenditures and in determining the extent and type of assistance to provide to surviving household members, issues we discuss in chapter 4. In all these instances, governments will unavoid-ably be involved in shaping social and legal attitudes toward the rights of the HIV-infected and the rights of those who are not infected.

The human rights rationale for a government role related to HIV/AIDS is dramatic and unambiguous in instances where sexual rela-tions are compelled by force. Here the government's overarching respon-sibility in protecting individuals from physical harm and forced exploi-tation is reinforced by the public interest in preventing the spread of

HIV. The universal obligation of governments to prevent rape and involuntary sexual servitude has been recognized in international human rights accords for decades. Although human rights advocates and protectors of traditional mores may argue whether the arranged marriage of a 14-year-old girl should be condemned or protected, all would agree to vigorously enforce prohibitions against rape and the sale of individuals, often adolescents, to houses of prostitution. Always abhorrent, rape and involuntary sexual servitude have become even more reprehensible in an age when its victims may be involuntarily exposed to risk of HIV infection. Governments that have been lax in prosecuting rape and forced prostitution must recognize that in the age of HIV/AIDS these crimes have become even more heinous than before.

Social Norms and Politics Make AIDS Challenging

DESPITE THE COMPELLING ARGUMENTS FOR GOVERNMENTS to confront AIDS, social norms and politics make designing and implementing effective AIDS policies uniquely challenging. The specific problems and their solutions will vary across countries. Nonetheless, four types of issues have commonly arisen:

- denial that HIV/AIDS may be a problem
- reluctance to help people who practice risky behavior to avoid infection
- preference for moralistic responses
- pressure to spend on treatment, at the expense of prevention.

Denial is typically seen at the earliest stages of the epidemic, when the long asymptomatic period of the disease makes its effects nearly invisible. An extreme form of denial includes unwillingness to acknowledge that extramarital sex and illicit drug use exist in the society. Some officials in societies with conservative social mores may be genuinely unaware of the extent of extramarital sex or illicit drug use; the more conservative the society, the greater the likelihood that such activities are clandestine. More typically, officials may be aware of such activities but lack information to evaluate their relevance to the AIDS threat. In such a situation, officials concerned about the possible negative reaction of their constituents may be reluctant to initiate the frank, public discussion that could provide the basis for formulation and implementation of an effective HIV prevention program.

In some instances, officials may acknowledge that HIV/AIDS poses a threat to society but nonetheless may be reluctant to advocate HIV prevention programs that focus directly on the people most likely to contract and spread the disease: sex workers, injecting drug users, male homosexuals and bisexuals with many partners, and others with high rates of sexual partner change. Although such approaches are most cost-effective—as we show in chapter 3—they may be impeded by two types of forces. On one hand, politicians and policymakers responding to the vast majority of their constituents, who do not engage in high-risk behavior, may feel little pressure to focus prevention programs on those most likely to contract and spread HIV. This is because few constituents are likely to understand the links between infection rates in people practicing high-risk behavior and their own risk of infection. On the other hand, to the extent that people engaged in high-risk activities have political influence, they or their advocates may oppose prevention efforts focused on themselves, out of concern that such programs will fuel discrimination. Given the lack of demand for prevention programs for those practicing high-risk behavior from the majority of their constituents, and resistance from those who would be the primary beneficiaries of such programs, officials may find it easier to initiate a general public information campaign, even if this is less likely to reach those most apt to contract and spread the disease.

Even if politicians and policymakers get past denial and the reluctance to target prevention interventions to those with the riskiest behavior, some interventions may well have broader social support than others. In many societies, encouraging abstinence from extramarital sexual relations or from injecting drug use would be widely viewed as morally correct, while providing free condoms for sex workers and their clients and offering clean needles to injecting drug users would be regarded by many people as facilitating immoral activity. Chapter 3 discusses why attempting to proscribe potentially risky behavior and to encourage more socially acceptable, low-risk behavior, while sometimes politically expedient, may have the unintended effect of exacerbating the spread of HIV. Societies and their governments should be aware of these costs when choosing how to confront the epidemic.

A final political obstacle to an effective government response arises only after people start getting sick and dying from AIDS. At this point, people who are infected with HIV and their families may be strongly motivated to lobby the government for subsidized treatment and care. We discuss the government response to the increased need for treatment and care in detail in chapter 4. Here it is sufficient to note that to the extent that such spending drains resources from effective HIV prevention, it will lead to more infections, sickness, and deaths.

Overview of the Book

THIS CHAPTER HAS PROVIDED BASIC INFORMATION ABOUT HIV that the remainder of the book will draw upon in analyzing how society in general, and governments in particular, can identify the public priorities in confronting the global epidemic of HIV/AIDS. We next analyze the epidemiology of HIV to identify some key principles that are essential to an effective response. This analysis concludes that acting as early as possible to prevent infections among people most likely to catch and spread the virus—people who have unprotected intercourse with many partners and people who share needles to inject drugs—would avert the largest number of secondary infections, not only among similar individuals but also in the general population (chapter 2). Are such measures possible? Which approaches are most cost-effective? How can governments improve upon their current efforts? Examining the experience of countries in confronting HIV/AIDS, we find that helping people most likely to spread HIV to protect themselves and others can indeed work and be extremely cost-effective. However, we also find that many governments have yet to implement programs with sufficient coverage of those most likely to contract and spread HIV or have failed to support these programs with broader social interventions, and so are missing valuable opportunities to prevent the spread of the epidemic (chapter 3).

What steps can government take to mitigate the impact of AIDS on infected individuals, the health sector, and surviving household members? Even when resources are very scarce, there are humane and affordable actions that governments can take to help people cope. However, these efforts should not draw resources from prevention, nor should government assistance be provided simply because of an AIDS diagnosis. Instead governments should integrate AIDS mitigation efforts with existing health care reform and poverty programs in ways that ensure that government assistance reaches those who need it most (chapter 4).

We next consider the strategic roles played by developing country governments, NGOs, bilateral donors, and multilateral institutions in financing and implementing AIDS policies within developing countries, and suggest ways in which these efforts could be improved. This discussion of partnerships concludes with an analysis of how public opinion and politics shape AIDS policy, and how developing country governments can work with other players to confront the epidemic (chapter 5). The book concludes with a summary of the main policy recommendations from the report for countries at various stages of the epidemic (chapter 6).

Appendix 1.1 Alternative Estimates of the Current and Future Magnitude of the HIV/AIDS Epidemic

EXTENSIVE SURVEILLANCE IN SOME COUNTRIES, COMBINED with spotty ad hoc surveys in others, allows an estimate of the extent of HIV infection for all countries in the world. Although more information is available about HIV infection than for any other important disease, the data in many countries are sparse or unrepresentative. Differences in judgment lead different experts to differing estimates of national infection, which aggregate to large differences in the estimated total number of HIV-infected people in the world. Since uncertainty also exists regarding the future trends in infections by country, differences among experts in the estimated current number of infections can translate to even larger differences in projections for the future.

The number of people in the world who were living with HIV infection in 1995 was variously estimated to be between 13 million (Murray and Lopez 1996) and 20 million (Global AIDS Policy Coalition, or GAPC). Two other estimates (UNAIDS 1996b and Bongaarts 1996) agree on the intermediate figure of 17 million. Figure 1.9 breaks down the global total by region for each of these four sources.

The figure makes clear that the four estimates are in rough agreement about the number of HIV-infected people in Latin America, the Middle East, North America, and Europe. For Africa and Asia, however, major discrepancies exist, greater than those that would be expected from the one-year difference in estimates. In Africa, UNAIDS and the GAPC agree on estimates that are 40 percent larger than those of Murray and

Figure 1.9 Number of Adults Living with HIV Infection, by Region: Comparison of Estimates, circa 1995
(millions)

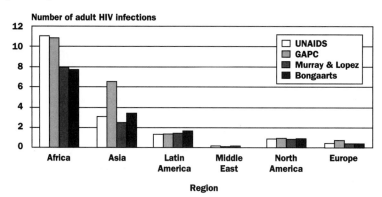

Source: The four projections are by UNAIDS (1996a) for 12/95, the Global AIDS Policy coalition (Mann and Tarantola 1996) for 1/96, Murray and Lopez (1996) for 12/94 and Bongaarts (1996) for 12/94.

Lopez and Bongaarts. In Asia, the GAPC is alone in predicting more than twice as many infections as any of the other three sources.

Much of the disparity in Asia is due to uncertainty surrounding the number of infections in India, which is almost exclusively derived from a few scattered surveys of urban high-risk populations. Extrapolating HIV infection from these small urban samples to a country of over 850 million is highly problematic. As Asia's second-largest and most populous nation, India has the potential to dominate the future course of the Asian epidemic and heavily influence infection estimates for Asia as a whole.

Figure 1.10 presents the estimates from three of the same sources for the future course of the epidemic, as reflected in the mortality rate from AIDS in five regions.[18] The disparity among HIV infection estimates extends to AIDS death rate estimates. While India was the main source of disagreement in estimates of current HIV infection, the countries of Eastern Europe and Central Asia (EECA) present an even greater challenge to those who attempt to predict the future course of the epidemic. Although Murray and Lopez and GAPC both project a vanishing AIDS death rate in EECA through the year 2020 (figure 1.10), more recent information suggests a larger and potentially explosive epidemic. HIV infection has spread with extraordinary speed among intravenous drug users in the former Yugoslav Republic of Macedonia, Poland, and

Figure 1.10 Current and Projected Future AIDS Death Rate per 1,000 People, by Region, 1990–2020: Comparison of Estimates

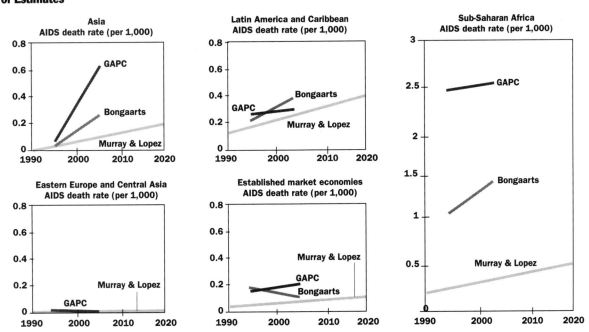

Source: The three sets of projections are by Murray and Lopez (1996), Bongaarts (1996) and the Global AIDS Policy Coalition (Mann and Tarantola 1996). Since the Global AIDS Policy Coalition (GAPC) does not explicitly project an adult death rate, the projection for 2005 used here is derived from their estimate of the number of new HIV infections in 1995 (table 1.5) by using Bongaarts' (1996) rule of thumb that AIDS deaths in any given year will approximately equal the number of people infected with HIV during the year ten years before.

Ukraine. For example, the percentage of HIV-infected injecting drug users in Nikolayev, a Ukrainian city on the Black Sea, rose from 1.7 percent in January 1995 to 56.5 percent eleven months later (AIDSCAP and others 1996). Furthermore, dramatic increases in sexually transmitted diseases in the region suggest an increasing vulnerability to HIV infection. Projections for EECA that do not take into account these recent outbreaks are likely to underestimate the severity of the HIV epidemic in those countries.

The most dramatic differences in the projected size of the AIDS epidemic are in Africa and Asia, where experts nevertheless agree that the impact will be large enough to measurably affect the growth and structure of the population.

Just as experts differ in their projections of the future number of deaths from AIDS, so they differ on the epidemic's impact on population levels and growth rates. Although negative population growth is not projected for any African country, both life expectancy and the dependency ratio will be adversely affected.[19] Differences in the projected impact of AIDS are sensitive to factors such as: the estimate of the infection rate in the base year, projection of future infection rates, the length of the asymptomatic period, the prenatal transmission rate, methodology, length of time from diagnosis of AIDS to death, the age and sex distribution of AIDS deaths, and the start year of the epidemic.

For a detailed discussion of differing projections for Africa, see Stover (*background paper*, 1996).

Notes

1. Garrett (1994) describes the first appearance of AIDS cases in several countries and the subsequent investigation leading to the 1984 discovery that HIV is the cause of AIDS.

2. Appendix 1.1 to this chapter compares the Murray and Lopez projections, relied on in the text, with the other, higher projections for the future course of the AIDS epidemic.

3. HIV can be isolated from the saliva of an infected person. Although there are a handful of cases of transmission through oral sex, there are no confirmed cases of transmission via saliva alone.

4. Murray and Lopez (1996) are unique in estimating current and future mortality disaggregated by disease. Their estimates of the current and future death rates from HIV/AIDS by region are smaller than those of Bongaarts (1996), especially in Africa, where Murray and Lopez estimate half the death rate in 2020 that Bongaarts estimates in 2005. Mann and Tarantola (1996) present much higher estimates than Bongaarts. See appendix 1.1 at the end of this chapter for a comparison of various estimates.

5. For comparison, infectious disease currently accounts for about 6 percent of the disease burden in devel-

oped countries, by either measure (Bobadilla and others 1993).

6. As of the end of 1993, an estimated 4.2 percent of worldwide TB infections were thought to be attributable to the AIDS epidemic, with the share expected to rise to 13.8 percent by the end of the century (Dolin, Raviglione, and Kochi 1993). In developing countries with severe HIV epidemics, the share is even larger. For example, 39 percent of adult TB infections in Abidjan, Côte d'Ivoire, can be attributed to HIV (De Cock 1993). In Africa, 19.5 percent of all TB deaths in 1990 were attributable to HIV, and this was projected to increase to 29 percent by 2000 (Dolin, Raviglione, and Kochi 1993). Murray and Lopez (1996) exclude all HIV-positive people from their count of TB deaths, even if the individual suffered from TB at the time of death.

7. As this book was being finalized in the spring of 1997, the World Health Organization (WHO) announced that the new "directly observed treatment strategy" for TB (the DOTS approach) has been so effective that the global number of TB cases is projected to remain flat rather than to increase. A revision of the Murray and Lopez projections, which takes into account this new development, would decrease the number of TB deaths, in-

cluding TB deaths among HIV-negative people, attributable to HIV. However, because the overall infectious disease burden would also decline, the relative importance of projected future deaths due directly to HIV/AIDS would increase. The same applies to projections for adult deaths caused by infectious diseases, discussed below.

8. A similar figure constructed from DALYs would yield roughly the same conclusions.

9. Because Murray and Lopez (1996) had not seen the very recent data on STDs and HIV in Eastern European and Central Asian countries described in chapter 2 of this book, they projected zero adult deaths from AIDS in those countries in the year 2020.

10. Individual-level data do not always show a negative relationship between individual or household income and HIV infection. Chapter 3 discusses the contradictory individual-level studies of this topic and possible ways to reconcile them with the aggregate finding reported here.

11. Each panel of figure 1.7 presents the relationship between one of the societal variables and HIV infection after purging the effects of the other seven explanatory variables. The figures were constructed by the *avplot* command in the 1997 STATA software package. See Over (*background paper,* 1997) for details and further results.

12. The index of poverty used in the upper-right panel of figure 1.7 is called the Gini coefficient and is defined on a scale from 0 to 1: 0 represents the perfectly equal distribution in which each person has exactly the same income and 1 represents the other extreme, absolute inequality such that all income is received by one person.

13. Since the dependent variable in these regressions is transformed into a "logit" as described in Over (*background paper,* 1997), changes in the independent variables are associated with specific changes in these logits, which correspond to changes in the infection rate. All the results control for the age of the epidemic, which is statistically significant. In the average country, prevalence among

urban residents at low risk is estimated to increase at the rate of 2.7 percentage points per year.

14. The European Commission has sponsored the development of a "toolkit" to assist planners in evaluating the potential links between their projects and the HIV/AIDS epidemic and incorporating those links into project design (European Commission 1997).

15. In 1993 and 1994, the United Nations, the World Bank, and the U.S. Bureau of the Census released country-specific demographic projections for Sub-Saharan Africa, which for the first time reflected the impact of the AIDS epidemic on population growth. The United Nations and the U.S. Bureau of the Census updated their estimates in 1996. Stover (*background paper,* 1997) analyzes the sources of the discrepancies among these alternative projections.

16. Estimates of the macroeconomic impact of AIDS include Over (1992); Kambou, Devarajan and Over (1992); Cuddington (1993); and Bloom and Mahal (1997). Ainsworth and Over (1994) review the literature.

17. This argument is made in detail in Kremer (*background papers,* 1996 a,b) and in Over (1997). The same arguments apply when needle sharing is the mode of transmission.

18. Murray and Lopez (1996) present projections to the year 2020 based on the background paper by Low-Beer and Berkeley (1996). Bongaarts (1996) presents projections by region for the years 1995 and 2005. Since Mann and Tarantola (1996) do not present their projections in table form, the death rates by region were computed from their tables of new AIDS cases, January 1, 1995, through December 31, 1995 (to get the death rate for 1995), and new cases of HIV infection, January 1, 1995, through December 31, 1995 (to get the death rate from AIDS in 2005).

19. Asian countries with lower birth rates than African nations are at risk of experiencing negative population growth.

CHAPTER 2

Strategic Lessons from the Epidemiology of HIV

H IV IS PRESENT IN ALMOST ALL COUNTRIES, BUT the speed with which the virus has spread has varied tremendously. In some countries HIV has infected only a tiny percentage of the population so far and its effects are all but invisible; in others, the virus has spread so widely that few families have been spared the tragedy of AIDS illness and death. Figure 2.1 shows the trend in the percentage of adults infected in various regions of the world.

Only some of the differences across and within regions can be explained by differences in the timing of the introduction of the virus. Consider the following:

- As of 1996, infection rates were still rising in all developing regions. In contrast, infection rates appear to have stabilized in North America and Western Europe at relatively low levels, even though the virus was introduced at nearly the same time as in Africa and Latin America.
- HIV has only recently been introduced in Eastern Europe and the former Soviet Union (not shown) but, as noted in chapter 1, the number of new infections is rising exponentially.
- In Thailand and parts of India, HIV infection among sex workers has climbed precipitously; however, it has so far remained low among sex workers in Indonesia and the Philippines.
- In Yunnan Province, China, and Manipur State, India, more than two-thirds of injecting drug users are infected, but in nearby Nepal the infection rate among injecting drug users has remained very low.

Figure 2.1 Estimated Trends in the Percentage of Adults Infected with HIV, by World Region

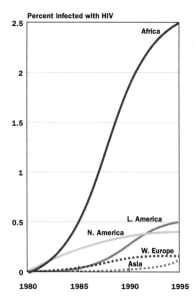

Source: Bongaarts 1996, figure 2. Used by permission

HIV is most widespread in Africa and rising in all developing regions.

■ The level of infection among pregnant women in urban areas of the Democratic Republic of the Congo (formerly Zaire) has leveled off at 4 to 5 percent, but in Botswana and Zimbabwe rates are six times as high and still climbing.

■ Infection levels are *declining* among Thai military recruits and young adults in Uganda.

What explains the different trajectories of the epidemic and what are the implications for policies to prevent HIV? In reviewing how HIV spreads in populations and the behavioral and biological factors behind the epidemic, this chapter identifies some important epidemiological principles that provide a foundation for the discussion in chapter 3 of government policies to prevent HIV. In the first part of the chapter, we review the determinants of the spread of HIV in populations. HIV does not strike individuals at random: both biology and individual behavior affect its spread. Most of the variation in trajectories of the HIV/AIDS epidemic across world regions can be explained by differences in behavior across societies and across groups within societies, which are in turn influenced by many of the economic and cultural factors described in chapter 1. In the absence of a cure or a vaccine, the key to arresting the spread of HIV is changing behavior. The epidemiology of HIV/AIDS suggests two important objectives for public programs to slow the spread of HIV, discussed in the second part of the chapter: governments should act as soon as possible and, irrespective of the stage of the epidemic, they should ensure prevention of infection among those with the riskiest behavior, who are most likely to contract and spread HIV. The chapter concludes with an overview of the level and distribution of HIV in developing countries. The extent to which HIV has saturated subpopulations with high-risk behavior and spread outward to those with low-risk behavior has important implications for government priorities in preventing HIV, discussed in chapter 3.

HIV Incidence and Prevalence, and AIDS Mortality

THE RATE OF SPREAD OF HIV AND CURRENT LEVELS OF infection are measured by *incidence* and *prevalence.*

- The *incidence* of HIV is the number of new cases, that is, the number of people who become infected during a specified period of time, usually over a twelve-month interval.
- The *prevalence* of HIV is the number of people currently infected with HIV at a given point in time. Because there is no cure for HIV/AIDS, HIV prevalence reflects the cumulative numbers of infections from the past and the mortality rate of those infected.

Incidence and prevalence of HIV and AIDS are often expressed as a rate—for example, in terms of the number of infections per 1,000 adults.

Figure 2.2 shows the relationship between HIV incidence, HIV prevalence, and AIDS mortality in a simulated epidemic for a typical Sub-Saharan African country. At the beginning of the epidemic, HIV prevalence grows rapidly and AIDS mortality is not yet evident because of the long asymptomatic period of most of those infected. Years later, when the first cases of AIDS appear, large numbers of people are already infected with HIV. Incidence may still be climbing but growth in prevalence may slow because of rising HIV/AIDS mortality or saturation of the population. As long as incidence exceeds mortality, the prevalence of HIV will continue to rise. Prevalence will peak in the year in which incidence exactly equals the rising mortality rate. Whether prevalence then levels off, declines, or resumes climbing toward a

Figure 2.2 HIV Incidence, Prevalence, and AIDS Deaths

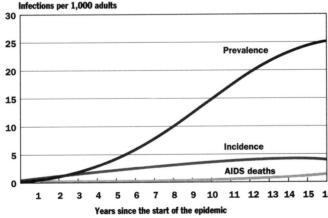

Infections per 1,000 adults

Source: Based on 1995 WHO data.

By the time people start dying of AIDS, HIV prevalence may already be very high.

new peak will depend on whether the number of new infections—incidence—is equal to, less than, or greater than the number of deaths of people with HIV/AIDS. In the absence of a cure, the key to reducing future HIV prevalence is by preventing new cases—that is, lowering incidence.

Stable or declining prevalence does not necessarily signal the end of the epidemic. Eventually, HIV prevalence will level off in all populations; in some it will stabilize at a high level, and in others at a low level. The factors affecting the height of the plateau in HIV prevalence are discussed below. However, a plateau simply indicates that there is an equilibrium in which the number of new infections exactly offsets AIDS mortality. In populations where prevalence is declining, mortality is occurring at a faster rate than new infections. The number of new infections may still be quite high, coexisting with high mortality.

The relationship between HIV incidence and prevalence and the lag in the appearance of AIDS cases have important implications for public policy:

- *Early intervention is critical to prevent an AIDS epidemic that can persist for decades.* Only a fraction of those infected with HIV are showing symptoms of AIDS at any given point in time. By the time that AIDS morbidity becomes a significant health issue, HIV may have spread widely in the population, making prevention efforts very difficult. Countries with few reported AIDS cases should not be complacent about launching prevention campaigns. While figure 2.2 alone is sufficient to make the case for early intervention, there are other compelling reasons that we shall return to later in this chapter.

- *The full impact of infection levels on mortality is delayed.* Even if all new HIV infections could be prevented, in the absence of a cure, AIDS deaths would continue for years because of the population already infected and the long asymptomatic period between HIV infection and AIDS. Countries where HIV prevalence is high are only beginning to experience the profound mortality impact of the epidemic, which will last for decades even with the best prevention efforts. These consequences are laid out in chapter 4 and reinforce the case for intervening as soon as possible to prevent HIV.

Biology and Behavior Affect the Spread of HIV

NOT ALL INFECTIOUS AGENTS INTRODUCED INTO A POPULA-
tion will be self-sustaining. If each infected person transmits
the infection, on average, to less than one other person over
his or her lifetime, then the infection will eventually disappear; if to
more than one other person, then the infection will expand. The *repro-*
ductive rate of a sexually transmitted disease is the average number of
susceptible people infected by an infected person over his or her lifetime (May and Anderson 1987, Thomas and Tucker 1996).[1] If each
person infected with a disease transmits it to exactly one other person,
then the reproductive rate is 1. In populations in which HIV has a
reproductive rate of less than 1, the epidemic will not be self-sustaining. Thus, the greater the reproductive rate of HIV, the more rapidly
the epidemic will spread.

What factors, then, determine the HIV reproductive rate in various
populations? We have seen in chapter 1 that the most common mode of
transmission of HIV is through sexual contact. Three main factors have
a large influence on the reproductive rate of all sexually transmitted diseases (STDs), including HIV:

- the amount of time a person remains infectious
- the risk of transmission per sexual contact
- the rate of acquisition of new partners.[2]

These factors are similar for transmission through contaminated injecting equipment, except that the risk of transmission per contact refers
to the risk per injection, and the number of partners refers to number of
people with whom injecting equipment is shared. The broad points in
the following discussion therefore apply to transmission through sharing
of contaminated needles, as well as through sexual contact.

Each of these three factors is in turn influenced by the biology of the
virus and by individual behavior. Biology plays an important role in the
amount of time a person remains infectious and in the risk of transmission per contact. But individual behavior also has a strong influence on
the risk of transmission per contact—for example, through decisions on
condom use, disinfecting shared needles, and seeking treatment for
other STDs. And individual behavior has a direct relation with the rate

of partner acquisition. Until medical science discovers either a cure or a vaccine, the most important avenue for reducing the spread of HIV will continue to be changing individual behavior.

The Duration of Infectiousness

The lack of a cure and the long duration of infectiousness are the main characteristics that distinguish HIV from most other STDs. The long duration of HIV infectiousness increases the likelihood that an infected individual will pass the infection to others. Further, because a person with HIV typically remains asymptomatic for years, an infected individual and his or her sexual (or injecting) partners are often unaware of the risk of transmission. Thus, the long duration of asymptomatic HIV infection potentially puts many more partners at risk than is the case for other STDs.

The impact of recently developed drugs that extend the lives of people with HIV may lengthen the infectious period. However, if these drugs significantly reduce the viral load, they might reduce the risk of infection per contact. Be that as it may, unless both dramatic medical advances and significant reductions in the costs occur, these new drugs are unlikely to have a significant impact on the duration of infectiousness in developing countries, since few developing countries have the financial or human resources to provide them. That leaves two primary mechanisms for prevention—reductions in the risk of infection per contact and reductions in the acquisition of new partners.

The Risk of Infection per Contact

The average risk of infection with HIV per sexual exposure is much smaller than that for other sexually transmitted diseases; however, because of the long period of infectiousness and numerous cofactors that enhance HIV transmission, the chance that an HIV-positive person who does not take precautions will eventually infect others can be quite high.

The most extensive studies of the risk of HIV transmission per exposure have been conducted in industrial countries. Because of generally superior health levels and the ready availability of treatment for other STDs, the average risk of HIV infection per sexual contact in industrial countries is quite small (table 2.1). For example, the average chance that an infected male will sexually transmit HIV to an uninfected female

Table 2.1 Probability of HIV-1 Infection per Exposure[a]

Mode of transmission	Infections per 100 exposures
Male-to-female, unprotected vaginal sex	0.1–0.2
Female-to-male, unprotected vaginal sex[b]	0.033–0.1
Male-to-male, unprotected anal sex	0.5–3.0
Needle stick	0.3
Mother-to-child transmission	13–48
Exposure to contaminated blood products	90–100

a. In the absence of cofactors, such as other STDs and variations in infectiousness over the incubation period.

b. Calculated as 1/3 to 1/2 times the rate for male to female.

Sources: Dabis and others 1993; DeGruttola and others 1989; Dunn and others 1992; European Study Group 1992; Haverkos and Battjes 1992; Mastro and de Vincenzi 1996; Padian, Shiboski, and Jewell 1991; Tokars and others 1993.

partner by unprotected vaginal sex is estimated at between 1 and 2 per 1,000 exposures. The risk of transmission from an infected female to an uninfected male partner through unprotected vaginal sex is one-third to one-half as great (Haverkos and Battjes 1992).[3] Thus, women are believed to have a somewhat greater probability of becoming infected from an infected male partner than the reverse. Anal sex carries the highest risk, especially for the receptive partner. The risk of transmission in unprotected anal intercourse, based on a study of men, is estimated to be between 5 and 30 per 1,000 exposures for the receptive partner. However, all of these figures very likely underestimate—perhaps severely—the average transmission probability per sexual act. They are generally based on studies of transmission within discordant couples—couples in which one partner is HIV-positive and the other is HIV-negative. Couples that are discordant for a very brief period are not captured in these samples; thus, the most infectious individuals are likely to be excluded. These studies also fail to capture couples in which neither partner tests positive for HIV, but one recently has become infected. Below we review evidence that this may be the period of highest infectiousness. If true, then studies of discordant couples are measuring HIV transmission during a less infectious period (Mastro and de Vincenzi 1996).

A transmission rate *per partnership,* not taking into account the length of partnership, might be a more realistic measure of the risks of sexual transmission of HIV within relationships.[4] A review of studies of per-partner transmission rates among heterosexuals in the United States and

Western Europe found an average transmission probability of about 23 percent from men to women, and of about half that rate (12 percent) from women to men (Mastro and de Vincenzi 1996). Yet even these rates are probably lower than those faced in developing countries because many people in developing countries are infected with other STDs that enhance HIV transmission, an issue we discuss below.

The "per contact" risk of HIV transmission with a commercial or casual partner in developing countries is thus likely to be substantially higher than the figures in table 2.1. Notwithstanding these considerations, the average infectiousness of HIV is believed to be substantially less than that for other STDs. In the case of gonorrhea, for example, the probability that an infected woman will transmit the disease to an uninfected male partner during intercourse is 20 to 30 percent per exposure, while the probability that an infected male will transmit the disease to his female partner is 50 to 70 percent (Hethcote and Yorke 1984).

Although sexual intercourse is the primary means of HIV transmission in both developed and developing countries, other modes of transmission carry a higher probability of spreading the infection. The chance that a mother will transmit the virus to her infant is variously estimated at 13 to 48 percent. Transmission probabilities through sharing contaminated injecting equipment between infected and uninfected injecting drug users are variable, depending on the specific injecting practices and which equipment is shared. The probability of transmission through an accidental needle stick in a medical setting, when the needle has been exposed to HIV-infected blood, is only about 1 in 250, or 0.3 percent. The transmission rate for transfusion of contaminated blood is nearly 100 percent.

The risk of infection per contact is not a constant; it can be influenced by a variety of factors, some of which may tend to exacerbate the epidemic. We discuss the most important of these below.

Risk may be highest soon after infection. Recent studies suggest that infectivity can vary dramatically according to the stage of HIV infection. The two peaks of infectivity are thought to coincide with the periods of highest viral load—the first and highest within the first few months of infection (before the production of antibodies to the virus) and the second, which is thought to be lower, at the very end of the asymptomatic period, as the body loses its battle with HIV (Pinkerton and Abramson 1996). Studies of homosexual men suggest that an individual faces a 10 to 30 percent chance of becoming infected during a single act of unprotected receptive anal intercourse if his partner is at the early, acute stage of the infection (Jacquez and others 1994). In the middle stage, the likelihood of infection drops to between 0.01 and 0.1 percent, but at the end stage it rises again to between 0.1 and 1.0 percent. The variation in probability of heterosexual transmission in early and late periods of infection has not been estimated but could have important implications for the size of the epidemic (box 2.1).

Greater risk of transmission immediately after infection may be one reason that the epidemic has taken off so rapidly in some developing countries. In Thailand, the average female-to-male risk of sexual transmission was estimated to be 3 to 6 infections per 100 exposures—much higher than the rates in table 2.1—perhaps because more people in Thailand were in the earliest, most infectious stage of the disease (Mastro and others 1994).[5] In addition, the likelihood of HIV transmission also differs by the type of the virus. HIV-1 is more easily transmitted and has a shorter incubation period than HIV-2 (De Cock and Brun-Vezinet 1996). HIV-1 has many subtypes with specific geographic distributions. However, there is no conclusive epidemiological evidence to date that any of these subtypes are more or less infective than others (Anderson and others 1996, Expert Group 1997).

Untreated STDs raise the risk of HIV infection per sexual exposure. STDs are far more common in developing countries than in industrial countries (table 2.2). Studies in both industrial and developing countries have found that people with current or past STDs are 2 to 9 times more likely to be infected with HIV.[6] However, because HIV and other STDs are both highly correlated with risky sexual behavior—high rates of partner change in particular—it is difficult to disentangle the extent to which conventional STDs actually enhance the transmission of HIV.

Nonetheless, there are compelling biological reasons for believing that untreated ulcerative STDs such as herpes, syphilis, and chancroid greatly increase the risk of HIV transmission per exposure: the lesions caused by these diseases provide a ready portal for transmission of HIV, whether they are on the HIV-infected or the uninfected partner. Enhanced HIV transmission in the presence of nonulcerative STDs such as gonorrhea, chlamydia, or trichomoniasis is also biologically plausible but the epidemiological evidence to support it has been weaker, mainly for methodological reasons (Laga and others 1993). For example, a recent study in Malawi found that the amount of HIV virus in the semen of

Box 2.1 The Impact of an Early Peak in Infectivity

MEDICAL RESEARCHERS ARE STILL UNSURE ABOUT the precise pattern of infectivity of HIV over its long incubation period. However, Roy Anderson (1996) has demonstrated that if HIV is the most infectious early on, as some medical researchers suspect, HIV incidence will climb more rapidly and reach a higher peak prevalence than if it is equally infectious early and late in the incubation period, or if it is more infectious late in the incubation period. In Anderson's simulation, epidemics with all three patterns of infectiousness eventually converge to the same HIV prevalence level. However, if the virus is more infective early in the incubation period, many more people are cumulatively infected. If people are most infectious immediately after becoming infected and before they develop antibodies to the virus, they would test negative for HIV precisely when they are most infectious. HIV could spread very rapidly among those with high rates of partner change during this brief, highly infectious period.

Box Figure 2.1 Shape of the Epidemic Curve under Alternative Assumptions about Infectiousness

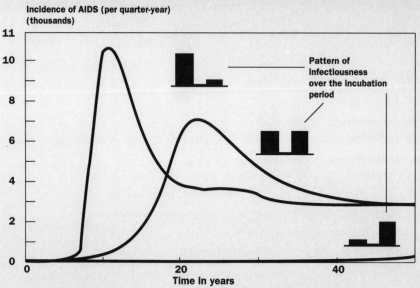

Source: Adapted from Anderson (1996), figure 4-5. Reprinted by permission of Oxford University Press.

HIV-positive men with urethritis was eight times higher than in a control group of HIV-positive men without it, and these concentrations diminished significantly when the urethritis was treated with antibiotics (Cohen and others 1997). Women are more likely than men to have STD infections without any apparent symptoms, and therefore many infections go untreated. Half of all women with gonorrhea, for example,

Table 2.2 Estimated Prevalence and Annual Incidence of Curable STDs among Adults 15–49, by Region

Region	Prevalence (millions)	Prevalence per 1,000	Incidence (millions)	Incidence per 1,000 per year
Industrial countries				
North America	8	52	14	91
Australasia	0.6	52	1	91
Western Europe	10	45	16	77
Developing countries				
Sub-Saharan Africa	53	208	65	254
South Asia	120	128	150	160
Latin America and Caribbean	24	95	36	145
Eastern Europe and Central Asia	12	75	18	112
Northern Africa and Middle East	6.5	40	10	60
East Asia and Pacific	16	19	23	28
Total	*250*	*85*	*333*	*113*

Note: Includes syphilis, gonorrhea, chlamydia, trichomoniasis. International differences in STD prevalence and incidence reflect differences in sexual behavior and in health care–seeking behavior.

Source: WHO/GPA 1995.

have no symptoms, compared with only 5 percent of men (Hethcote and Yorke 1984). Thus, if nonulcerative STDs do facilitate HIV transmission, they are likely to differentially raise the transmission probabilities to and from women, since a higher percentage of women are likely to have asymptomatic STD infections that go untreated. A recent review found that, in eleven African countries, from 5 to 17 percent of pregnant women tested positive for syphilis; in Jamaica the rate was 5 percent, and in Haiti more than 10 percent (Van Dam, Dallabetta, and Piot 1997).

Whatever the precise nature of the link between HIV and other STDs, there is evidence that treating symptomatic STDs reduces HIV transmission. In the early 1990s a randomized controlled trial in rural areas of Mwanza region, Tanzania, found that treatment of symptomatic classic STDs lowered the incidence of HIV among adults by more than 40 percent (Grosskurth and others 1995a). The extent to which this result can be generalized to other countries is likely to depend on many country-specific factors, including the underlying prevalence of HIV and STDs, the types of STDs that are prevalent, the quality of treatment services, and existing levels of STD treatment prior to the intervention. At the outset of the Mwanza study, HIV prevalence among adults 15 to 54 was already high—4 percent (Grosskurth and others 1995b). Simulations of the HIV epidemic in rural Uganda indicate that the proportion of HIV infections for which STDs were a cofactor was highest early in that epidemic (Robinson and others 1997). This suggests that the effectiveness of STD treatment in slowing HIV incidence in Mwanza might have been even greater had it occurred much earlier in the epidemic.

Male circumcision may be a factor. Some researchers have found a correlation between HIV infection and lack of circumcision among men and believe that this may account in part for the rapid spread of HIV in Sub-Saharan Africa.[7] Ethnographic studies suggest that men are least likely to be circumcised in central, eastern, and southern Africa, along a north–south swath through the Rift Valley (Bongaarts and others 1989). This also happens to be the area with the highest rates of HIV infection in urban areas. In 1989, in five countries where more than three-quarters of men were not circumcised, the urban prevalence of HIV was roughly 16 percent. In contrast, the average level of urban HIV infection was only 1 percent in 20 other countries where more than 90 percent of men were estimated to be circumcised.

One reason why uncircumcised men could be at higher risk of contracting HIV and passing it to others is that they are at higher risk of developing ulcerative STDs, particularly chancroid. Poorer genital hygiene among uncircumcised men may also play a role, particularly in low-income and unsanitary settings. A study in Kenya found that, even among men without chancroid, uncircumcised men were more likely to sero-convert (29 percent) than those who were circumcised (2.5 percent) (Plummer and others 1991). However, the amount of increased risk of HIV infection from lack of circumcision alone has not been established and whether or not such a risk exists is still debated. This is because circumcision is highly correlated with many other factors besides chancroid. In particular, ethnicity and religion are strong determinants of whether or not men are circumcised. It is therefore difficult to disentangle the effect of male circumcision (or lack of it) from that of other cultural norms that affect sexual behavior.

Clearly, even if male circumcision is protective against acquiring and spreading HIV, it is not sufficient to prevent infection. High proportions of men in West Africa are circumcised, yet HIV has nonetheless spread rapidly there. More than three-quarters of U.S.-born men are circumcised, but that has not prevented a sexually transmitted HIV epidemic in the United States (Laumann, Masi, and Zuckerman 1997). Conversely, in Western Europe and South America, circumcision is uncommon, yet the HIV epidemics in those areas have not reached the scale of the one in eastern and central Africa (de Vincenzi and Mertens 1994).

Behavior affects the probability of transmission. Although the basic transmission probabilities for HIV per exposure are founded in the biology of the virus, fortunately they can be substantially reduced through behavioral change. Using latex condoms and obtaining treatment for conventional STDs can lower the probability of transmission through sexual contact. Sterilization of injection equipment can dramatically reduce transmission among injecting drug users and among patients in medical facilities. And mother-to-child transmission can be reduced through both medical treatment and behavioral changes. The prospects for changing behavior to reduce HIV transmission are discussed in chapter 3.

The Rate of Partner Change

While transmission probabilities have an important influence on the reproductive rate of HIV, the rate of sexual partner change probably accounts for the greatest differences in the rate across groups and countries. Similarly, the rate at which injecting drug users change partners with whom they share unsterilized injecting equipment strongly influences the HIV reproductive rate among them. Finally, in medical settings, the rate of reuse of unsterilized injecting equipment for multiple patients is analytically equivalent to the rate of partner change (box 2.2). In all three situations, the higher the rate of partner change is, the greater is the likelihood that the virus will pass from infected to uninfected people.

In contrast, although the probability of becoming infected per exposure is higher for people receiving transfusions of infected blood and for the children of HIV-positive mothers, these groups are unlikely to infect many others. The rate of partner change among transfusion recipients, for example, is quite low, on average. Because the HIV reproductive rate for these modes of transmission is probably less than 1, if the virus were spread only by transfusion or from mother to child, the epidemic would most likely not be sustained. In the absence of condom use or sterilization of shared injecting equipment, rapid rates of partner change sustain the epidemic.

Both the *average* rate of partner change in a population and the *variation* of the rate across individuals have an impact on the spread of HIV

Box 2.2 HIV Transmission through Medical Injections

IN MANY PARTS OF THE WORLD, MEDICAL PRAC-titioners or patients prefer injections over oral med-ication—often because of patients' or medical practi-tioners' perceptions that injections are more effective than oral medication. Studies in countries as diverse as India, Kenya, Nigeria, Uganda, and Vietnam have confirmed the popularity of injections among pa-tients (Birungi and Whyte 1993, Reeler 1990). They are often favored by both public and private practi-tioners, including traditional healers, pharmacists, and professional injectors. As a result, behavioral sur-veys conducted in 1989–90 found that from one-third to more than half of adults in eight of nine countries received medical injections in the past year (Ferry 1995). Among those receiving injections, the average was between three and six per year.

In the resource-scarce settings of developing countries, injecting equipment—including suppos-edly disposable equipment—may be reused on many patients without proper sterilization. A study of three dispensaries in Burkina Faso, for example, found that for every 1,000 injections, the dispen-saries used from 14 to 250 syringes and from 70 to 700 needles (Wyatt 1993). If such equipment is not properly sterilized between uses, HIV and other blood-borne diseases can be transmitted between patients, in the same way that they are transmitted among injecting drug users. Transmission of hepati-tis, syphilis, malaria, and polio through unsterile injection equipment has been documented in devel-oping countries, and unsterile injection equipment is a major cause of abscess (Wyatt 1993). Patients in countries hard-hit by AIDS are often well aware of the risks of HIV transmission by medical injection; surveys in two communities in rural Uganda found that 63–83 percent of households kept their own needles and syringes at home to avoid having to share with others in medical situations (Birungi, Asiimwe, and Whyte 1994).

The number or share of HIV infections transmit-ted through unsterile medical injections is not known and, given the frequency of injections in many countries, might be difficult to document. The age distribution of HIV cases and the correla-tion with other known risk factors strongly suggest that sexual transmission is nevertheless the main mode of transmission in most countries. However, in countries where HIV infection is widespread and where sterilization practices are poor, the risk of HIV transmission by medical injection is real. This risk can be minimized by improving sterilization practices and by minimizing the use of injections. Because of the public and professional popularity of injections in many countries, minimizing their unnecessary use may require substantial public education.

in populations. Other factors being constant, the higher the average rate of partner change is, the higher is the reproductive rate of HIV. How-ever, in a population in which a few people have very high rates of part-ner change and many people have very low rates, HIV and other STDs will spread more quickly than if the same average number of partners were distributed more equally across the entire population (Anderson and May 1988, Over and Piot 1993).

Surveys of sexual behavior suggest that there is in fact quite significant variation in the rate of partner change across subgroups in a single population.[8] To take one example, figure 2.3 shows the distribution of men and women 15 to 49 in Rio de Janeiro, Brazil, who had at least one nonregular sexual partner in the previous twelve months, according to the total number of nonregular partners. There are two "peaks" in the distribution of people by their number of nonregular partners—one large peak among those with no nonregular partners or only a few, and another small peak among those with very many partners. Roughly half of men (56 percent) and 90 percent of women reported *no* nonregular partner, meaning that they either did not have any sexual partner or that they had sex only with their spouse or with some other regular partner. Those who did have nonregular partners usually acknowledged only a few. For example, roughly 12 percent of men and 6 percent of women reported only one nonregular partner in the past twelve months. On the other hand, a small percentage of men—nearly 2 percent—reported having 20

Surveys of sexual behavior show that most people have no casual partners or only a few, but a few people have very many partners.

Figure 2.3 Distribution of Men and Women, Ages 15 to 49, with at Least One Casual Sexual Partner, by Number of Nonregular Partners in the Past Year, Rio de Janeiro, Brazil, 1990

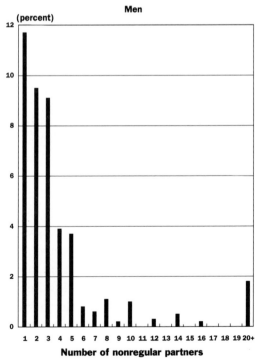

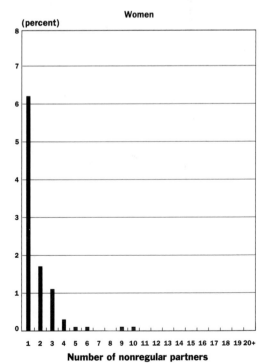

Note: 56 percent had no nonregular partners.

Note: 90 percent had no nonregular partners.

Source: Background paper, Deheneffe, Caraël, and Noumbissi 1996.

or more partners in the same period. The "two-peaked" distribution of people according to their nonregular sexual partners is typical of those found from sexual behavior surveys in other countries (*background paper*, Deheneffe, Caraël, and Noumbissi 1996) .

The variation, or *heterogeneity*, in sexual behavior is even more striking in Thailand, where, in 1990, 28 percent of men 15 to 49 had a nonregular partner in the past year and almost 4 percent had 20 or more partners, while only 2 percent of women reported *any* nonregular partner.[9] Sampling methods for these surveys are typically less successful in capturing sex workers, who are usually a small percentage of the population but an important component in the second peak of the bimodal distribution. High rates of partner change in a very small subgroup may be sufficient to sustain an STD or HIV epidemic that can gradually spread to the rest of the population.

Mixing patterns. The path of the epidemic within the overall population depends on the degree and pattern of mixing among people with high-risk behavior, and the mixing between people with high-risk behavior and people with low-risk behavior. By "high-risk behavior," we mean unprotected sexual intercourse with multiple partners or sharing of unsterilized injecting equipment. People with high-risk behavior are very likely to become infected and to unknowingly pass HIV to others. People with low-risk behavior—who have few partners, who consistently use condoms, who do not inject drugs, or (if they do) do not share injecting equipment—are less likely to pass HIV to others. However, they are nevertheless at risk of becoming infected through transfusion of contaminated blood or sexual mixing with people who practice high-risk behavior. And young children are at risk of becoming infected from their mothers perinatally but are highly unlikely to spread HIV.

In a sexually-transmitted HIV epidemic, the speed at which HIV spreads from people with a large number of partners to those with very few partners depends on the extent of mixing between people with different levels of sexual activity. If people with large numbers of partners have intercourse only with others who are similarly active (known as *assortative* sexual mixing), then HIV will tend to rise rapidly within those groups but only very slowly and to a limited extent in the rest of the population. As a result, the epidemic will achieve lower peak levels of infection in the entire population than if those with large numbers of partners also have sex with those who have fewer partners (known as random or *disassortative* mixing) (Anderson 1996; Anderson, Gupta, and Ng 1990).

Mixing patterns explain why HIV does not spread through a population at a uniform rate. Rather, it spreads in a series of smaller epidemics that race through overlapping subpopulations whose behavior puts them at various degrees of risk, then outward to those with less-risky behavior with whom they mix.

Sex workers whose clients do not use condoms, injecting drug users who share unsterilized injecting equipment, and others with very high rates of partner change are typically the first to be infected in an HIV epidemic. HIV prevalence in these groups can rise very rapidly. Figure 2.4 shows rapid increases in HIV prevalence among sex workers in several cities in developing countries. Some of the differences in the rate of increase across cities can be attributed to differences in the timing of the introduction of the virus. However, other factors also appear to be at work. HIV prevalence among sex workers in Santo Domingo, Dominican Republic, has risen more slowly; this is believed to be due to very high rates of condom use among brothel-based sex workers (Peggy McEvoy, personal communication).

HIV tends to move even more rapidly among injecting drug users who share injecting equipment than among sex workers because the risk of transmission per contact is much greater. In countries where those who inject drugs commonly share injecting equipment, HIV can infect

Figure 2.4 Increasing Prevalence of HIV among Sex Workers, Seven Cities in Developing Countries, 1985–95

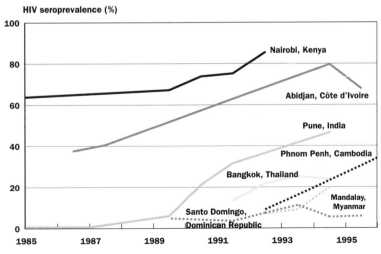

HIV prevalence has reached high levels and continues to climb among female sex workers in many developing countries

Source: U.S. Bureau of the Census (database), 1997.

the majority of users in a matter of months, as has been the case in parts of Asia and in Ukraine (figure 2.5).

Figure 2.6 shows the incidence of AIDS in various population groups in six areas of Latin America. In Brazil, the first wave of the epidemic was among men who have sex with men.[10] This was followed a few years later by an epidemic among injecting drug users, most of whom are also men. Later, the disease spread to sex workers and the female partners of bisexual men and injecting drug users.

The timing and pattern of the epidemic waves can be very different, even within a single region. In the Andean Area, Mexico, and the Southern Cone, the epidemic first struck men who have sex with men, as it did in Brazil. In the Caribbean and the Central American Isthmus, heterosexual transmission quickly outpaced transmission in other groups. In Thailand, HIV spread first among men who have sex with men and injecting drug users, and then spread among sex workers and their clients. Researchers have since determined that the epidemics among injecting drug users and sex workers in Thailand were largely independent, spreading two distinct variants of the virus (Ou and others 1993).

Figure 2.5 Rapid Diffusion of HIV among Injecting Drug Users, Asia and Ukraine, Various Years

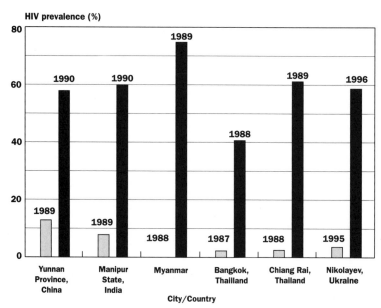

Once HIV is introduced among injecting drug users who share needles, HIV prevalence can go from practically zero to near saturation levels in just a few months.

Source: Stimson 1996; for Ukraine, UNAIDS 1996d.

Figure 2.6 Annual Incidence of AIDS Cases in Latin America and the Caribbean according to Risk Factor, 1982–95

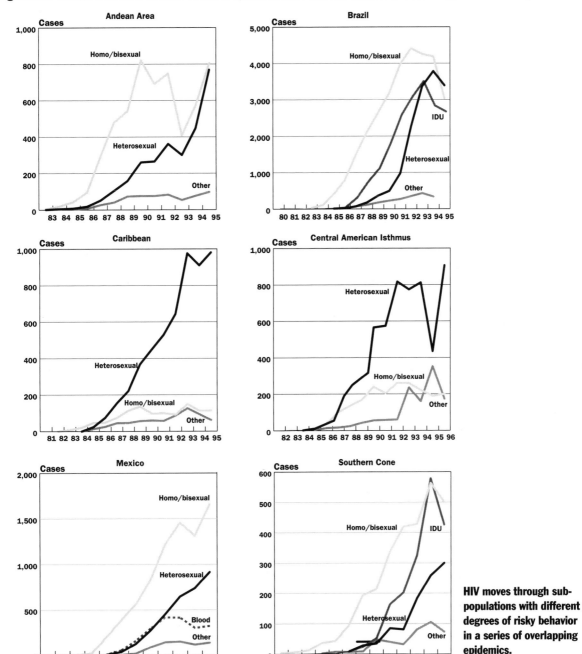

HIV moves through sub-
populations with different
degrees of risky behavior
in a series of overlapping
epidemics.

IDU Injecting drug users.
Andean Area: Bolivia, Colombia, Ecuador, Peru, Venezuela.
Southern Cone: Argentina, Chile, Paraguay, Uruguay.

Source: PAHO data, 1996.

Subpopulations that practice high-risk behavior are constantly in flux (Weniger and Berkley 1996). Over time, some individuals abandon high-risk behaviors or die, while others initiate high-risk behavior, adding to the high-risk subpopulation. High-risk behavior usually varies over an individual's life cycle. Sexual activity is often highest among young, unmarried adults, particularly men. As men and women marry and age, levels of casual sex typically decline. The age profile of the percentage of men and women engaging in sex with a nonregular partner in figure 2.7 demonstrates this well, although it also reflects to some extent temporal changes in social norms. As we shall see in chapter 3, changing socioeconomic factors can also induce people to adopt or abandon risky behavior. The dynamic nature of the subpopulation practicing high-risk behavior at any point in time prevents their HIV-infection rate from reaching 100 percent.

If there is very little mixing between people with different degrees of risky behavior, the overall epidemic may have multiple peaks. Incidence may rise and fall several times as first one group and then another becomes nearly saturated by the virus. Thus, a sustained decline in incidence in a specific group of people with risky behavior does not necessarily signal the end of the epidemic in the entire population (Anderson 1996; Anderson, Gupta, and Ng 1990).

Concurrent partnerships. Partnerships that overlap in time are concurrent partnerships. Examples of concurrent partnerships include: partnerships between married men or women and commercial or casual sexual partners; people engaging in long-term relationships with more than one casual partner; and polygyny, the practice of having more than one wife. In two populations in which individuals have the same average number of partners in a given period, HIV and other STDs will spread more rapidly in the population in which partnerships are concurrent than in the population in which partnerships occur sequentially (*background paper*, Morris 1996). This is because when partnerships are sequential, the virus cannot spread to a new susceptible person until the dissolution of one relationship and the start of another. In a concurrent partnership, it can infect more susceptible people in a shorter period of time.

Recent research has focused on the role of "bridge populations" in the spread of HIV. These are partnerships that link people in groups that otherwise might not mix, such as partnerships between people with high-risk behavior and those with very low-risk behavior (*background paper*, Morris 1996; Morris and others 1996). For example, men who

Figure 2.7 Probability of a Casual Sex Partner over a 12-Month Period, by Age and Gender

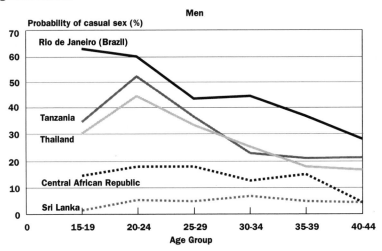

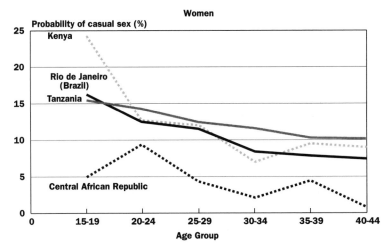

> Levels of sexual activity vary over an individual's life cycle. Sexual activity is often highest among young, unmarried adults.

Note: These results control for schooling, urban residence (where applicable), and occupation.
Source: Background paper, Deheneffe, Caraël, and Noumbissi 1996.

have unprotected intercourse with sex workers and have either a wife or steady girlfriend may transmit HIV to monogamous women who would not otherwise be at risk. The extent of such mixing in Thailand was recently captured by a survey of sexual behavior among low-income men and long-haul truckers in three provinces (table 2.3). The bridge population comprises a very large percentage of both groups—about 17 percent of low-income men and 25 percent of the truckers. Sexually active

Table 2.3 Distribution of Two Samples of Thai Men, by Type of Sexual Network, 1992

Network type	Low-income men		Truckers	
	%	N	%	N
No bridge	*83.2*	*817*	*74.9*	*245*
No partners	16.0	157	3.1	10
Wife only	45.3	445	59.6	195
Other only	6.4	63	0.6	2
Wife + other	2.5	25	3.4	11
Sex worker only	12.9	127	8.3	27
Bridge	*16.8*	*165*	*25.1*	*82*
Wife + sex worker	7.9	78	15.3	50
Other + sex worker	6.7	66	4.3	14
Wife + other + sex worker	2.1	21	5.5	18
Total	*100.0*	*982*	*100.0*	*327*

N Number of men.
Note: Reference period is in the past 6 months.
Source: Background paper, Morris 1996.

injecting drug users are another potential bridge population. In Manipur state, India, within two years of the first reported HIV case among drug injectors, 6 percent of the noninjecting sexual partners of injecting drug users were infected (Sarkar and others 1993).

The impact of heterogeneity, mixing, and concurrency: A simulation. The combined effect of heterogeneity in sexual behavior, mixing between groups of different behaviors, and concurrent partnerships can have a profound impact on the potential course of the HIV/AIDS epidemic, absent any intervention. Figure 2.8 shows simulations of heterosexual HIV epidemics in four imaginary populations with different underlying patterns of sexual behavior (*background paper,* Van Vliet and others 1997). The four simulated populations are identical in every respect, except in their patterns of sexual behavior, as follows:

- *Commercial and casual sex:* In this population, some men engage in commercial and casual sexual relationships before and after marriage; some women have relations with casual partners before marriage but all women are monogamous after marriage. The commercial and casual partnerships can be concurrent with each other and with marriage. Sex workers comprise about a one-fifth of 1 percent of all women.

Figure 2.8 The Impact of Different Baseline Patterns of Sexual Behavior on a Heterosexual HIV/AIDS Epidemic: STDSIM Results

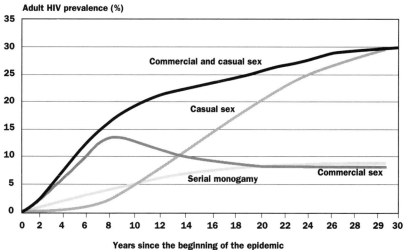

Simulations show that HIV spreads most rapidly in populations with concurrent commercial, casual, and marital sex, and least rapidly if there are no concurrent partnerships.

Source: Background paper, Van Vliet and others 1997.

- *Commercial sex:* Some men engage in commercial sex before and after marriage with a small group of female sex workers comprising only one-quarter of 1 percent of the female population. Neither men nor women have casual sexual relationships before or after marriage. Commercial and marital relationships can be concurrent.
- *Casual sex:* Some men and women have casual sexual relationships before and after marriage. There are no sex workers in this population, but casual partnerships can be concurrent with each other and with marriage.
- *Serial monogamy:* Some men have commercial and casual sexual relationships before marriage and some women have casual relationships before marriage. After marriage, neither men nor women have extramarital relationships. All relationships before marriage are serial; this is the only one of the four populations with no concurrent relationships. Sex workers are 0.05 percent of the female population.

Using the STDSIM simulation model described in box 2.3, it is possible to see how the course of the epidemic would differ in these four

Box 2.3 STDSIM: Modeling Behavior and Sexually Transmitted Diseases

THE SIMULATIONS IN THIS CHAPTER AND IN chapter 3 were conducted by researchers at the Erasmus University, Rotterdam, the Netherlands, using STDSIM, a computer model they developed to estimate the transmission of and impact of preventive interventions on HIV and four classic STDs—gonorrhea, chlamydia, syphilis, and chancroid (*background paper*, Van Vliet and others 1997). The STDSIM model can describe the spread of HIV and other STDs in populations with different demographic characteristics, sexual behavior, health care settings, and feasible interventions.

Each STDSIM simulation shows the cumulative outcome of interactions between a large number of hypothetical individuals. Each individual has a life history and specific characteristics, some of which remain constant while others change. In this model, individuals start and end sexual relationships, contract STDs, and, except for HIV, are cured. Rela-

tionships between men and women are explicitly modeled, with parameters for the frequency of intercourse and the duration of each relationship. When a simulated person becomes infected, the program considers that his or her partner may also become infected.

For example, the life history of a man and a woman who enter into a steady relationship might evolve as follows. Before the marriage, both have had other sexual relationships. After the marriage, the man continues casual relationships with other women and becomes infected with an STD, which he transmits to his wife. The man seeks treatment for his STD but quite soon he is reinfected by his wife, who is unaware that she is infected. This process might repeat itself until both partners seek treatment for their STDs. The STDSIM model is described in detail in Van der Ploeg and others (1997).

imaginary populations with no specific intervention or behavior change.[11] Baseline levels of condom use are assumed to be 5 percent among casual partners and 20 percent among sex workers. HIV is introduced in the population in year zero.

The results in figure 2.8 show the trends in HIV prevalence for the entire population, including members with high- and low-risk behavior. The first curve, at the top of the figure, shows the path of the epidemic in the population in which people have concurrent relationships and commercial and casual sex. Thirty years into the epidemic, HIV begins to show signs of leveling off, but at a very high level—30 percent. In the population in which there is only commercial sex and marital sex, HIV prevalence peaks eight years after the virus is introduced at about 13 percent, then declines to an equilibrium prevalence of about 8 percent 20 years into the epidemic. Prevalence declines even in the absence of any behavior change because the people with the riskiest behavior, infected at the outset, begin to die. Most new cases of HIV infection at that point occur among people who have recently adopted high-risk behavior. In the population in which there is casual but no commercial sex, the epidemic progresses more slowly, reaching a prevalence rate of only about 3 percent eight years after the start of the epidemic. However, prevalence continues to climb, reaches 30 percent, and is still increasing 30 years after HIV is introduced. Finally, in the population that practices serial monogamy with commercial, casual, and marital partners but no concurrent partnerships ("serial monogamy," the lowest curve), HIV prevalence rises more quickly than in the population with only casual sex, but levels off at about 9 percent. Thus, even without behavior change to prevent HIV, the course of an HIV epidemic can be quite different across populations, depending on the heterogeneity of behavior, the extent of mixing, and the degree of concurrency in partnerships.

Implications for Public Policy

THE PRECEDING DESCRIPTION OF HOW HIV/AIDS SPREADS through populations, particularly the role that variations in individual behavior play in determining the course of the epidemic, have important implications for government policies to prevent HIV.

Act as Soon as Possible

Governments that invest in effective prevention before HIV/AIDS becomes evident can avert suffering, save lives, and avoid potentially massive expenditures on AIDS treatment and care. To demonstrate the importance of changing behavior as early as possible in the epidemic, figure 2.9 shows the impact of reducing HIV transmission by a moderate increase in condom use in the hypothetical population in which HIV is spread by both commercial and casual sex.[12] The top line shows the course of the epidemic without any intervention or spontaneous behavior change; the lowest line shows the impact of raising condom use from 5 percent to 20 percent among men with commercial and casual sexual partners, three years after the start of the epidemic; the middle line shows the impact of the same increase in condom use 15 years after the start of the epidemic. An intervention that succeeded in raising condom use relatively early in the epidemic would reduce peak HIV prevalence from 30 percent to 20 percent. Increased condom use that occurs later prevents HIV prevalence from rising, leaving it constant at 22 percent. By the end

Figure 2.9 The Impact of Increased Condom Use by Men on Adult HIV Prevalence, Early and Late in the Epidemic

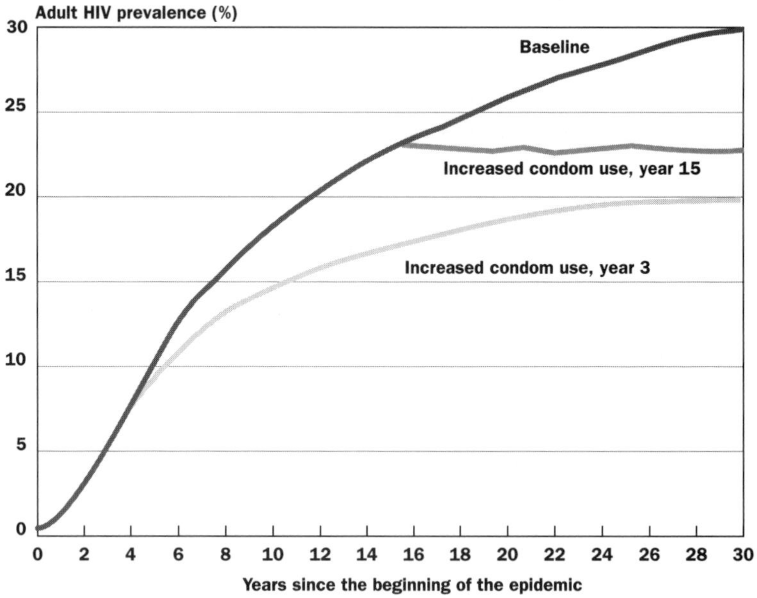

Early interventions to change behavior will reduce peak prevalence and save many lives.

Note: Baseline is the STDSIM model in which HIV is spread by commercial and casual sex and 5 percent of men use condoms. In the simulation, condom use rises to 20 percent.
Source: Background paper, Van Vliet and others 1997.

of the 30-year simulation, the earlier intervention has prevented more than twice as many infections and three times as many deaths as the later intervention (*background paper*, Van Vliet and others 1997). Note, however, that the level of behavior change modeled in year 3 of the epidemic, when prevalence is already high (5 percent), is not early enough to prevent substantial infection of the population.

There are several reasons why early intervention to change high-risk behavior is preferable to later action. Early in an epidemic, HIV spreads exponentially. Because very few people are infected, the probability is greater that unprotected sex or needle sharing involving an infected person and a random partner will result in a new infection. Further, if the viral load is highest in the first months of infection, as scientists suspect, early in the epidemic a very large proportion of those infected will be highly infectious. Later in the epidemic, unprotected sex between an infected person and a random partner is relatively less likely to result in transmission—both because of the lower average infectiousness and because there is a greater likelihood that a random partner is already infected.[13] Another reason for acting early is that interventions cannot be put in place instantly; a period of trial and error may be necessary to discover which approaches work best in particular settings. Finally, as will be shown in chapter 4, from a budgetary perspective it is far less costly to prevent HIV infection than to treat people with AIDS.

As we shall see later in this chapter, governments representing nearly half of the population of developing countries—2.3 billion people—are in areas with nascent HIV/AIDS epidemics and are still in a position to act early and decisively to prevent a widespread epidemic. Among those countries are China, certain states of India, Indonesia, the Philippines, Eastern Europe, North Africa, and the Middle East. Yet even in countries where HIV is already prevalent among those with risky behavior, there are still opportunities for prompt intervention to prevent the spread of HIV into the next wave of people susceptible to infection.

Despite the compelling case for acting early, governments have often been slow to respond. This is understandable in those countries where HIV struck first, since very little was known in the 1970s and early 1980s about how HIV was transmitted and how its spread could have been prevented. In particular, the existence and implications of the long asymptomatic period of HIV infection were not well understood. Countries where the epidemic hit later have had the opportunity to benefit from increased knowledge about the disease, and governments are becoming quicker to launch national AIDS prevention programs, no doubt due in part to the efforts of international programs like the Global Programme on AIDS and UNAIDS (box 2.4). Yet even today, some policymakers dismiss the significance of a relatively small number of reported AIDS cases, not recognizing that such a situation is precisely when prevention is needed. As recently as 1994, a senior health official in one populous developing country chided reporters that "to make a few thousand reported AIDS cases into an epidemic is just absurd." Tragically, such attitudes persist, despite the now widely understood fact that reported AIDS cases are but the tiny, visible portion of a deadly, mostly invisible, and potentially explosive HIV epidemic. Officials who ignore the signs of a pending epidemic miss an opportunity to mount an early, less costly, and highly effective response.

There are many reasons why policymakers may be slow to confront HIV/AIDS. In virtually every country, the HIV/AIDS epidemic has been preceded by a period of denial. AIDS is perceived to be a problem

Box 2.4 Are Governments Intervening Early Enough?

THE FIRST COUNTRIES HIT BY THE HIV/AIDS EPI-demic in the late 1970s and early 1980s were caught by surprise: the viral source of AIDS, the long asymptomatic period of HIV infection, and the ways that HIV is spread were not well understood. Governments in these countries were not able to launch prevention programs until years after the introduction of HIV. In many African countries HIV had already spread to the general population, and the numbers of AIDS cases were already rising at alarming rates. Countries where HIV was introduced later have had an opportunity to learn from this tragic experience and to act early to prevent a widespread epidemic. Have they done so?

The answer appears to be yes. Box figure 2.4 shows the amount of time between the first reported AIDS case and the start of a government national AIDS program for 103 countries for which such data are available. In nearly a third of the countries, the first case of AIDS was reported before 1985. For this group, which includes most of the industrial countries, national AIDS control programs were launched an average of almost five years after the first reported case. More than a third (37 percent) of the countries reported their first AIDS case in 1985–86; these countries launched AIDS prevention programs an average of 18 months later. Among countries that had their first reported AIDS case in 1990–94, national AIDS programs were launched an average of one year before the first reported AIDS case. Vietnam, for example, launched its program in 1990, three years before their first reported AIDS case.

In fact, the improved response time of country AIDS prevention programs was largely associated with the mobilization of the international community in the mid-1980s to launch international anti-AIDS programs. This movement was spearheaded by WHO and its Global Programme on AIDS,

Box Figure 2.4 Lag between the First Reported AIDS Case and Initiation of a National AIDS Control Program, 103 Industrial and Developing Countries

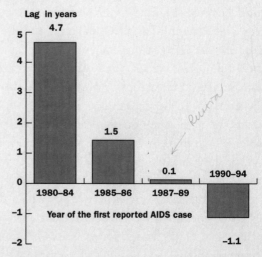

Source: Authors' calculations, based on data from UNAIDS Country Support 1996 and Mann and Tarantola 1996, table D-5.

which helped countries to develop national AIDS programs and action plans. Fully 60 percent of the countries in box figure 2.4 launched their national AIDS program in 1986–88.

Although the launching of a national AIDS program is a welcome indication that policymakers recognize the importance of prevention, organizing and implementing an effective program takes time. Further, an early response does not necessarily foreshadow the quality of the response, the coverage of programs, or the degree of political commitment, which in large part determine a program's effectiveness. Chapter 3 will examine the available cross-country data on the quality of the national response.

of foreigners, an imported problem confined to tourists or to people who have lived abroad. Injecting drug use and unprotected sex with multiple partners are said to be problems only in other countries, not in one's own country. The lack of reliable data on HIV prevalence and risk behavior makes the potential for an epidemic easy to ignore. Even when officials and politicians understand that the conditions for an HIV epidemic exist in their country, they may be reluctant to raise the issue publicly or to initiate programs to prevent HIV among people engaged in high-risk behavior, who are often stigmatized. Or, they may believe that the limited resources available would be better spent fighting more common health problems, such as tuberculosis and malaria. Finally, officials and business leaders have sometimes attempted to conceal information about the extent of the epidemic because they fear it will discourage tourism or investment.

Although some of these obstacles cannot be dismissed lightly, failure to overcome them and act early to prevent the spread of HIV can be extremely costly. In rich and poor countries alike, denial and other social and political barriers have delayed interventions that might have significantly reduced the epidemic, saving money and, most important, lives.

The Multiplier Effect of Changing the Highest-Risk Behavior

We have seen that an HIV epidemic can only persist if the reproductive rate of HIV is greater than 1 in at least some subgroup of the population, and that the shape of the epidemic is largely influenced by the degree of mixing between people with different degrees of risky behavior. Other things being equal, those with the highest rates of partner change are most likely to contract and spread HIV (box 2.5). It follows that preventing HIV infection in someone with a high rate of partner change—sexual or injecting—will indirectly avert many more future infections than preventing infection in a person who practices low-risk behavior and is thus less likely to infect others. Therefore, for HIV, as for other sexually transmitted diseases, the most efficient strategy for reducing the spread of the disease is to prevent transmission among those with the highest rates of partner change (Hethcote and Yorke 1984, Over and Piot 1993). Prevention of infection among those with the highest rates of partner change has a "multiplier" effect in terms of preventing many more subsequent, secondary infections, most of them among individuals with low-risk behavior (box 2.6).

Box 2.5 Who is Most Likely To Contract and Spread HIV?

NOT EVERYONE IN THE POPULATION IS EQUALLY likely to become infected with HIV and to transmit it to others. People with large numbers of sexual and needle-sharing partners who do not take precautions to prevent HIV infection by using condoms or clean injecting equipment are very likely to become infected. These unsafe practices then will create many opportunities for HIV to spread through the population. On the other hand, people who have no sexual or injecting partners, or very few partners, or who always take precautions to avoid contracting and spreading HIV, are very unlikely to become infected or to infect others.

The distribution of individuals according to the likelihood that they will contract and spread HIV varies across populations and, within any given pop-

ulation, over time. In general, over the course of a year, most people have no casual sexual partners or at most one or two; their reproductive rate of HIV will tend to be quite low. Some people, however, have several partners and a somewhat higher reproductive rate of HIV, and a much smaller number of people have very many partners and a very high reproductive rate of HIV, much greater than one. *The upper portion of this distribution—that is, people who have the greatest numbers of partners and do not use condoms or clean injecting equipment—are those most likely to contract and spread HIV.* Although these individuals are a very small percentage of the total population, enabling them to adopt safer behavior, thereby protecting themselves and others, is essential to curbing the epidemic.

How much more effective is preventing transmission among people with the highest rates of partner change? The top panel of figure 2.10 compares the impact of slowing transmission by raising condom use in three different subpopulations in a heterosexual epidemic: women in steady relationships, men engaging in commercial and casual sex, and sex workers. The top line shows the baseline increase in prevalence in the absence of any change in condom use and is identical to the baseline epidemic in figure 2.9. The second line shows the impact of instantaneously raising condom use from zero to 20 percent among women in steady relationships. Because these women do not have a high rate of partner change, higher condom use has almost no effect until HIV prevalence reaches about 15 percent, and only a small effect thereafter. The third line shows the impact of instantaneously increasing the rate of condom use from 5 to 20 percent among the 40 percent of men who have a commercial or casual partner annually.[14] This intervention, like the previous one, has very little effect for several years, until prevalence reaches about 13 percent. The fourth and lowest line shows the impact of increasing from 20 to 90 percent the rate of condom use among sex workers, a mere 0.20 percent of the adult female population.[15] Raising condom use

Box 2.6 The Multiplier Effect of Reducing HIV Transmission among Sex Workers in Nairobi

A SIMPLE CALCULATION ILLUSTRATES THE POWER of reducing HIV transmission among those with the highest rates of partner change. A highly successful HIV prevention program in Nairobi, Kenya, provided free condoms and STD treatment to 500 sex workers, 80 percent of whom were already infected with HIV (Moses and others 1991). The women had an average of four partners per day. Following the interventions, condom use rose from 10 to 80 percent, averting an estimated 10,200 new HIV infections per year. One-third of the prevented cases were among clients of the sex workers and two-thirds among the clients' other partners, including their wives.

Suppose, instead, that the program had raised condom use to 80 percent among 500 men chosen at random from the low-income community in which the sex workers worked. How many infections would have been averted? At that time, about 10 percent of the male population was infected with HIV, and the average man had four partners *per year*. Using the same assumptions about transmission rates, condom effectiveness, and secondary infections, raising condom use to 80 percent among the men would have prevented only 88 new HIV infections per year among the men's partners (S. Moses, personal communication). These calculations demonstrate how prevention programs can be more effective if they raise condom use among people with the highest rate of partner change. Raising condom use in commercial sex transactions in a sustainable way involves changing behavior among both sex workers and their clients.

Box Figure 2.6 Infections Averted per Year by Raising Condom Use to 80 Percent in Two Populations in Nairobi

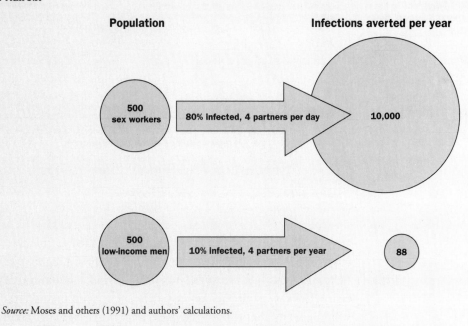

Source: Moses and others (1991) and authors' calculations.

Figure 2.10 The Impact of Increased Condom Use by Various Subpopulations on Adult HIV Prevalence in the Entire Population

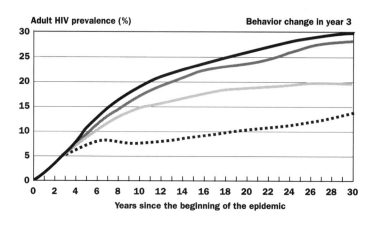

Condom interventions among those with the highest rates of partner change are highly effective in stopping the increase in HIV early in this simulated heterosexual epidemic in which HIV is spread by commercial and casual sex.

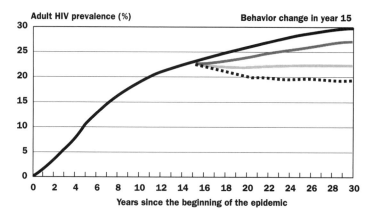

Increased condom use among those with the riskiest behavior is also highly effective in a later-stage epidemic, but behavior change among other groups of the population will be necessary to lower HIV prevalence more rapidly.

Note: Baseline is the model in which HIV is spread by commercial and casual sex, with no behavior change.

Source: Background paper, Van Vliet and others 1997.

among the small number of sex workers and their clients has a much larger impact on prevalence, both because they have a higher rate of partner change and because we have assumed higher condom use. If 90 percent of sex workers maintained consistent condom use, prevalence would rise to about 14 percent after 30 years, while if condom use among men with commercial and casual partners can be maintained at 20 percent—a change in behavior involving many times as many people in a given year—prevalence would stabilize at about 20 percent.[16]

These benefits of preventing transmission among those with the largest number of partners during the early stages of the epidemic have been well documented (Garnett and Anderson 1995; Stover and Way 1995; *background paper*, Van Vliet and others 1997). However, as HIV spreads, an increasing number of new cases occur among people who are not themselves engaging in risky behavior, but who nevertheless become infected by their partners. In fact, in advanced epidemics the majority of those infected may practice low-risk behavior. When the overwhelming number of new cases is among people with low-risk behavior, does it still make sense to try to prevent HIV transmission among the relatively small segment of the population that continues to engage in the highest-risk behavior?

The answer is yes. In the bottom panel of figure 2.10, we show a simulation of the impact of raising condom use in the same three subpopulations shown in the top panel, but applied much later—fifteen years after the start of the epidemic. The long delay and higher prevalence rates notwithstanding, high condom use among those with the highest rate of partner change—sex workers, in this simulation—is still more effective in lowering prevalence in the whole population than is moderate condom use among men engaged in commercial and casual sex, who have fewer partners. Either of these strategies is far more effective than raising condom use among women in steady relationships. One can imagine other sexual behavior regimes where condom use among sex workers might make less of an absolute difference than condom use in casual relationships, or where these levels of condom use in any one of the subpopulations or even in all of them combined are not sufficient to reverse the course of the epidemic; some of these other scenarios will be discussed in chapter 3. *Even so, interventions that prevent transmission among those with the highest rates of partner change will prevent more secondary cases per primary case averted than interventions that change only the behavior of the population that practices low-risk behavior.* It follows that, if interventions that reduce risky behavior among those with the highest rates of partner change are not much more expensive per primary case averted than programs for low-risk populations, then they can be very cost-effective.

The Level and Distribution of HIV Infection in Developing Countries

DEVELOPING COUNTRIES FIND THEMSELVES AT DIFFERENT stages of the HIV/AIDS epidemic due largely to differences in the timing of the introduction of HIV and in sexual and drug-injecting behavior. In this section, we classify countries according to the level and distribution of HIV infection, which, as we shall see, has a significant influence on the cost-effectiveness of preventive interventions (chapter 3), and on the scope of interventions to mitigate the impact of the epidemic (chapter 4).

Despite the obvious value of a country typology, low-quality and inadequate data and our still-incomplete knowledge of the disease itself make any evaluation of a country's status very tentative. Because collecting incidence data is very costly, HIV monitoring systems collect data on prevalence. However, prevalence data are rarely collected for representative samples of the population. Our typology is therefore necessarily based on prevalence among frequently monitored groups with presumed high-risk behavior—sex workers, injecting drug users, homosexual and bisexual men, STD patients, and the military—and one frequently monitored group assumed to be at lower risk—pregnant women attending antenatal clinics. Most of these groups present significant sampling problems. For sex workers, injecting drug users, and homosexual and bisexual men, it is often impossible to identify a representative sample. Even the prevalence data for pregnant women, which may be systematically collected, are not usually nationally representative but are confined to women in urban areas who attend certain clinics. Further, because women only become pregnant if they are sexually active and most pregnant women are from younger age groups, women at antenatal clinics cannot be a proxy for the general population. Thus, much of the information about HIV prevalence comes from ad hoc samples, and in some cases the samples are very small.

Notwithstanding these and additional problems that we discuss below, by using available data from research studies and epidemiological surveillance, countries can be classified according to two broad criteria: first, the extent of HIV infection among groups of people often found to engage in high-risk behavior,[17] and, second, whether the infection has spread to populations assumed to practice lower-risk behavior. The typology includes three stages of the HIV/AIDS epidemic:

- *Nascent:* HIV prevalence is less than 5 percent in all known sub-populations presumed to practice high-risk behavior for which information is available.
- *Concentrated:* HIV prevalence has surpassed 5 percent in *one or more* subpopulations presumed to practice high-risk behavior, but prevalence among women attending urban antenatal clinics is still less than 5 percent.
- *Generalized:* HIV has spread far beyond the original subpopulations with high-risk behavior, which are now heavily infected. Prevalence among women attending urban antenatal clinics is 5 percent or more.

Due to a lack of data, this typology does not address several important factors. In particular, it does not distinguish between countries based on incidence, the rate of new infections. As we have seen, because HIV cannot be cured and lasts many years, prevalence can rise even if incidence is declining. Prevalence data do not reveal whether the number of new infections is increasing, declining, or level, either in specific subpopulations or among the entire population.[18] Moreover, prevalence data are available only for a few specific subpopulations. Prevalence can stabilize in one or more of these subpopulations even as it spreads rapidly through others that are not monitored. These shortcomings highlight the importance to governments of collecting additional data on HIV incidence and prevalence so that policymakers can formulate an effective response (box 2.7).

Even with much better data on incidence and prevalence, we would still lack sufficient behavioral information to confidently predict the course of the epidemic in a specific country. Surveys of sexual behavior by the World Health Organization's Global Programme on AIDS in the late 1980s and early 1990s were among the first attempts to measure behavioral risk factors for HIV infection in developing countries (Cleland and Ferry 1995). WHO also studied drug-injecting behavior and the risk of HIV infection in thirteen cities in industrial and developing countries in 1989 (WHO, Program on Substance Abuse 1994). These and more recent studies have increased our knowledge about risk factors in developing countries. Even so, almost two decades into the epidemic only a few geographical areas and a handful of developing countries have been covered. Policymakers in most countries simply do not know how many people engage in commercial sex, casual sex, or injecting drug use; how frequently they do so; or the extent to which they take actions to re-

Box 2.7 Monitoring the Spread of HIV

HIV/AIDS PROGRAM PLANNERS NEED TIMELY data on trends in HIV prevalence among certain sub-populations to design, implement, and monitor the impact of HIV/AIDS interventions. The World Health Organization and UNAIDS recommend that countries conduct periodic surveys of HIV prevalence among populations likely to engage in high-risk behavior, such as patients at STD clinics and people enrolled at drug treatment centers (AIDSCAP and others 1996, Chin 1990, Sato 1996). Residual blood collected for other tests performed at these facilities is tested for HIV after removing all information that would allow individuals to be identified, to monitor levels and trends in HIV prevalence while maintaining the confidentiality of test results. This is called the "unlinked, anonymous" method of HIV surveillance. In countries where HIV prevalence has reached significant levels in populations with high-risk behavior, surveillance should be extended to populations that practice low-risk behavior, such as women attending antenatal clinics.

Monitoring HIV prevalence of these "sentinel" populations—often referred to as "sentinel surveillance"—is preferred because large-scale surveys of HIV and STD prevalence in the general population are expensive and often do not capture enough people with high-risk behavior to detect trends. Furthermore, for ethical reasons, HIV testing in these larger surveys can only be done with the informed consent of participants. This will bias the results if people who choose not to participate are more or less likely to engage in risky behavior than those who do participate. At any rate, HIV is unlikely to show up in a major way in the general population until fairly late in the epidemic. Important trends can be detected by monitoring sentinel populations, with the understanding that these groups are not representative of the general population and the results are not adequate to project the number of current and future AIDS cases. Sentinel surveillance should be augmented by an AIDS case-reporting system, which will help in estimating the start date of the epidemic and the impact on mortality and the health system. HIV and AIDS epidemiological surveillance is only one component of the essential information that government has a key role and comparative advantage in providing; other types of information are discussed in chapter 3.

duce their risk of contracting HIV. Basic information on the levels of condom use and partner change among the general population is also unavailable. Without this information, it is simply not possible to accurately predict the course of the epidemic.

For these reasons, we cannot predict with certainty in which low-prevalence countries HIV infection will take off or at what level prevalence will stabilize. Some countries where HIV is currently regarded as a minor problem may turn out to resemble eastern Africa, where the virus spread rapidly through groups with high-risk behavior and widely into the general population. In other countries HIV may infiltrate groups with high-risk behavior but never evolve into a generalized epidemic, even without intervention. Or, knowledge of HIV may cause people to adopt less risky behavior, with or without government urging. Countries with a high incidence and prevalence of STDs other than HIV are likely to be particularly vulnerable to a large and rapidly spreading HIV epidemic, because STDs and HIV are spread by the same behavior and STDs enhance HIV transmission. However, since most other STDs can be cured, countries with low STD prevalence may still have behavior patterns that are conducive to the rapid spread of HIV.

Lacking the information to predict the course of the epidemic, and given the terribly high human and financial costs of HIV/AIDS, *it would be prudent to assume the worst and act aggressively to minimize the epidemic as early as possible.* The remainder of this chapter uses this typology to outline the state of the epidemic as of mid-1996 in four developing regions. A list of countries and estimates of prevalence in different subpopulations are in table 1 of the statistical appendix to this report.

Africa

Roughly 90 percent of all HIV transmission in Sub-Saharan Africa is by heterosexual sex. HIV has spread rapidly among people with high-

In seventeen Africa countries, more than 20 percent of female sex workers in cities are infected with HIV.

Figure 2.11 HIV Infection in Urban Sex Workers in Sub-Saharan Africa, Various Years

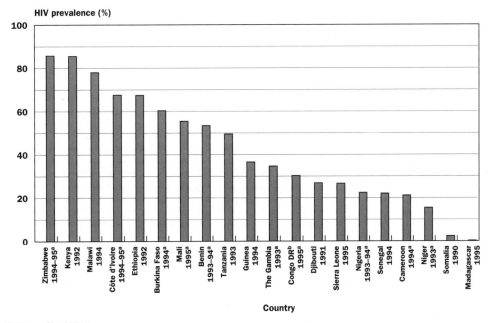

a. HIV-1 and/or HIV-2.
b. Formerly Zaire.
Source: U.S. Bureau of the Census (database), 1997.

Figure 2.12 HIV Seroprevalence among Pregnant Women in Selected Areas of Sub-Saharan Africa, 1985–95

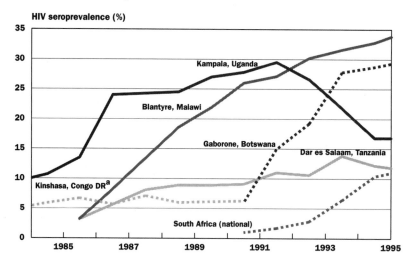

Infection rates among women attending antenatal clinics have grown rapidly to high levels in some countries, stabilized at lower levels in others, and appear to be declining in Kampala, Uganda.

a. Formerly Zaire.
Source: U.S. Bureau of the Census (database), 1997.

risk behavior and widely among those assumed to be at lower risk. Prevalence among urban sex workers exceeds 20 percent in seventeen countries, and is 50 percent or more in nine countries (figure 2.11). Infection rates among women attending antenatal clinics have grown rapidly to high levels in some areas, have stabilized at lower levels in others, and appear to be declining in Kampala, Uganda (figure 2.12). HIV has infected more than 5 percent of women attending urban antenatal clinics in nineteen countries, and in six countries more than 20 percent are infected. An estimated two-thirds of all new cases of mother-to-child transmission worldwide occur in Sub-Saharan Africa (UNAIDS 1996d).

The countries with generalized epidemics include most in eastern, southern, and central Africa, plus Côte d'Ivoire, Benin, Burkina Faso, and Guinea-Bissau in West Africa (figure 2.13). There is often considerable geographic variation in infection levels within countries. Nigeria, which has more than 100 million people and is the region's most populous country, has areas at all three stages of the epidemic. In more than half of Nigeria's states the epidemic is concentrated. HIV has spread most widely in Lagos, along the west coast, and in Delta, Plateau, Borno, and Jigawa states, located to the east and northeast. However, in three

Figure 2.13 HIV Infection in Africa and the Middle East

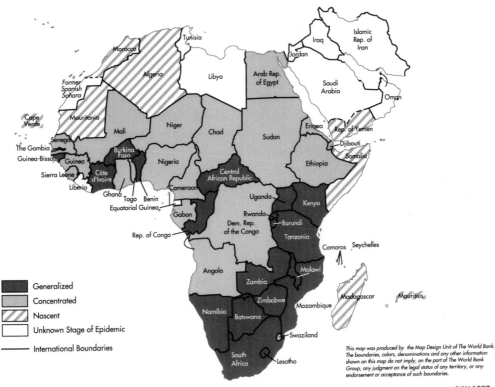

Generalized

Concentrated

Nascent

Unknown Stage of Epidemic

International Boundaries

This map was produced by the Map Design Unit of The World Bank. The boundaries, colors, denominations and any other information shown on this map do not imply, on the part of The World Bank Group, any judgment on the legal status of any territory, or any endorsement or acceptance of such boundaries.

JULY 1997

states—Edo, Niger, and Oyo—the epidemic is still nascent with low prevalence levels, even among subpopulations with high-risk behavior. HIV was detected early in the Democratic Republic of the Congo (formerly Zaire), but in contrast to many eastern and southern African countries, prevalence has stabilized at less than 5 percent on average in urban antenatal clinics (Piot 1994; statistical appendix, table 1). In Uganda, one of the hardest-hit countries in Africa, HIV prevalence among young people has declined (box 2.8).

Most Sub-Saharan African countries face the dual challenge of lowering HIV prevalence—which can happen only over many years—and of coping with the impact of existing high prevalence on the health system and society. Their domestic budgetary resources to accomplish this are quite limited. Countries with nascent epidemics in Sub-Saharan Africa—Cape Verde, Madagascar, Mauritania, Mauritius, and Somalia—have a unique opportunity to intervene early and aggressively to pre-empt a

Box 2.8 Declining Seroprevalence in Uganda

RECENT STUDIES FROM UGANDA SHOW A DECLINE in HIV seroprevalence rates, particularly among young adults. Among adults in Masaka district, Uganda, overall HIV seroprevalence declined by less than 1 percentage point between 1989 and 1994, from 8.2 percent to 7.6 percent (Mulder and others 1995). However, the decline was greater among the young—from 3.4 to 1.0 percent among males and from 9.9 to 7.3 percent among females ages 13 to 24 (box figure 2.8). The greatest declines were seen in males ages 20 to 24 and females ages 13 to 19. Yet HIV seroprevalence increased among males and females over 25 years old. In neighboring Rakai District, HIV prevalence in adults 15 to 59 declined from 23.4 in 1990 to 20.9 percent in 1992 (Serwadda and others 1995). Once again, the youngest group showed the largest decline: for ages 13 to 24, prevalence declined from 17.3 to 12.6 percent.

HIV prevalence is also declining among pregnant women attending antenatal clinics in Uganda. In the main referral hospital in Kampala, Uganda's capital and largest city, HIV seroprevalence among pregnant women fell from 28 percent to 16 percent between 1989 and 1993 (Bagenda and others 1995). All age groups under age 38 experienced a decline in prevalence, but the decline was greatest for those under 19. Similar declines were observed among pregnant women at antenatal clinics in other urban areas of Uganda between 1991 and 1994 (Asiimwe-Okiror and others 1995).

The overall decline in HIV prevalence among the two rural populations of adults in Masaka and Rakai districts can be accounted for almost completely by rising mortality, with no change in incidence (Serwadda and others 1995). However, according to nationwide surveys conducted in 1989 and 1995,

Box Figure 2.8 HIV Seroprevalence in Rural Masaka, Uganda, 1989 and 1994

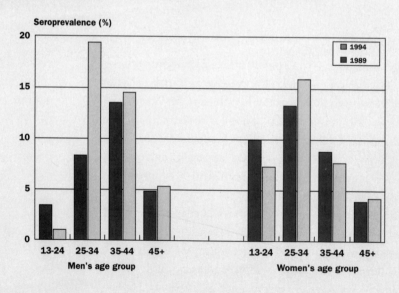

Source: Mulder and others 1995.

Box 2.8 *(continued)*

there have been important changes in sexual behavior that could explain the apparent decline in incidence of HIV among young adults observed in antenatal clinics, particularly in urban areas (Asiimwe-Okiror and others 1997, Stoneburner and Carballo 1997). The percentage of young adults age 15 to 19 who have ever had sexual intercourse declined from 69 to 44 percent among men and from 74 to 54 percent among women. Condom use has risen substantially among all age groups, and the percentage with a casual partner has declined, particularly among the young.

Is the decline in prevalence in young adults the result of policy? Interventions by both nongovern-mental organizations (NGOs) and the Ugandan government have focused on reducing the frequency of sexual partner change, distributing and promoting condoms, and controlling sexually transmitted infections (Mulder and others 1995). However, knowledge of the routes of HIV is widespread in Uganda, and many young people have personal experience of a loved one dying from AIDS. In the absence of a carefully designed evaluation, assessing the contribution of specific interventions is extremely difficult. Given the importance of the issue to other countries with generalized epidemics, international support for such an evaluation is warranted.

full-scale epidemic. High STD prevalence in Madagascar makes it vulnerable to the rapid spread of HIV (box 2.9). The epidemic in most of North Africa and the Middle East is also nascent, although there is evidence of rapidly climbing HIV infection among injecting drug users in Bahrain and Egypt, as has been the case in Asia. A large number of North African and Middle Eastern countries could not be classified because of lack of data.

Latin America and the Caribbean

More than half of the countries in Latin America and the Caribbean have concentrated epidemics (figure 2.14). These include the most populous countries in the region—Brazil and Mexico. Six countries have nascent epidemics, two (Guyana and Haiti) have generalized epidemics, and two (Bolivia and Panama) have insufficient information to be classified.

Injecting drug use and sex between men have played a major role in transmission in many countries in Latin America. Roughly one-quarter of all HIV infections in Brazil (24 percent, 1992) and a third in Argentina (39 percent, 1991) have been attributed to transmission through injecting drug use, which is an important source of transmission in Uruguay as well (Bastos 1995, Libonatti and others 1993). The epi-

Figure 2.14 HIV Infection in Latin America and the Caribbean

Generalized
Concentrated
Nascent
Unknown Stage of Epidemic
International Boundaries

This map was produced by the Map Design Unit of The World Bank. The boundaries, colors, denominations and any other information shown on this map do not imply, on the part of The World Bank Group, any judgment on the legal status of any territory, or any endorsement or acceptance of such boundaries.

demic is well established among homosexual and bisexual men in Argentina, Brazil, Colombia, Mexico, and Peru and has infected significant numbers of sex workers in Argentina, Brazil, the Dominican Republic, Guyana, Honduras, Jamaica, and Trinidad and Tobago. Although the data are spotty, the relatively high prevalence of HIV among injecting drug users, homosexual and bisexual men, and sex workers in Latin America suggests that in many of these countries the virus is poised to spread to the low-risk sexual partners of people who engage in high-risk behavior.

In the Caribbean and parts of Central America, HIV is spread mostly through heterosexual transmission. Male and female cases are roughly equal in Haiti; the epidemic has spread broadly to 8 percent of pregnant

Box 2.9 The Case For Early Intervention in Madagascar

AS A COUNTRY WITH RELATIVELY FEW CASES OF HIV infection, Madagascar is in a good position to intervene early to stop the rapid spread of HIV. Fewer than half a percent of STD patients and sex workers were infected with HIV in 1992, but the potential exists for an explosive HIV epidemic in Madagascar because of the high prevalence of other STDs (Behets and others 1996). As of 1995, almost one-third of sex workers and about one in eight pregnant women in urban areas were infected with syphilis. According to one projection model, by 2015 HIV seroprevalence among Malagasy adults could be either 3 percent, using the course of Thailand's epidemic as a model, or 15 percent, using the course of Kenya's epidemic as a model.

Certain behavioral risk factors common to STD and HIV transmission are common in Madagascar. In 1992, about one in ten men reported recent sexual contact with a prostitute (Andriamahenina 1995). Stable concurrent partnerships outside of marriage were reported by 11 percent of pregnant women, nearly one-third of STD patients and one-quarter of prostitutes. Condom use is low in Madagascar. Only one-third of sex workers reported always using condoms with stable extramarital partners and only 5 to 8 percent of STD patients reported always using condoms (Andriamahenina 1995). Among all women aged 15 to 49 surveyed in 1992 by the Madagascar Demographic and Health Survey, only 0.5 percent used condoms (Refeno and others 1994).

The government of Madagascar launched its national AIDS control program with the encouragement of and financial support from WHO/GPA in 1988, three years before its first reported AIDS case. The country already had a system of STD surveillance prior to the introduction of HIV/AIDS, consisting of fifteen STD treatment clinics in major towns. However, the system is poorly equipped and needs major revamping. To address the gaps in knowledge about HIV prevalence, the government conducted and financed, through a World Bank loan, its first survey of STD/HIV prevalence and risky behavior among high-risk groups in 1994–95. As we shall see in chapter 3, strengthening epidemiological surveillance, raising levels of awareness of HIV/AIDS, and vigorously promoting condom use and STD treatment among individuals with the riskiest behavior should be the priority for Madagascar, given the levels and distribution of infection that prevail. Fortunately, condoms will soon have a higher profile in the country, as USAID has funded a new condom social marketing program in major cities as of the end of 1996.

women, and there is significant mother-to-child transmission. More than 70 percent of AIDS cases in the Dominican Republic is attributed to heterosexual transmission; the ratio of male-to-female cases now stands at 2 to 1 and is declining (ONUSIDA 1997). HIV prevalence among pregnant women in that country has risen to a national average of 2.8 percent, and in some areas has reached 8 percent. Following a similar path, 1 percent of pregnant women in Honduras are also infected with HIV. In Guyana, which is in South America but faces on the Caribbean, nearly 7 percent of women attending antenatal clinics were infected, as of 1992.

Asia

In most Asian countries for which there is information, the epidemic has reached a concentrated stage either nationwide or at least in some states or provinces (figure 2.15). This includes regions of the world's two most populous countries, China and India, most of Indochina, and Malaysia. In the remaining Asian countries for which there is information, the epidemic is nascent; infection among those presumed to practice high-risk behavior is less than 5 percent.

Patterns of infection in east, south, and southeast Asia have been greatly influenced by the proximity of many countries to the "Golden Triangle" of heroin production, located at the border between Lao PDR, Myanmar, and Thailand, and to its distribution routes (*background paper*, Riehman 1996). HIV infection was first detected among those who inject drugs in Bangkok in 1987; during the next year it spread rapidly among injecting drug users in the Thai capital (Stimson 1994). The pattern was quickly repeated among injecting drug users in northern Thailand and along the border areas between southern Thailand and northern Malaysia. In 1989, HIV infection was identified in Myanmar, Yunnan Province in China, and in Manipur State in India. HIV was detected among injecting drug users in Singapore in 1990.

Injecting drug use has been the main transmission mode in China, where the most highly infected province, Yunnan, is adjacent to international drug routes. Male injecting drug users in Yunnan account for 78 percent of HIV infections in China (Zheng 1996). In other Chinese provinces, infection rates are thought to be low, even among those who practice high-risk behavior (Yu and others 1996). Economic reforms that have helped to reduce the number of people in poverty in China by more than half since the late 1970s have also resulted in large increases in internal migration that could generate conditions conducive to the spread of HIV. Studies have estimated that nearly 100 million people, roughly one in twelve people in China, have moved either temporarily or permanently from their registered residence (Nolan 1993, Peng 1994). Much of the movement involves migration within provinces, but an estimated 20 million migrants have moved from poor areas of western China to eastern provinces (Nolan 1993). Most migrants are young, single, and male, but many women have also migrated; some have reportedly become involved in prostitution. STDs, which were all but eliminated in China in the 1960s, are rising rapidly (Cohen and others 1996, Kang

Figure 2.15 HIV Infection in Asia

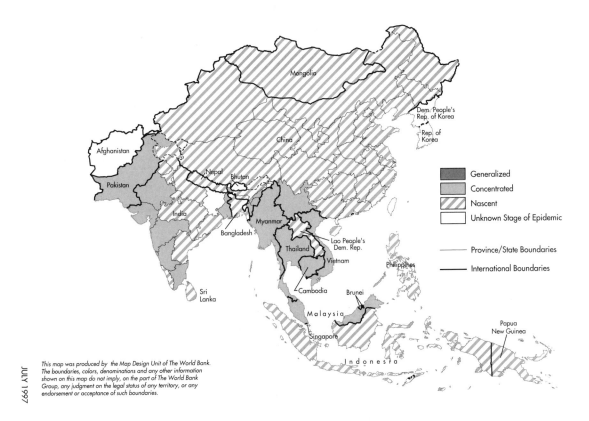

This map was produced by the Map Design Unit of The World Bank. The boundaries, colors, denominations and any other information shown on this map do not imply, on the part of The World Bank Group, any judgment on the legal status of any territory, or any endorsement or acceptance of such boundaries.

JULY 1997

1995). Early preventive interventions for migrants and sex workers in areas receiving migrants could reduce the likelihood of an epidemic of HIV and other STDs among these mobile groups.

Among the nations of South Asia, the epidemic is believed to be spreading most rapidly in India and Pakistan. In India, HIV is widespread among injecting drug users in the northeastern states of Manipur and Mizoram and is spreading to their sexual partners; prevalence in antenatal clinics in Manipur has reached 2 percent. HIV is well established among sex workers and STD patients in much of southern India, including populous Maharashtra and Tamil Nadu states (Jain, John, and Keusch 1994). In the city of Mumbai (formerly Bombay), HIV prevalence among pregnant women has reached 1 to 2 percent. In Pakistan, the infection rate among injecting drug users in Lahore was 12 percent; as of 1995, HIV infection among women attending antenatal clinics was still extremely low. Transmission by those who inject drugs also may be a factor near a second major heroin-producing area, the "Golden Crescent," where Pakistan's Northwest Frontier meets the Badakhshan area of Afghanistan and the Baluchistan area of Iran (*background paper*, Riehman 1996). However, there are no recent data on HIV prevalence among drug users or other groups in these areas. In Nepal, prevalence so far remains very low among injecting drug users in Katmandu, partly because of interventions discussed in the next chapter. Bangladesh's HIV epidemic is still nascent, but, without behavior change, HIV could spread quickly among a population of brothel-based sex workers and their clients.

In most of southeast Asia, with the significant exceptions of Indonesia, Lao PDR, the Philippines, and Papua New Guinea, the HIV epidemic is at the concentrated stage. Injecting drug use has played a central role in the launching of HIV, often in conjunction with commercial sex, but heterosexual transmission is now the predominant mode of transmission. HIV is firmly established among injecting drug users and sex workers in Cambodia, Myanmar, and Thailand, and 1 to 3 percent of pregnant women are HIV-positive in those countries. In Thailand, HIV prevalence peaked at 4 percent among military conscripts in 1993, but has recently been declining following a national campaign to reduce sexual transmission of HIV through greater condom use and a reduction in commercial sex. In Cambodia, however, infection levels in the military have reached nearly 7 percent. In Malaysia and Vietnam, more than three-quarters of HIV infections are attributed to transmission through injecting drug use (Hien 1995, Kin 1995). Yet sexual transmission in Malaysia is clearly on the rise; nearly 40 percent of HIV/AIDS cases seen at the University of Malaya Medical Center since 1986 were thought to be due to heterosexual transmission (Ismail 1996). In contrast, although HIV has been detected sporadically for some time among sex workers in the Philippines and Indonesia, it has not spread rapidly, even within that group; as of mid-1996 these two populous countries remained at the nascent stage (Jalal and others 1994, Tan and Dayrit 1994).

Eastern Europe and the Former Soviet Union

The rapid social change and economic dislocation that has accompanied the collapse of socialism in Eastern Europe and the former Soviet Union (FSU) have created a situation in which the potential for an HIV epidemic looms large. The available data on HIV prevalence suggest that most countries in the region are still in the nascent stage (figure 2.16).

However, reliable information on HIV prevalence by subpopulation is scarce in all but a handful of countries—almost two-thirds of all of the countries in this region cannot be classified based on the available information.

Ukraine has a concentrated HIV/AIDS epidemic, based on high HIV prevalence among injecting drug users; between January and August 1995, HIV prevalence among those who inject drugs rose to 13.0 percent, from just 1.4 percent. Just five months later, more than half of the injecting drug users in the Ukrainian city of Nikolayev were infected (UNAIDS 1996d). A survey of new injecting users in Poland in 1995

Figure 2.16 HIV Infection in Eastern Europe and Central Asia

This map was produced by the Map Design Unit of The World Bank. The boundaries, colors, denominations and any other information shown on this map do not imply, on the part of The World Bank Group, any judgment on the legal status of any territory, or any endorsement or acceptance of such boundaries.

Figure 2.17 Reported Cases of Gonorrhea in Eastern Europe, 1986–94

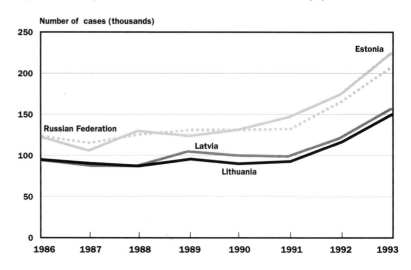

The rapid increase in STD cases in Eastern Europe signals heightened vulnerability to an HIV/AIDS epidemic.

Source: WHO/GPA 1995.

found HIV prevalence of 4.7 percent; a few years earlier, among longer-term injectors in the city of Warsaw the rate was 45 percent (WHO/EC Collaborating Centre 1996b). Given that those who inject drugs often travel to neighboring countries, it is reasonable to expect a similarly rapid take-off among injecting drug users in the Russian Federation and Belarus (Bourdeaux 1996).

In Romania, HIV initially spread primarily among children; over 90 percent of AIDS cases in 1990 were among children under 13 years of age. It was erroneously believed that transfusing blood among children would provide important nutrients and boost the immune system (Hersh and others 1991). Instead, it spread HIV among them. The practice has since been abandoned.

A key signal of the potential for an HIV epidemic in this region is the dramatic increase in STDs experienced by most countries since the collapse of the Soviet Union. The number of gonorrhea cases has nearly doubled between 1990 and 1994 in four Eastern European countries (figure 2.17). In Ukraine, the number of syphilis cases increased more than tenfold between 1991 and 1995 (AIDSCAP and others 1996).

Regardless of the stage of the epidemic, there are compelling reasons for governments to find ways to encourage people who practice risky behavior to adopt safer practices as soon as possible. Fortunately, in most areas of the developing world it is not too late to avert a generalized epidemic. Half of the population of developing countries—2.3 billion people—live in areas where the HIV/AIDS epidemic is still nascent. Another third of the population of developing countries lives in areas where the epidemic is already concentrated but has yet to become generalized. In all these areas, action now to help people with the highest-risk behavior protect themselves and others from HIV infection can save millions of lives and avoid massive future expenditures on AIDS treatment and care. Even in areas where the epidemic is already generalized, action to prevent infection among those most likely to contract and spread HIV can still make a substantial difference.

What steps can be taken to help people who engage in high-risk sexual activity or injecting drug use to protect themselves and others from HIV infection? The next chapter discusses two broad complementary approaches: altering the perceived costs and benefits of individual choices, and changing the social environment that shapes and constrains these choices.

Notes

1. Thomas and Tucker (1996) point out that the use of the word "rate" is not technically correct, since it is in fact a number, not a rate per unit of time. The reproductive rate was first applied to HIV transmission by May and Anderson (1987). Previous authors had referred to this as the "infectee number" (Hethcote 1976, Nold 1978).

2. The number of exposures per partner and the type of sexual act also affect the spread of HIV through the population, but empirical evidence suggests that the rate of partner change is far more important (Anderson, Gupta, and Ng 1990).

3. However, a review of sixteen studies that compared female-to-male and male-to-female transmission found female-to-male rates as low as 5 percent and male-to-

female transmission rates as high as 140 percent (Haverkos and Battjes 1992).

4. Transmission rates per partnership are not much affected by the length of the partnership or number of exposures. This is because in a relatively long-term partnership, either the HIV-positive partner is truly infectious and transmission will occur relatively quickly, or the HIV-positive partner is relatively uninfectious and transmission will not occur despite many contacts over a long period.

5. High infection rates with other STDs may also have contributed to high transmission rates in the study.

6. For example, Hook and others 1992, Laga and others 1993, Lazzarin and others 1991, Mastro and oth-

ers 1994, Plummer and others 1991, Quinn and others 1990.

7. For example, Bongaarts and others 1989; Caldwell and Caldwell 1996; Conant 1995; de Vincenzi and Mertens 1994; Moses and others 1990, 1995; Simonsen and others 1988.

8. Like all data on sexual activity, such surveys are subject to a variety of errors. Since they ask about private behavior, misreporting may skew results. In many cultures men may exaggerate the number of their sexual partners, while women may do the opposite. People with high rates of partner change may not be able to recall their number of partners accurately, and sex workers may not be captured by surveys using standard sampling techniques. Even if we assume a significant degree of error, however, the variation in the rate of partner change within populations is striking.

9. By 1993, the percent of men 15–49 with nonregular sexual partners in the past twelve months had declined to 15 percent in Thailand (Thongthai and Guest 1995).

10. Men who have sex with men include self-identified homosexual and bisexual men as well as heterosexual men who have sexual relations with other men.

11. The impact of behavior change on the epidemic will be introduced later in this chapter and in chapter 3.

12. The baseline is the same as the top line in figure 2.8 for an epidemic fueled by commercial and casual sex, with concurrent partnerships.

13. While this "pre-emption" effect slows the epidemic's rate of spread at the population level as the epidemic matures, the probability that an uninfected individual will encounter at random an infected partner has greatly increased because a large share of the population is infected.

14. In this model, 40 percent of men have a casual partner or a sex worker in a year. However, the individuals are not the same over time. It is assumed that men who use condoms with casual partners also use them with sex workers.

15. We choose to model a higher level of condom use among sex workers because it has been shown to be more feasible than similarly high levels of condom use among other subpopulations.

16. Note that in this hypothetical population with concurrent and casual sex, behavior change in any one of these groups is not sufficient to reduce HIV prevalence to zero. That would require simultaneous increases in condom use in more than one group.

17. We recognize that the characteristics of certain subpopulations, like military service or sexual orientation, are imperfect predictors of risky behavior. While members of these subpopulations may practice riskier behavior *on average*, in some countries they practice low-risk behavior, as evidenced by lower rates of partner change, high rates of condom use, or limited sharing of injecting equipment.

18. An exception would be adolescents, who can be presumed to be uninfected as they enter adulthood. HIV prevalence among adolescents at a point in time is likely to reflect recent infection.

CHAPTER 3

Efficient and Equitable Strategies for Preventing HIV/AIDS

WHILE RESEARCHERS CONTINUE THE search for a cure or a vaccine that is affordable in developing countries, the greatest hope for combating HIV/AIDS in the foreseeable future is that of helping people to choose safer behavior so that they will be less likely to contract and spread HIV. But *can* government affect the very private and intimate behaviors that spread HIV/AIDS? And if it can, what are the priority actions that governments should take to prevent the epidemic, so as to maximize the impact of scarce resources?

This chapter identifies government priorities in the prevention of HIV/AIDS. We find that public policy *can* affect the behaviors that spread HIV. There *are* effective interventions, and governments have many ways to influence private behavior. In the first two sections of the chapter, we focus on two complementary approaches. The first aims to influence individual choices directly within the existing economic and social context by changing the costs and benefits of various types of behavior, making safer behavior a more attractive choice. However, individual behavior is shaped and often constrained by the economic and social context; some individuals' choices are very constrained. A second, complementary approach is to change the economic and social conditions that make it difficult or impossible for some people to protect themselves from HIV. The benefits of this approach typically extend far beyond HIV prevention. It is therefore often difficult to assess the cost-effectiveness of measures to affect the economic and social environment, but they are likely to improve the efficacy of direct interventions to individuals.

Which activities are most justified for government action and, among them, which should be priorities? There are three activities in which governments have an indispensable role in ensuring the efficiency and equity of prevention programs: providing public goods related to prevention, such as the collection and dissemination of information about the epidemic; reducing the negative externalities of risky behavior by promoting safer behavior among people who are most likely to contract and spread the virus; and promoting equity by ensuring that the most destitute are not denied access to the means to protect themselves from HIV. These activities will reduce the spread of HIV most quickly and will benefit everyone in society, including those with low-risk behavior and the poor. Private individuals would not invest in them in sufficient quantity to slow the epidemic. In addition to satisfying these criteria of public economics, government-sponsored preventive activities should be cost-effective. We review some important factors that influence the cost-effectiveness of government actions and propose a set of public priorities for preventing HIV according to the stage of the HIV/AIDS epidemic.

In the final section of the chapter, we review the available evidence on the extent to which national governments have succeeded in following the broad prevention strategy outlined in this chapter. We find that while governments, in collaboration with private actors, have implemented many worthwhile programs, some of which have had demonstrable impact, there is a need for renewed effort to implement the activities that are likely to be most cost-effective from the government perspective in preventing the epidemic. Perhaps the major impediment to more effective, efficient, and equitable prevention is a lack of political commitment, first, to collect and disseminate the data on HIV prevalence, risk behavior, the effectiveness of programs, and their costs, and, second, to work constructively with those at highest risk of contracting and spreading HIV to prevent infection. These are important issues that national governments, international donors, and nongovernmental organizations need to act on urgently; they are discussed at greater length in chapter 5.

Before we begin, a caveat: this chapter reviews evidence on the impact of policies and suggests priority actions for governments. However, it is not intended as a blueprint for implementing these programs in the field, or as a primer on "best practices" for specific interventions or pro-

Box 3.1 Best Practices in HIV Prevention and Treatment Online

UNAIDS IS PUTTING UP-TO-DATE INFORMATION ON best practices in HIV prevention, treatment, and AIDS impact online on its website: www.unaids.org. Information will be available on more than 40 topics, among them community mobilization, male and female condoms, counseling and testing, HIV education, epidemiology, human rights, STDs, and HIV vaccines. The collection of materials on each topic will normally have five components:

- *UNAIDS point of view.* A short advocacy document for journalists and community leaders that cites the key facts and figures, addresses myths and misconceptions, and sets out ways of dealing with the topic.

- *Technical update.* A short technical overview of the topic for managers of HIV/AIDS programs and projects, summarizing the main problems involved as well as the "best practice" responses.
- *Best practice case studies.* Examples of "best practice" on the topic in specific countries.
- *Presentation graphics.* A selection of slides or overheads and talking points for presentations.
- *Key materials.* A maximum of ten reports, articles, books, compact discs, or videos that represent the most up-to-date, authoritative thinking on the topic.

grams. This huge task is beyond the scope of this volume and the expertise of its authors. For this, UNAIDS is assembling an extensive collection of resources to which practitioners may refer (box 3.1).[1] This chapter also focuses on changing behavior to prevent HIV. Programs to ensure a safe blood supply, while preventing some infections, will not be sufficient to prevent an HIV epidemic spread mainly by sexual or drug-injecting behavior. The same is true for medical interventions to prevent transmission of HIV from mothers to their children. Blood safety and treatment to prevent mother-to-child transmission are discussed in the context of the impact of HIV on the health sector in chapter 4.

Influencing Individual Choices

MOST THINGS IN LIFE ENTAIL SOME RISK, YET PEOPLE WILLingly take risks when they perceive that the benefits of an action outweigh the costs. For example, drivers speed and pedestrians dart across busy streets, despite the increased risk of injury or death. People start smoking although they know it might lead to lung cancer and heart disease. Sometimes risk enhances pleasure.

Mountain climbers scale the Himalayas, their thrill perhaps intensified by the danger they face. All of these decisions reflect individual preferences and an assessment of costs, benefits, and risks (Philipson and Posner 1993). Sex and injected drugs offer very intense if short-lived pleasure. Do individuals also weigh costs, benefits, and risks when deciding whether and how to engage in these activities?

Fortunately for efforts to slow the epidemic, the answer is yes. A substantial body of economic research, much of it in developing countries, has shown that actual and perceived costs and benefits, some of which can be affected by government policies, significantly influence private decisions about marriage, childbearing, and contraceptive use.[2] It is therefore reasonable to assume that sexual behavior that spreads HIV can also be influenced by public policy. Similarly, studies have shown that under the right conditions drug users change their injecting behavior to reduce the likelihood of being infected with HIV. This section discusses four types of policies that can alter the perceived "costs" and benefits of various activities so that individuals will avoid behavior that is very likely to spread HIV: providing various types of information, reducing the costs of using condoms, reducing the costs of using sterile injecting equipment, and attempting to raise the costs of engaging in commercial sex and injecting drug use.[3]

Knowledge of HIV Reduces Risky Behavior—But Not Enough

Knowledge about the levels of HIV infection in the population, how HIV is transmitted, and how to avoid contracting it can induce some people who engage in high-risk behavior to adopt safer sexual and injecting practices or to refrain from casual or commercial sex and injecting drug use altogether. In Thailand, for example, the public revelation in 1989 that 44 percent of sex workers in the northern city of Chiang Mai were infected with HIV is believed to have contributed to the growing use of condoms in commercial sex, even before large-scale government condom programs began (Porapakkham and others 1996). Similarly, there is evidence that condom use rose in the United States in the 1980s, in part independently of the impact of prevention programs (box 3.2).

Moreover, there is considerable evidence that people engaged in high-risk activities are more motivated to learn about HIV than others, since they are more likely to become infected and suffer the consequences.

Box 3.2 Behavior Responds to Increased Risks: AIDS Incidence and Increased Condom Use in the United States

A STUDY IN THE UNITED STATES HAS SHOWN THAT young American adults increased condom use in the 1980s in response to the growing perceived risks of HIV, net of the impact of state HIV prevention programs (Ahituv, Hotz, and Philipson 1995).[1] This suggests that there will be some spontaneous behavioral response to the rising costs of unprotected sex imposed by the AIDS epidemic.

In 1984, the first of the six years covered by the study, there had been relatively few AIDS cases, and there were few differences in condom use across census regions of the United States. As the cumulative number of AIDS cases climbed differentially, so did condom use: the larger the per capita cumulative number of AIDS cases was, the larger was the increase in condom use. Across all states for the six-year period, condom use among 25- to 27-year-olds doubled from 8 percent to more than 16 percent; for African Americans in this age group condom use almost tripled, from 7 to 19 percent. The responsiveness of condom use to per capita cumulative AIDS cases was greatest among sexually active single men and single men in urban areas, the two groups in the sample most likely to be at high-risk of HIV infection. Condom use among married men, who presumably had a much lower risk of acquiring HIV, did not vary according to the per capita cumulative AIDS cases in the state.

What caused these changes in behavior? If the states with the largest AIDS epidemics responded by establishing stronger AIDS prevention programs, disentangling the impact of growing per capita cumulative AIDS cases and state-run prevention programs on condom use would be extremely difficult. In fact, the study did not find a strong correlation between the extent of the AIDS epidemic and the strength of a state's HIV prevention efforts. To further isolate the effect of the per capita cumulative AIDS cases on condom use, the study controlled for such factors as age, gender, race and ethnicity, marital status, urban residence, education, and parents' income, as well as for state-specific controls that proxied for AIDS program expenditures.

The strong correlation between per capita cumulative AIDS cases and condom use persisted, even after controlling for these variables. Furthermore, with the passage of time the responsiveness of condom use to per capita cumulative AIDS cases increased. The authors estimated that 32 to 65 percent of the actual increase in condom use could be attributed to increases in the cumulative number of AIDS cases—an objective measure of heightened risk—as opposed to other state- and individual-level factors.

---

[1]The study used a sample of 8,956 respondents for the 1984–90 waves of the National Longitudinal Survey of Youth (NLSY79), which began in 1979 with interviews of 12,000 people from the 1958–65 birth cohorts.

Surveys conducted by the Global Programme on AIDS in 1989–90 found that respondents who had engaged in risky behavior knew more about modes of transmission and the severity of AIDS than those who had never engaged in risky behavior (Ingham 1995). Similarly, Demographic and Health Surveys (DHS) in seven African countries found that individuals with characteristics correlated with higher-risk sexual behavior were more likely than others to be aware that condoms prevent transmission of HIV (box 3.3). A study in Imphal, the capital of Manipur State in northeastern India, found that injecting drug users, more than 80 percent of whom were infected with HIV, were far more

Box 3.3 Who Knows How Much about Preventing HIV/AIDS?

WE WOULD EXPECT THAT PEOPLE WHOSE BEHAVIOR puts them at higher risk of acquiring or spreading HIV would have greater incentive to find out how HIV is transmitted and how it can be prevented. If this were indeed the case, people who are more likely to engage in high-risk behavior would tend to know more about HIV prevention than those whose behavior does not put them at risk.

Data from seven countries in Sub-Saharan Africa support this hypothesis. Box figure 3.3 shows the percentage of men and women with various background characteristics who knew that HIV transmission could be prevented by condoms. The data are from a pooled sample of DHS data on adults from Burkina Faso, the Central African Republic, Côte d'Ivoire, Senegal, Tanzania, Uganda, and Zimbabwe.

Those groups with the lowest level of knowledge about the protective effect of condoms (far left in the figure) were married men and women in their late 40s who lived in rural areas, had little or no schooling, and no nonregular sexual partners in a recent time period. They were also the least likely to need this knowledge: HIV infection levels are lower in rural areas than in urban areas in Sub-Saharan Africa, and monogamous couples would not need to use condoms to avoid infection. Lack of education might also be responsible for low levels of knowledge; DHS data show a strong correlation between education level and knowledge of condoms as a means of preventing HIV (*background paper*, Filmer 1997).

At the other extreme are the two bars on the right of the figure, which correspond to the groups of men and women with the highest levels of knowledge about condoms as a means of protecting against HIV. They were younger, single, lived in urban areas, and recently had a nonregular sexual partner, all factors that would also put them at higher

risk of HIV infection. Their higher levels of schooling also probably contributed to greater awareness. However, knowledge of HIV prevention among individuals likely to be at highest risk is still well short of 100 percent. This suggests that additional efforts are needed to provide basic information about prevention to those who are most likely to contract and spread the virus.

Box Figure 3.3 Percentage of Adults Who Know That Condoms Are a Means of Protection against HIV Transmission, by Individual Characteristics, Seven Sub-Saharan African Countries

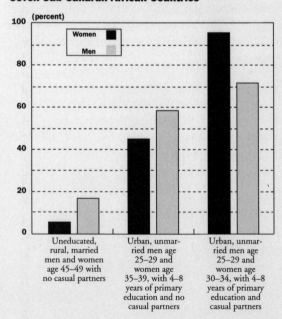

Source: Pooled DHS data for Burkina Faso, the Central African Republic, Côte d'Ivoire, Senegal, Tanzania, Uganda, and Zimbabwe.

knowledgeable about HIV transmission than was a comparison group of college students (Sarkar and others 1996).

While some of those with high-risk behavior may have become knowledgeable as a result of targeted prevention programs, the fact that they are more knowledgeable than others and yet persist in risky behavior underscores the point that simply increasing awareness of HIV will not change high-risk behavior enough to end the epidemic. Indeed, if this were the case, the now-widespread knowledge that HIV is transmitted through unprotected sexual intercourse and shared needles would have marked the beginning of the end of the epidemic.

Researchers have identified a number of interrelated factors that affect the way in which people use knowledge to assess the costs and benefits of risky behavior and internalize them. These factors include the extent to which: they understand how HIV infection would affect them personally; they perceive their own behavior to be risky; and they have the skills necessary to negotiate safer behavior with partners and to resist social pressures. Interventions that address these issues can result in substantially more behavior change than knowledge alone (Choi and Coates 1994; Holtgrave and others 1995; Oakley, Fullerton, and Holland 1995). These interventions can take many forms, from public information campaigns using mass media to training and education programs conducted face-to-face (box 3.4). Some of these approaches are likely to be more effective than others. For example, a lecture coupled with skills training that included role playing, psychodrama, and group discussions was more effective in raising condom use among homosexual men in the United States than was a small group lecture alone (Valdiserri and others 1989). Approaches that change behavior by improving the information base are enhanced by other policies, discussed below, that lower the costs of safer behavior.

Nevertheless, many knowledgeable people who have the skills to reduce their risk and have internalized the dangers will continue to engage in risky behavior. Given the devastating personal cost of contracting HIV, why isn't the danger of being infected sufficient to cause people to forgo all risky activities? One reason is that the costs of reducing risk are clear and immediate while the benefits are uncertain and distant. Whether people are willing to incur these up-front costs to reduce the risk of future sickness and premature death from AIDS depends on their assessment of the probability of infection from a specific act and the discount rate applied to future years of healthy life. Assessing the probabil-

---

Box 3.4 What Is "Information"?

AT THE OPERATIONAL LEVEL OF HIV PREVENTION programs, information sometimes refers to the first of three components in "information, education, and communication" activities, or IEC. Billboards, pamphlets, and public service messages on radio and television are examples of this narrow definition of information. However, this report uses information in the much broader sense commonly understood by economists and many others to include all types of knowledge, regardless of how it is acquired or shared. Thus, in this report, the phrase "providing information" encompasses such diverse activities as:

- making available basic knowledge, such as the facts about HIV transmission and how to protect oneself
- training in skills and motivation, such as how to negotiate condom use or to sterilize injecting equipment

- education, such as reproductive health and HIV/AIDS education in schools
- counseling to help people assess their own risks and take appropriate action.

Some ways of providing information are more effective than others in helping people to adopt safer behavior. Identifying the ways in which information can be provided most effectively at the lowest possible cost is an important operational question, but is beyond the scope of this report. While information, broadly defined and efficiently provided, can almost certainly change the behavior of some people to an extent, it is unlikely to be sufficient to change behavior enough to stop the HIV/AIDS epidemic, especially among those most likely to contract and spread the virus.

---

ity of infection is not easy, since most people do not know the HIV infection status of their partners or even the extent of their partners' current or past risky behavior. It is hardly surprising that, for people in the thrall of passion, addiction, or difficult economic circumstances, the immediate and certain costs of reducing their risk of contracting HIV will sometimes appear greater than the uncertain future benefits. Moreover, even if people making decisions about their private behavior take into full account the potential costs to themselves, they are likely to greatly undervalue the costs of their behavior to society in terms of secondary infections that unintentionally may result.

For these reasons, although better information on the risks of HIV will induce people engaging in high-risk behavior to reduce their risk somewhat, they will still engage in riskier behavior than society would prefer in the interest of curbing the AIDS epidemic.[4] Additional measures will be necessary to change the costs and benefits of risky behavior and its alternatives in ways that encourage safer behavior.

Lowering the Costs of Condom Use

Condoms are highly effective in preventing HIV transmission, both directly and by reducing the transmission of other STDs (Pinkerton and Abramson 1997). However, even people who are fully aware of the risks of HIV and of the protective benefits of condoms may not use them. The costs of condom use include not only the price of the condom but also the potential inconvenience and embarrassment of obtaining and using one, and, for some people, reduced sexual pleasure. Policies that lower these costs, by lowering the price of condoms, improving their availability, and increasing their social acceptability, would be expected to increase condom use and reduce HIV transmission.

Condom social marketing programs aim to achieve all of these objectives: they sell condoms at low, heavily subsidized prices; they make condoms readily available by selling them at nontraditional outlets, such as pharmacies, drug stores, truck stops, bars, and hotels; and they increase the social acceptability of condoms with advertising campaigns and grass roots activities, such as street theater, that show condom use as normal, healthy, and even fun. These programs are often targeted at low-income households, where the price of condoms is more likely to inhibit demand. Following the introduction of condom social marketing programs, sales have increased dramatically (figure 3.1). In many of the countries shown in figure 3.1, condoms were virtually unavailable or unknown prior to the program. Most condom social marketing programs are subsidized by international donors and many, such as the program in India, receive government subsidies. Some governments have also lowered condom prices by reducing import tariffs and sales taxes on condoms and latex rubber.

The potential impact of condom social marketing coupled with reductions in taxes and tariffs can be seen in Brazil. Before the social marketing campaign began, condoms cost between $0.75 and $1 each, and market volume was stagnant at about 45 million pieces; a tariff on imported condoms kept prices high and sales low. In 1991, subsidized *Prudence* brand condoms, priced at about $0.20 each, were launched with an intensive information campaign, and the tariff on imported condoms was gradually reduced. By 1995, total market volume had more than tripled to an estimated 168 million condoms (Clemente and others 1996). Both the domestic suppliers and the socially marketed condoms experienced sales growth, and *Prudence* condoms claimed about 11 percent of the greatly expanded market.[5]

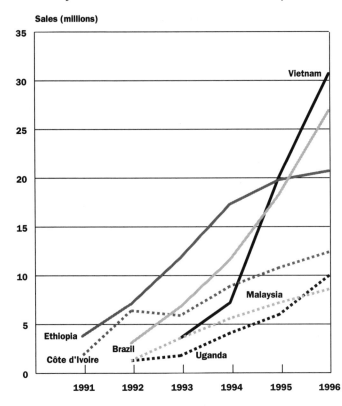

Figure 3.1 Socially Marketed Condom Sales in Six Countries, 1991–96

Condom social marketing programs raise condom sales by offering condoms at subsidized prices, by increasing their availability, and by popularizing their use.

Source: Statistical appendix, table 3.

Some governments, NGOs, and condom social marketing programs also have launched condom promotion programs focused specifically on commercial sex. The social marketing program in Cambodia, for example, includes special activities to reach sex workers and their clients at bars and hotels. Some of these programs distribute condoms to sex workers for free. Several have succeeded in raising condom use among sex workers and have shown a clear impact on the incidence of HIV. Box 3.5 describes the success of one such program involving sex workers in the Congo DR (formerly Zaire); similar success has been achieved in a program involving prostitutes in Nairobi, Kenya (Ngugi and others 1988; see box 2.6). Thailand launched a nationwide campaign that included condom distribution in brothels and a mass advertising campaign promoting condom use in commercial sex, with the goal of achieving 100 percent condom use in commercial sex. Condom use has risen dra-

Box 3.5 Preventing HIV among Kinshasa Sex Workers

AN STD PREVENTION PROGRAM IN KINSHASA, Congo DR (formerly Zaire), that offered 500 HIV-negative sex workers free condoms, STD tests, treatment, counseling, and group discussions about prevention succeeded in increasing condom use and reducing the incidence of STDs, including HIV (Laga and others 1994).

Prior to the program, only 11 percent of the sex workers used condoms, and those who did used them occasionally. Three months into the program, more than half of the women reported that they were using condoms consistently—that is, they were engaging in not more than one unprotected act per week. By the end of the three-year project, consistent condom use had risen to more than two-thirds of the participating women. The main obstacle cited for not achieving 100 percent condom use was refusal by male clients to use them.

In the course of the project, the incidence of HIV-1 dropped from 11.7 to 4.4 per 100 woman-years of observation (box figure 3.5). There was also a decline in the incidence of three treatable STDs—

gonorrhea, trichomoniasis, and genital ulcer disease—which probably contributed to the reduction in HIV transmission. Women who used condoms consistently and attended the clinic regularly had lower HIV-incidence than those whose condom use and visits to the clinic were less frequent.

Two important lessons from the success of this program are the complementarity of condom and STD treatment interventions with sex workers and the need to raise the willingness of clients to use condoms to ensure success. While increased condom use was probably most directly responsible for the reduction in HIV and STD incidence and a less costly component of the project, the STD treatment component was also important for ensuring the health of the sex workers and maintaining their participation in the project. Not taken into account in the evaluation is the fact that midway through the project, the social marketing program for *Prudence* condoms was launched in Kinshasa, and likely had an important impact on the willingness of clients to agree to condom use (Marie Laga, personal communication).

Box Figure 3.5 Incidence of HIV-1 and Other STDs among HIV-Negative Sex Workers over Three Years

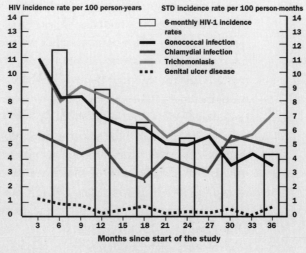

Source: Laga and others 1994. © The Lancet Ltd. 1994. Used by permission.

matically, to over 90 percent of commercial sex acts, and HIV prevalence has declined among several groups with high rates of partner change (Nelson and others 1996, Rojanapithayakorn and Hanenberg 1996).

Policies to increase condom use among people who have unprotected sex with many partners are a potentially powerful means of reducing the size of the HIV epidemic and thereby of providing significant benefits to the entire society. These programs, however, can be politically controversial. Providing subsidized condoms to people with many sex partners may be viewed by the public as condoning immoral behavior. Better understanding of the potential benefits of such programs—and the costs to everyone of failing to adopt them—is an important first step in increasing their acceptance by society generally and by constituencies that might otherwise oppose them.

Lowering the Costs of Safe Injecting Behavior

As we have seen in chapter 2, once HIV is introduced to a population of injecting drug users, the risk of infection to those who share injecting equipment is extremely high—higher and more immediate than for any other group engaging in high-risk behavior. Since HIV greatly increases the personal cost of injecting with shared equipment, we would expect to see substantial behavioral change, provided that there are low-cost ways for injecting drug users to avoid becoming infected.

Not all injecting drug users are addicted, but for those who are, quitting is rarely a low-cost option.[6] It can be extremely painful, can take a long time, and is often not successful: within a year or two of completing drug treatment, 70 to 80 percent of people treated for heroin use typically resume injecting (Golz 1993, McCoy and others 1997).[7] This is not to suggest that treatment programs are without value, even from the limited perspective of confronting HIV/AIDS. Often they are the only available channel for reaching injecting drug users with HIV prevention programs. And although treatment often does not result in permanent cessation of drug use, it can sometimes result in less risky behavior—lower injecting rates and less needle sharing—among those who resume injecting (Blix and Gronbladh 1988; Metzger 1997; Rezza, Oliva, and Sasse 1988). Overall, however, there is little evidence to suggest that treatment programs alone will be an effective means of containing the HIV/AIDS epidemic among injecting drug users.[8] Thus, although providing information on the risks of HIV transmission and lowering the costs of drug treatment programs may induce some inject-ing drug users to stop injecting altogether and others to adopt safer prac-tices, many are likely to continue risky injecting behavior unless the costs of safer injecting behavior—primarily the costs of obtaining sterile equipment or bleach—are sufficiently low.

Unfortunately, in many countries safer injecting behavior can be very costly to the injecting drug user. Research suggests that difficulty in ob-taining sterile injecting equipment is the most important reason that in-jecting drug users share equipment (National Research Council 1989, Vlahov 1997). Problems in obtaining sterile equipment are likely to be especially severe in developing countries, where low incomes and general scarcity of sterile injecting equipment and even bleach leave many in-jecting drug users with no access to clean equipment. In Myanmar, for example, many addicts go to professional injectors who sell the drugs, injecting one person after another with a single needle attached to an eyedropper or plastic tubing (Oppenheimer 1995). Injecting drug users in Ho Chi Minh City, Vietnam, reportedly obtain injections at "shoot-ing booths" ("Ho Chi Minh City . . ." 1996). A survey of one town in Manipur State in northeastern India found that virtually all who inject (97 percent) share equipment: in most cases, eyedroppers attached to sy-ringes (Sarkar and others 1996). Often, even if a sterile syringe can be obtained at a price that an injecting drug user is able to pay, possessing one can lead to arrest. In such instances, an injecting drug user must weigh the risk of HIV infection from sharing equipment against the risks of prison if caught with a syringe.

One inexpensive way for governments to lower the cost of safer in-jecting behavior is to remove legal barriers to the purchase and posses-sion of sterile injecting equipment. When armed with knowledge of the risks and provided with legal, easy access to sterile equipment, injecting drug users in many settings have rapidly adopted safer behavior. After needles were made available from pharmacies without a prescription in 1992 in the U.S. state of Connecticut, the percentage of injecting drug users who shared needles declined from 71 to 15 percent within three years (Span 1996). In Bangkok, Thailand, over 90 percent of injecting drug users surveyed reported changing their behavior to reduce their risk of HIV infection; 80 percent said they were obtaining sterile injecting equipment from pharmacies instead of sharing equipment with others (Choopanya and others 1991).[9] HIV prevalence among injecting drug users in Bangkok, which had been climbing rapidly, stabilized at around

40 percent, about half the level that was reached in neighboring countries where syringes and information about HIV were less readily available (Weniger and others 1991).

In some locales, governments or NGOs have moved sooner and more actively to minimize the harm from injecting drugs in the face of the HIV epidemic. Such "harm reduction" programs have often held seroprevalence among injecting drug users at remarkably low levels. In the five cities in table 3.1, harm reduction programs have maintained HIV prevalence among injecting drug users at less than 5 percent for at least five years, while in neighboring cities infection rates among injecting drug users soared to 50 percent or more (Ch'ien 1994; Des Jarlais and others 1995; Lee, Lim, and Lee 1993; Poshyachinda 1993; Wong, Lee, and Lim 1993). Each of the cities cited began programs early, before HIV had widely infected injecting drug users. The programs included information on safe injecting, needle exchange programs (NEPs), bleach programs, and referral for drug treatment. Besides containing the epidemic, these programs also raised the demand for drug treatment in all five cities. However, among all of the cities, only Sydney, Australia, launched a massive expansion of drug treatment programs.

Table 3.1 Prevention Programs in Cities That Have Kept HIV Infection among Injecting Drug Users below 5 Percent

Program site	Began early	Needle exchange	Legal purchase of injecting equipment	Bleach distribution	Community outreach	Expanded drug treatment	Extensive voluntary HIV counseling & testing	IDUs reporting change in behavior (%)
Glasgow (Scotland)	√	√	√		√			84
Lund (Sweden)	√	√			√		√	82
Sydney (Australia)	√	√	√	√	√	√		84
Tacoma (U.S.)	√	√		√	√			73
Toronto (Canada)	√	√		√	√			87

IDU Injecting drug user

Source: Des Jarlais and others 1995, table 3 and text.

Needle exchange programs, by providing new sterile injecting equipment in exchange for used syringes, minimize needle sharing and remove contaminated needles from circulation. Many NEPs also supply free condoms to help prevent the spread of HIV to the sexual partners of injecting drug users, as well as education about reducing high-risk behavior and referral for drug treatment.[10] NEPs have been widely adopted in Australia, New Zealand, and many Western European countries with great success in reducing transmission of HIV as well as other blood-borne infections, like hepatitis B and hepatitis C. A recent study compared trends in HIV prevalence among injecting users in a sample of 81 cities in North America, Europe, and Asia and the South Pacific (Hurley, Jolley, and Kaldor 1997). In the 52 cities without NEPs, HIV prevalence among injecting drug users increased by 5.9 percent per year, on average, while it declined by an almost equal percent annually (5.8 percent) in the 29 cities that had NEPs. Nonetheless, political opposition to needle exchange programs remains a potent force. In the United States, these programs are rare and distribution of syringes is often illegal. A recent study estimated that a national NEP in the United States would have prevented from 4,000 to 10,000 HIV infections between 1987 and 1995 among drug users, their sexual partners, and children; if such a program had been implemented in 1996, it could still have prevented from 5,000 to 11,000 infections over the next five years (Lurie and Drucker 1997).

Can harm reduction strategies also work in low-income countries? The answer appears to be yes. Although there is less experience with harm reduction strategies in developing countries, the remarkable success of a needle exchange program in Nepal suggests that such programs should be attempted more widely. Launched in 1992, the program has helped to hold HIV prevalence among Katmandu's 1,500 injecting drug users at less than 2 percent, even as HIV has soared in other Asian countries along the drug trade routes (box 3.6).

Where resources for needle exchange are lacking, bleach distribution may offer a less expensive alternative. Bleach is not only cheaper, it is also less controversial than needle exchanges, and, when used correctly, highly effective in killing HIV on infected equipment (Siegel, Weinstein, and Fineberg 1991). Furthermore, improved availability of bleach is virtually the only option for harm reduction among injecting drug users in prisons, since prison authorities are unlikely to distribute needles because they can be used as weapons. However, bleach is not always

Box 3.6 Harm Reduction among Injecting Drug Users in Nepal

NEPAL MOVED EARLY TO PREVENT THE SPREAD OF HIV among injecting drug users, with dramatic results. In 1992 a nongovernmental organization called the Lifesaving and Lifegiving Society began to provide education, condoms, bleach, needle exchange, and primary health care to about 650 of Katmandu's 1,500 injecting drug users. The group has collaborated with the Ministries of Home Affairs and Health and law enforcement agencies, while offering injecting drug users confidential and nonjudgmental assistance (Peak, Maharjan, and Crofts 1994).

Among the drug users participating in the program, the mean frequency of injecting fell from 24 injections per week, shortly before the start of the program, to 17 injections per week in 1994. The number of unsafe injections dropped by half, the number of people with whom equipment was shared fell 21 percent, and the number of times equipment was shared decreased 29 percent. Most telling is the fact that HIV prevalence among injecting drug users in Katmandu has remained low, less than 2 percent, even as it has soared in nearby countries (Maharjan and others 1994). Whether HIV prevalence remains low in the years to come will depend on continued harm reduction efforts among injecting drug users.

Mills and others (1993) estimated that the total costs for the program after one year of operation were $7,333, and that the average cost per client contact was $3.21 per contact. Since the virus often spreads from injecting drug users to their partners and their children, this expenditure represents an investment not only in protecting the injecting drug users themselves, but also in preventing a much wider epidemic.

readily available. For example, before 1991, bleach was all but unknown in Manipur State in India (Sarkar and others 1996); a bleach distribution program launched that year in the city of Churachandpur increased the percentage of injecting drug users using bleach to sterilize syringes from 31 to 72 percent. Bleach programs produce the greatest gains in life expectancy per uninfected injector when they are implemented early, while levels of infection among injecting drug users are still 2 percent or less (Siegel, Weinstein, and Fineberg 1991).

Programs that reduce the cost of safer injecting procedures often arouse fears that they will encourage people to initiate drug use or discourage addicts from seeking treatment. If efforts to reduce the cost of safer injection behavior did encourage illicit drug use, this would have to be weighed against the benefits of reduced HIV transmission. Fortunately, there is substantial evidence indicating that this is not the case (National Research Council 1989). Two evaluations of NEPs in six industrial countries failed to find any evidence that the programs raised the number of injecting drug users or that they increased the number of improperly discarded needles (Lurie and others 1993; Normand, Vlahov, and Moses 1995; U.S. GAO 1993).[11]

Raising the Cost of Risky Behavior: The Uncertain Effect of Enforcement

We have seen that lowering the costs of safer sex and safer injecting behavior can encourage less-risky behavior and reduce the transmission of HIV. But both prostitution and use of illegal and addictive drugs impose substantial negative externalities on society in terms of sexually transmitted and blood-borne diseases, crime, the costs of law enforcement, and incarceration. It is not surprising, therefore, that programs that attempt to promote safer behavior among sex workers, injecting drug users, and other people who engage in high-risk behavior may be politically unpopular, especially if they are perceived to be condoning prostitution or drug addiction. An alternative approach that often has broad popular appeal is to discourage such activities through punitive measures and stepped up enforcement of existing laws. From the standpoint of economic theory, such attempts to raise the cost of potentially risky behavior could discourage it, provided that costs can be raised sufficiently. However, because sexual activity and injecting drug use are private activities, enforcing prohibitions is costly and difficult in practice (Minon and Zwiebel 1995). Moreover, as we discuss below, such attempts may have unintended consequences that may exacerbate the epidemic.

Raising the costs of commercial sex. Attempts to curtail commercial sex have only rarely been effective. Indeed, numerous studies have shown that prohibition and punishment of commercial sex causes sex workers to shift their place of business and to change the way they solicit clients to avoid prosecution. Singapore, for example, attempted to eradicate prostitution by closing "red-light" districts in commercial areas; brothels soon appeared in residential areas (Ong 1993). Similar attempts in the Philippines have driven sex workers underground (Brown and Xenos 1994). These efforts rearrange the problems associated with commercial sex, but do not eliminate them. Worse, people who elude law enforcement and continue to engage in commercial sex are likely to become more difficult to reach with public health interventions to encourage safer behavior.

Enforcement problems aside, punishing commercial sex is not very efficient in preventing HIV, since the virus is transmitted not by commercial sex *per se* but by unprotected intercourse with multiple partners, regardless of whether one party pays the other. Commercial sex can be either high-risk or low-risk, depending on the activity and whether or not a condom is used. Thus, even if commercial sex could be eliminated by these measures, HIV could persist and spread through casual sex networks, although perhaps less rapidly. Moreover, it is reasonable to expect that the more effective is the curtailment of commercial sex, the more casual sex networks will expand.

These problems notwithstanding, it is possible for prohibition and punishment of commercial sex to slow the epidemic, provided that high-risk commercial sex transactions are reduced sufficiently and that non-commercial high-risk sex activities do not increase by an offsetting amount. China is one of the few countries in history that seems to have dramatically reduced prostitution and STDs in this manner, for about 20 years beginning in the early 1950s (Cohen and others 1996). However, this was achieved not as an end in itself, but in the context of massive social and political revolution and extensive government control of individual freedoms and the economy. The economic controls proved to be incompatible with growth and have since been relaxed; the internal migration that has ensued has created conditions conducive to the reappearance of prostitution, casual sex, and the spread of STDs.

Few, if any, countries are likely to be willing or able to impose the extensive social controls and bear the high costs experienced by China in the 1950s simply to control HIV. Short of such extensive efforts, attempts to eradicate commercial sex will cause some sex workers to find other types of work and will discourage some clients. Those who persist, however, will become more clandestine and thus more difficult to reach with information about HIV and policies that encourage condom use (*background paper*, Ahlburg and Jensen 1996).

An alternative to prohibiting and punishing commercial sex is to legalize and regulate it. Although this approach sometimes arouses considerable political opposition, it can make sex workers and their clients easier to monitor and to reach with information, condoms, and STD treatment. For example, in legal brothels in Australia condoms are used almost universally and STD rates are low (Feachem 1995). The public health advantages of regulated prostitution can also be seen in the historical experience of pre-independence India (box 3.7).

But regulated commercial sex faces the problems one would expect in any attempt to create a monopoly where there is a ready alternative supply. The likely result is a legal, high-priced market, presumably with lower transmission of HIV, and an unregulated, low-priced market, which authorities cannot easily monitor and where transmission of HIV

> ## Box 3.7 Health Benefits of Regulating Prostitution in Pre-Independence India
>
> REGULATED PROSTITUTION OFFERS MANY OPPORtunities to monitor the health of prostitutes and their clients, to treat disease, and to prevent infections. However, these measures are often politically difficult to implement. Historical experience with regulated prostitution in India under British rule illustrates both the public health benefits of regulated prostitution and the power of public opposition to undermine such efforts.
>
> The Contagious Diseases and Cantonment Act of India was enacted in 1864 to control the spread of STDs among the occupying British forces (Farwell 1989). The act regulated "first class" Indian prostitutes and brothels frequented by British sol-
>
> diers, mandating periodic medical inspection of prostitutes for infection. The act worked well in lowering the transmission of STDs. However, public opposition to these measures in Britain and India weakened enforcement. As a result, hospital admissions for STDs rose among the troops, peaking in the 1890s.
>
> A second Cantonments Act passed in 1899 gave the military forces more authority in curbing the spread of disease. Once prostitution could again be regulated, hospital admissions for STDs among the British soldiers fell dramatically—from 536 per thousand in 1895 to only 67 per thousand in 1909.

will presumably be higher (*background paper*, Ahlburg and Jensen 1996). In Singapore, for example, where brothels have been regulated since the government abandoned efforts to stamp out prostitution, a significant informal commercial sex sector persists. Moreover, mandatory health screening of prostitutes, even in the regulated brothels, is not always enforceable (Ong 1993). A similar pattern can be seen in Australia, where the 1986 Prostitution Regulation Act reduced the number of brothels in Melbourne by 65 percent. Result: the price of sex in brothels rose, while the number of streetwalkers and escorts increased (Hatty 1993).

In summary, prohibiting and punishing commercial sex is unlikely to be an effective approach to reducing HIV transmission. Although fewer people would presumably engage in risky behavior, those who do so despite the threat of prosecution will be more difficult to reach with public health interventions. Depending on the balance between these two outcomes, draconian measures to curtail commercial sex could actually worsen an HIV epidemic. The net impact of legalization and regulation is also difficult to predict and will depend on how much regulation raises the price of sex in the legalized sector and the extent to which programs are successful in encouraging safer sex in the illegal, unregulated sector.

Raising the costs of drug use. Arguments for and against raising the costs of drug use through prohibition and punishment are broadly similar to those for prohibiting commercial sex. Again, it is important to note that neither drug use *per se* nor injecting *per se* spreads HIV; rather, it is the sharing of unsterilized equipment that does so. Many who use illicit drugs do not inject, and often those who inject do not share needles. So efforts to eliminate illegal drug use, while they might be rationalized on other grounds, are potentially a very inefficient and costly way of reducing the unsafe injecting behavior that spreads HIV.

Nonetheless, the already substantial political appeal of drug enforcement efforts is likely to be strengthened by the close association between use of injected drugs and HIV. It is therefore useful to consider the likely impact on HIV of the two main strategies for eradicating illegal drugs: restricting supply, by attempting to halt the drug trade, and reducing demand, by punishing drug users or forcing them into treatment.

The most politically popular way to attempt to reduce drug use is to reduce the availability of drugs. However, drug interdiction may simply rearrange the problem or make it worse. For example:

■ Addicts may switch to other substances. In India, when the government tried to restrict the heroin trade, the price of heroin rose and addicts switched to synthetic opiates; injecting behavior was unchanged (Pal and others 1990).

■ Users may shift from smoking to injecting, which requires a smaller dose to produce euphoria but greatly raises the risk of HIV. For example, efforts to control opium smoking in Bangkok, Calcutta, and other areas in India were followed by an increase in heroin injection (Des Jarlais and others 1992; Sarkar and others 1993).

■ The drug trade may shift to other areas where people not previously exposed to drugs may begin injecting. For example, as a result of efforts to halt the drug trade in other regions, West Africa has emerged as an important transit point for cocaine from South America and heroin from Southeast Asia bound for Europe and North America. Similarly, stepped-up enforcement in Nigeria shifted the drug trade to Côte d'Ivoire, Zambia, and Zimbabwe (Stimson 1993).

If restricting the supply of injectable drugs does not effectively reduce risky injecting behavior and may actually increase it, what about attempting to reduce demand? Because most injecting drug users are chemically dependent, prohibition and threats of punishment are notoriously ineffective in reducing their demand for drugs. A survey of 450 injecting drug users in Manipur State in India, where addicts are imprisoned, found that only 2 percent regarded the threat of imprisonment as a reason to stop injecting drugs; half of the addicts surveyed had previously been in prison (Sarkar and others 1993). And far from reducing HIV transmission, imprisonment may have the opposite effect. Unable to obtain syringes, prisoners who inject drugs frequently resort to shared, improvised equipment, such as ballpoint pens, which would be very difficult to sterilize, even if bleach were available. Mandatory drug treatment is likely to be even less successful in ending drug use than voluntary treatment, since patients entering such programs presumably have very little desire to change their behavior.

In summary, efforts to raise the cost of injecting drugs through drug interdiction or the punishment of injecting drug users may increase rather than decrease risky injection behavior. Although the data on the impact of such efforts on HIV incidence are fragmentary, the available evidence suggests that harm reduction programs, including information about HIV, sterile injection equipment or bleach kits, and referral for voluntary treatment programs will be more effective and less costly in reducing risky injection behavior than drug interdiction or incarceration of addicts. This is particularly true given the strong evidence, discussed above, that injecting drug users *do* respond to this information by reducing risky injection behavior that spreads HIV.

Easing Social Constraints to Safe Behavior

A SECOND APPROACH TO REDUCING THE SPREAD OF HIV AIMS to change the social and economic factors that shape—and sometimes constrain—individual choices about risky behavior. Measures pursued by this approach have many other benefits besides reducing the HIV epidemic and they are already on the agenda of most developing country governments. The benefits are sometimes more difficult to quantify because of their broad impact, which extends far beyond HIV prevention. However, these measures are highly complementary to policies that directly affect the costs and benefits of risky behavior. Included in this approach are measures that alter social norms, raise the status of women, and reduce poverty.

Altering Social Norms

Some social norms discourage behaviors that transmit HIV, while others may be conducive to high-risk behavior or may discourage people from adopting safer behavior. HIV is likely to spread more widely where multiple, concurrent partnerships are the norm. In urban areas of parts of Sub-Saharan Africa, for example, traditional polygyny has evolved into many forms of formal and informal marriages and consensual unions, often concurrent and long term. The resulting sexual networks are highly conducive to the rapid spread of HIV (Caldwell, Caldwell, and Orubuloye 1989; National Research Council 1996). Social norms and peer pressure that encourage men to use the services of prostitutes or that venerate men with many female "conquests," while placing a high value on female chastity create the conditions for an explosive HIV epidemic.

Recent surveys reveal dramatic differences in the premarital sexual experiences of unmarried youth in different countries, even within the same region. The difference between what is acceptable for women and for men is also striking. In Rio de Janeiro (Brazil), for example, 61 percent of never-married men 15 to 19 had sexual intercourse in the past 12 months, compared with only 9 percent of never-married women (Caraël 1995). In Manila (the Philippines) and Thailand, 15 and 29 percent, respectively, of men 15 to 19 had sexual intercourse in the 12 months before the survey, but for women the share was 0 to 1 percent. There is also substantial diversity across countries in Sub-Saharan Africa; for example, in the Central African Republic, Côte d'Ivoire, Guinea-Bissau, and Kenya, young men and women are both very likely to have had sexual intercourse, while in Burundi and Togo, the proportion of young men and women who have had sexual intercourse is very small. The challenge for policymakers in countries with social patterns that are very conducive to the spread of HIV is to encourage safer behavior, without stigmatizing those who engage in unsafe behavior in ways that make them more difficult to reach with public health interventions.

Norms on marriage and childbearing can also affect the spread and prevention of HIV. The tradition of a man selecting a bride who is five or ten years younger than himself spreads HIV from one generation to the next (*background paper*, Morris 1996; Ssengonzi and others 1995). In eastern Africa, the traditional custom of levirate marriage was once a major contributor to the spread of HIV. According to this custom, a woman whose husband has died must marry his brother or, at a minimum, have sexual relations with the brother. In part because of the risk of contracting HIV, the practice is now in decline, but still widespread. Throughout Sub-Saharan Africa, mothers improve the survival of young children by breastfeeding as long as two years after birth. However, in some societies there is a taboo against marital sex while the mother is breastfeeding, leaving husbands to seek sexual gratification elsewhere. Finally, efforts to encourage condom use among married couples for HIV prevention will be especially difficult in societies where people want large families and a woman's social status and economic well-being are heavily dependent on the number of children she has. This is largely the case in Sub-Saharan Africa, where the benefits of increased condom use among married couples in suppressing the HIV/AIDS epidemic would be greatest (Bankole and Westoff 1995).

While sexual conservatism may be one of the best protections against HIV at the societal level, adherence to norms is never complete. An HIV epidemic can nevertheless occur, and conservatism can result in stigma of those who become infected or who are in social groups associated with high-risk behavior, making it more difficult to support safer behavior. Sometimes religious and political leaders stigmatize condom use as immoral, erecting additional social barriers and costs to safer and more responsible behavior. Thus, even though broadly based sexual conservatism may be helpful, treating HIV and the behavior that spreads it as a moral issue rather than a public health issue can hinder efforts to contain the epidemic.

Improving the Status of Women

In most societies, the lower social and economic status of women reduces their ability to insist upon male sexual fidelity and to negotiate safe sex. These problems can be particularly acute in societies where women's inheritance rights, property and child custody rights in divorce, and even the right to own land and other property are limited. In some instances, a wife's mere suggestion that her husband use a condom can provoke physical abuse. Even where the situation is not this stark, women's lower literacy, lower incomes, and low economic independence relative to men's give them less access to prevention information, fewer resources to purchase condoms and treat conventional STDs, and less ability to leave a relationship that puts them at risk of contracting HIV. For all these reasons, many more women than men face situations in which they are unable to choose behavior that would protect them from HIV.

Women who sell sexual services often face particularly constrained choices. Unemployment, divorce, desertion, and the breakdown of the extended family are among the factors that can lead women to offer sex for money (Plange 1990). One-half of sex workers interviewed in Calcutta cited extreme poverty as the reason for entering sex work, and 22 percent cited "family disturbances" (Chakraborty and others 1994). Often one of the few jobs available to unsupported, single women with limited education, prostitution can be quite lucrative compared with available alternatives. In The Gambia, for example, sex workers earn three times as much per day as women in informal sector jobs and as much as senior civil servants (Pickering and Wilkins 1993). Low-priced sex workers in Bali, Indonesia, earn more per week than the average civil servant does in a month (Wirawan, Fajans, and Ford 1993). In metropolitan Bangkok and a northeastern province of Thailand, the average take-home pay of sex workers is more than twice that of the average woman of the same age in other jobs (Bloom and others forthcoming). Controlling for their age and education, female sex workers earned over 50 percent more than they would have in other jobs for which they were qualified.

Women may also engage in commercial sex to fulfill family obligations. Throughout the developing world, poor families seek to insure themselves against economic risks by diversifying economic activities and through networks of family members spread over broad geographic areas. Adult children migrate to urban areas, abroad, or to specific areas to seek lucrative jobs that will allow them to save and remit earnings to their families. This motivation, coupled with the high returns for prostitution, explains much of the supply of sex workers, particularly in Asia (Archavanitkul and Guest 1994; Wawer and others 1996a). Fewer than one in ten sex workers in Bali, Indonesia, are from Bali (Wirawan, Fajans, and Ford 1993). Nepali women make up half of the population of prostitutes in Mumbai (Bombay) brothels (Human Rights Watch/Asia 1995). In Indonesia, Nepal, and Thailand, migration networks between particular villages that supply young women and specific areas in urban commercial sex districts are well established (Archavanitkul and Guest 1994; Human Rights Watch/Asia 1995; Jones, Sulistyaningsih, and Hull 1994).

Ending legal restrictions on women's rights, encouraging social equality, and raising economic opportunities for women not only make it easier for women to avoid HIV; they are also important in fostering development. Policies that specifically help women include increasing girls' school enrollment; guaranteeing equal employment opportunities; outlawing and severely punishing slavery, rape, wife abuse, and child prostitution; and guaranteeing inheritance, property, and child custody rights. A growing economy is also an important ingredient for increased economic opportunities for women. Of course, improving the status of women will also open new opportunities to choose risky behavior; a few women may do so. However, it is reasonable to expect that the vast majority of women, given expanded choices, would seize the opportunity to choose safe behavior and avoid HIV infection.

Reducing Poverty

Poverty and low socioeconomic status also constrain the decisions people make about risky behavior. Those with low incomes, for example, may not be able to afford to treat STDs or to buy condoms. Poor families may see commercial sex as a lucrative occupation for young and poorly educated daughters. People with less education may have less access to information about the dangers of high-risk behavior or be less able to understand prevention messages. This explains why those most likely to contract STDs and other infectious diseases within a society are the poor and uneducated. And it is supported by the cross-country analysis in chapter 1, which showed that developing countries with higher incomes have lower levels of HIV infection.

However, at the individual level within countries, the probability of HIV infection is often greater among men and women with higher incomes and schooling. Most of the evidence comes from studies conducted in eastern and central Africa in the late 1980s and early 1990s. For example, in a study of female outpatients and pediatric patients in Kigali, capital of Rwanda, women whose main partners had higher education and income were more likely to be infected than those whose partners had low education and income; infection rates showed similar patterns according to the partners' occupations (table 3.2). Among women attending family planning clinics in Dar es Salaam, Tanzania, those whose partners had more than twelve years of schooling were *five times* more likely to be infected than those whose partners had no schooling (Msamanga and others 1996). In Malawi, HIV seroprevalence was lowest for pregnant women whose partners had no education (5 percent) but rose to 16 percent among women whose partners had more than

Table 3.2 Percentage of Women Ages 19 to 37 Infected with HIV, by Their Partners' Socioeconomic Status, Kigali, Rwanda

Partner's characteristic	HIV-positive (percent)
Education (years)	
0–4	18
5–7	32
8–11	34
Monthly income (RWF)	
None	22
1–9,999	25
≥ 10,000	35
Occupation	
Farming	9
Military	22
Private sector	32
Civil servants	38

Note: Sample size: 1,458.
Source: Allen and others 1991.

seven years of education (Dallabetta and others 1993). In rural Rakai District, Uganda, household heads with any education were more likely to be infected than those with no schooling (*background paper,* Menon and others 1996b). In Kagera Region of Tanzania, the probability of dying from AIDS was higher for women with primary schooling or secondary schooling, compared with that of women with no schooling (*background paper,* Ainsworth and Semali 1997).

The gap in infection rates between the educated and uneducated in the early 1990s was greater in rural areas than in urban areas of Eastern Africa, for both men and women. In the town of Mwanza, Tanzania, for example, women were more likely than men to be infected, but there was no difference in HIV prevalence among men or women according to their schooling (table 3.3). However, in rural areas of Mwanza and in Rakai District, Uganda, there were sometimes marked differentials in prevalence by education, which were larger among women than among men.[12]

Finally, higher-income adults in Central and Eastern Africa were more likely to be infected than those with lower incomes. Workers with higher incomes were more likely to be infected in two businesses in Kinshasa, Congo DR (formerly Zaire), in the late 1980s (Ryder and others 1990). The workers in the (higher-paying) bank had higher HIV infection rates than those in the (lower-paying) textile mill and, within each

Table 3.3 Relation between Education and HIV Status, Men and Women, Mwanza Region, Tanzania, and Rakai District, Uganda

Study site	Level of schooling	HIV prevalence (percent)	
		Men	Women
Mwanza, Tanzania (urban)	Fewer than 4 years	9.6	15.3
	4 years or more	8.5	15.3
Mwanza, Tanzania (rural)	Fewer than 4 years	2.7	3.0
	4 years or more	4.2	5.9
Rakai District, Uganda (rural)	None	7.5	13.5
	Primary	17.6	29.8
	Secondary	19.7	40.7

Sources: Barongo and others 1992, Grosskurth and others 1995b, Serwadda and others 1992.

firm, managerial workers had higher HIV prevalence than manual workers. In Rakai District, Uganda, heads of household with higher-quality dwellings were half again as likely to be infected as those without, controlling for age, gender, marital status, education, and occupation (*background paper*, Menon and others 1996b).

Why would adults with higher socioeconomic status have higher HIV infection rates? First, men with higher education and income will find it easier to attract and support additional commercial and casual sexual partners. For example, analysis of data from the WHO/GPA sexual behavior surveys found that in five African sites, as well as in Thailand, Manila (the Philippines), and Rio de Janeiro (Brazil), the more education a man has, the more likely he is to have had a casual, nonregular sexual partner (*background paper*, Deheneffe, Caraël and Noumbissi 1996).[13] A second reason is that men and women with more education and higher incomes are likely to travel more and thus have more opportunities for a variety of sexual contacts.

Do these results imply that for HIV, unlike other infectious diseases, including other STDs, poverty reduction and rising education may actually *increase* the spread of HIV? This would seem to contradict the finding in chapter 1 that HIV infection rates are lower in countries with higher income and literacy.

This seeming discrepancy between findings at the individual and international levels can be explained by two factors. First, at the time many of these people became infected, in the early- and mid-1980s, awareness and knowledge of HIV prevention were low. Thus, the protective advantages that higher education and income would normally provide—a greater ability to learn about HIV prevention and more resources to purchase condoms or take other steps to avoid infection—did not come into play. Second, lacking this knowledge and failing to take protective measures, those with higher income with their greater number of partners were more exposed to HIV. Since HIV prevalence is cumulative, over a long enough period this would result in higher HIV prevalence among those with higher education and income than among the poor, with fewer partners.

If these explanations are correct, then as knowledge of how to avoid HIV infection becomes available, people with more education and higher incomes would be in a better position to learn about it and avoid infection. As a result, HIV incidence should decline more rapidly among those who are better-off, eventually reversing the positive relation between income and prevalence found in African studies.

The limited available evidence suggests that this is in fact occurring. In Brazil, for example, about three-quarters of those newly diagnosed with AIDS through 1985 and for whom educational data were available had a secondary or university education. By 1994 only about one-third of those with newly diagnosed AIDS had that much education (Parker 1996). In urban areas of Butare, Rwanda, HIV incidence was higher among women in low-income households (Bulterys and others 1994). This result is consistent with trends in industrial countries; in the United States, for example, new infections are more likely to occur among those with low socioeconomic status (Cowan, Brundage, and Pomerantz 1994; Krueger and others 1990). Further, DHS data confirm that in all of the developing countries studied, the more education men and women had, the greater was the likelihood that they were using condoms (figure 3.2).[14] In another study, Thai men with the highest permanent income and assets were more likely than other men to consistently use condoms with sex workers (Morris and others 1996). Among 21-year-old Thai military conscripts in the early 1990s, HIV incidence was lower for conscripts with more education (Carr and others 1994).[15]

Several studies of sex workers have found that women with higher incomes were more likely to use condoms and had lower levels of HIV infection. In three cities in São Paulo State, Brazil, for example, sex workers who charged higher prices had fewer clients, were more likely to have always used condoms in the past year, were less likely to have injected drugs, and consequently were less likely to be infected with HIV

Figure 3.2 Percentage of Men and Women Using Condoms with a Casual Partner, by Education, Eight Countries

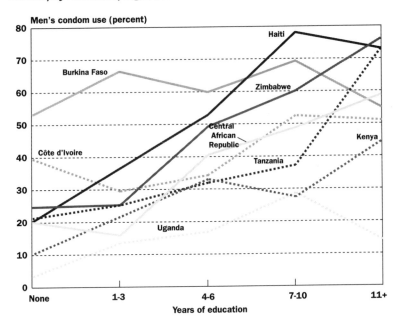

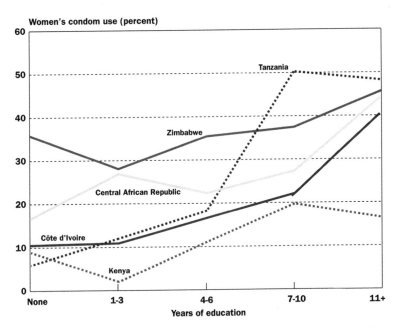

Men and women with more schooling are more likely to use condoms for casual sex.

Note: This figure shows the probability of condom use, holding constant age, urban residence, occupation, and assets. The reference period varied. See note 14 at the end of the chapter.
Source: Background paper, Filmer 1997, from DHS data.

and other STDs than were low-priced sex workers (Lurie and others 1995).

In conclusion, despite the links found between higher socioeconomic status and HIV infection in some areas, government policies to raise incomes and schooling and reduce poverty should lower the economic constraints to engaging in safer behavior and reduce the incidence of HIV over the long term. These policies are often pursued because they have far-reaching social benefits; they are also highly complementary to short-term interventions to encourage safer sexual and injecting behavior and will likely enhance the impact of these interventions.

Setting Government Priorities in Preventing HIV

LOWERING THE COSTS OF SAFER SEXUAL AND DRUG-INJECTing behavior for those who are most likely to contract and spread HIV can reduce risky behavior, and this in turn will have a powerful impact on the course of the epidemic. Given the many possible ways of attempting to achieve these ends, what are the highest priority programs from the perspective of government spending?

This section proposes a broad prevention strategy for governments at all stages of the epidemic to maximize the impact of limited government resources in curbing the spread of HIV. In keeping with the principles of public economics, governments should finance or directly implement interventions that are essential to stopping the spread of HIV but that private individuals or firms will not have sufficient incentive to finance on their own—namely, provision of public goods, reduction of the negative externalities of behavior that spreads HIV, and protection of the poor from HIV infection. Programs that address these issues will improve the efficiency and equity of government prevention efforts. In addition, following the principles of epidemiology from chapter 2, program effectiveness will be improved if governments act as soon as possible and if they succeed in preventing infection among those most likely to contract and spread HIV. These recommendations are not meant to limit the scope of government involvement if there are resources and public will to undertake even more. Rather, the intention is to identify the minimum set of activities that all governments should be engaged in to improve the efficiency and equity of prevention programs, and a rational order in which to expand them.

This strategy is a broad one, based on underlying principles of public economics, epidemiology, and cost-effectiveness. It remains for individual countries to identify the specific combination of programs, policies, and interventions to pursue this strategy in a cost-effective way. We discuss a number of factors that sometimes uniquely affect the cost-effectiveness of public activities to prevent HIV. However, even within these guiding principles, the cost-effectiveness of HIV prevention programs is likely to vary considerably according to the setting. Programmatic choices are necessarily country-specific because the costs and effectiveness of interventions, as well as the characteristics and accessibility of those most likely to contract and spread HIV, vary widely across settings.

Public Economics and Government Priorities

Government policies can potentially address three types of market failures that occur with respect to the prevention of HIV. The first problem involves the underprovision of public good—specifically, the lack of incentives for the private sector to collect and disseminate information crucial to the prevention of the epidemic. The second involves the negative externalities of high-risk behavior: people deciding whether or not to take steps to protect themselves from HIV are likely to consider the costs to themselves of becoming infected but may not consider the cost of secondary HIV infections that may result if they become infected. The third type of market failure involves equity: very poor people are less able to protect themselves against HIV than others.

Providing public goods. The collection and production of information about the prevention and control of HIV is almost entirely a public good, since it is impossible for a private agent to capture the resulting benefits. Essential information includes data on the levels and trends of HIV and STD infections, the prevalence of high-risk behavior, and the costs and benefits of prevention programs (box 3.8). Everyone benefits from such information, but there is insufficient incentive for it to be produced privately in sufficient quantity.

Reducing the negative externalities of behavior that spreads HIV. If people engaged in risky behavior were fully informed of the dangers of HIV infection and how it could be prevented, if they had the means to prevent it, and if they were the only ones to suffer consequences if they became infected, then the argument for government intervention to prevent their infection would be weak. However, unprotected intercourse with multiple partners and unsafe injecting

Box 3.8 What Data on HIV and Other STDs Should Governments Collect?

THE TYPES OF INFORMATION DESCRIBED BELOW are crucial if the government is to design and implement effective measures to reduce high-risk behavior.

- *Levels and trends of HIV and other STD infections.* Information on the levels and trends of HIV and other STD infections in the general population, in subpopulations with high- and low-risk behavior, and in specific geographical areas is necessary to monitor the spread of the epidemic and the impact of public policies. This information is typically gathered through unlinked anonymous testing of blood samples from groups of individuals—women attending antenatal clinics, blood donors, STD patients, and injecting drug users in treatment programs—as part of an epidemiological surveillance system (AIDSCAP and others 1996, Chin 1990, Sato 1996). Specialized surveys can track trends in HIV prevalence among subpopulations that engage in high-risk behavior. In countries where the HIV epidemic is nascent, levels and trends in the prevalence of other STDs can be indicative of patterns of sexual behavior that will spread HIV.

- *Prevalence of high-risk behavior and the characteristics of those who practice it.* Information on the prevalence and nature of risky behavior is indispensable for estimating the potential size and rate of spread of an HIV epidemic, and deciding when, where, and how to intervene. What are the high-risk behaviors in a given

country? Are they prevalent in the general population or confined to distinct, identifiable groups? These questions can be answered through representative surveys of the general population and smaller surveys of behavior in groups thought to be at high risk. Gathering such information can be very difficult: sexual behavior surveys are often controversial, and surveys of both sexual and drug-use behavior face logistical and accuracy problems. The information is nonetheless crucial to estimating the future course of the epidemic, the number of individuals at high risk, and the most effective approach to reducing risky behavior.

- *The costs of interventions and their impact on HIV incidence.* Much information has been collected on the impact of programs on knowledge and behavior, but regrettably few studies have documented the impact of these behaviors on the incidence of HIV; fewer still have attempted to estimate cost-effectiveness. Assessments of cost-effectiveness in terms of the impact on HIV incidence, rather than on behavioral change alone, are likely to highlight the crucial importance of reducing risky behavior among those most likely to contract and spread HIV. Such studies can be helpful not only in shaping more effective policies, but also in overcoming political objections to subsidizing safer behavior among people most likely to transmit HIV to others.

practices raise the risk of infection for everyone, even for those who do not engage in high-risk behavior.

As shown at the beginning of this chapter, people who practice risky behavior are likely to take some steps to reduce their risk of contracting HIV because of the enormous personal costs if they become infected. However, because they react primarily to the increased risk to themselves, their response will not reflect the full costs—that is, the negative externalities—of unintentionally spreading HIV to the rest of society. Many people who do not practice high-risk behavior would be willing to pay those who do practice it to adopt safer behavior in order to contain the epidemic and thus lower everyone's risk of contracting HIV. Government is the only mechanism through which this can happen. Thus, there is a strong justification for government to subsidize safer behavior among people who engage in high-risk behavior in order to protect everyone—either by preventing primary infections among those who engage in high-risk behavior or, when such individuals are already infected, by preventing the infection from spreading to others.

The principles of public economics and epidemiology are therefore in agreement that governments should give high priority to the prevention of infection among people most likely to contract and spread HIV. The extent of negative externalities of risky behavior can be measured by the number of secondary infections that a person would generate if infected with HIV. By this definition, the riskier an individual's behavior, the more secondary infections he or she is likely to generate, and thus the greater the negative externalities of his or her behavior. The reproductive rate of HIV among such individuals is likely to be much greater than 1, sustaining the epidemic. On the other hand, people who practice low-risk behavior—such as abstinence from sex or injecting drug use, sexual monogamy, consistent condom use, or injecting with sterile equipment—will generate few, if any secondary infections. They are vulnerable to the negative externalities generated by others' risky behavior.

Some readers may ask: Shouldn't government devote at least as many resources to preventing the spread of HIV among people who do not engage in high-risk behavior but who may nonetheless contract the virus? After all, such individuals usually constitute the vast majority of the population. In countries where government has a broad mandate to provide health services to everyone, and where resources are adequate, expanding government-financed preventive interventions to the low-risk population should not be ruled out. However, even in these cases, governments should first ensure that programs achieve adequate coverage of those most likely to contract and spread HIV, starting from those with the highest-risk behavior, since this is the most efficient way to protect everyone.

There are situations, however, in which interventions for the entire population are essential to the success of programs aimed at reaching those most likely to contract and spread HIV. For example, although information that helps people with low-risk behavior to further reduce their risk will have very little impact on the overall epidemic, they need to understand how HIV is transmitted and—more important—how it is *not* transmitted. Inaccurate information or even accurate information presented in ways that create feelings of fear and vulnerability can incite discrimination against those who are at high risk of contracting and spreading HIV and people who are HIV-positive or living with AIDS (Allard 1989). This is not only unwarranted and unjust, but it may also compromise the ability to launch the prevention programs most likely to slow the spread of HIV, as well as efforts to mitigate the epidemic discussed in chapter 4. Information that fuels fear of HIV-infected people has been a serious problem in nearly all countries where there has been extensive early publicity about the epidemic. To avoid this problem, governments should make certain that the general population understands that HIV cannot be caught from a handshake or other casual contact, and take steps to reduce stigma and protect those at risk of HIV from discrimination.

Helping poor people to avoid HIV. Government's first step should be to lower the costs of safe behavior for the poor by improving the functioning of markets, for example, by eliminating tariffs on condoms and restrictions on condom advertising. If prevention is still too costly for the poor, subsidies may also be warranted. In countries where particular groups face significant barriers to acquiring information—because of illiteracy, different languages, or lack of access to newspapers, radio, or television—subsidized information for the disadvantaged groups will improve their access to prevention. Besides subsidized information about HIV, the most important preventive measures are those that make it easier for poor people to obtain condoms, treatment for STDs, and access to safe blood. However, unless there is substantial overlap between the poor and those with high-risk behavior, subsidized HIV interventions for the poor will address equity but will not be sufficient to contain the epidemic, particularly in its early stages.

Cost-Effectiveness from a Government Perspective

Governments, like all entities that must operate within a fixed budget, seek "value for money" in deciding between alternative expenditures. Cost-effectiveness analysis is a tool for deciding among alternative courses of action when resources are scarce. It seeks to determine how to achieve the maximum effect within a given budget or, viewed differently, how to achieve a desired effect at the lowest possible cost. For HIV prevention and other health interventions, the effects or benefits per dollar spent can be calculated without regard to who pays for the intervention or who benefits. However, some health interventions that are cost-effective by this measure would be undertaken and financed by individuals themselves even without government involvement. (For example, many individuals will seek counseling, use condoms, or use sterile injecting equipment to avoid contracting and spreading HIV, whether or not these services are subsidized by the government.) Measures for which individuals would have paid on their own are not a priority for a government's limited funds. Rather, governments should first spend their limited resources on interventions that are cost-effective in the conventional sense *and* that would not occur in the absence of government involvement—that is, cost-effective interventions that provide public goods and reduce the negative externalities of high-risk behavior. This strategy will ultimately prevent the largest number of subsequent infections among all subgroups in the population, including the majority who do not engage in high-risk behavior.

Some governments have the mandate to intervene broadly in the health sector and to assume responsibility for providing curative and preventive health care for all citizens, irrespective of whether government spending is efficiency-enhancing. Would this mandate change the priorities discussed above? No, it would not. Unless the costs of changing behavior among those most likely to contract and spread HIV are extraordinarily high, promoting behavior change among them is likely to be the most cost-effective in terms of preventing infection in the general population.

Several factors that affect the cost-effectiveness of alternative publicly financed HIV/AIDS prevention programs do not necessarily enter into cost-effectiveness calculations for private individuals. There are also a number of issues specific to the cost-effectiveness of HIV interventions that need to be taken into account.

Public benefits include averted secondary infections. In calculating the benefits from prevention programs, it is important to consider not only the benefits received by the person directly affected, but also the number of secondary infections that are averted. Secondary infections measure, in effect, the extent to which negative externalities are being addressed by programs. Failure to include secondary infections among the benefits of public health programs will result in serious underestimates of the benefits of prevention in groups that are very likely to contract and spread HIV, since the number of people in these groups tends to be small relative to the total population but the number of secondary infections per person tends to be large.

Public prevention should augment, not replace, private prevention efforts. Public programs will be more cost-effective if they can raise additional demand for risk-reducing behavior without "crowding out" private sources of inputs. For example, in societies where condoms are already widely available at low cost, subsidized condom projects that are not specifically directed toward those most likely to contract and spread HIV may simply shift the financing of existing condom sales from private pockets to public ones, with no impact on the epidemic. Whether the resulting redistribution enhances equity depends on the income distribution of condom purchasers relative to the general population. As we saw earlier in this chapter, high-income sex workers and men with higher education are more likely to use condoms for casual and commercial sex. Thus, untargeted condom subsidies are likely to help those who are relatively better off rather than enhance equity, and are likely to substitute for condoms that were already being purchased privately rather than generate additional demand. The same argument holds for the potential substitutability of public for private services for STD treatment and safe blood. The effectiveness of public programs should therefore be measured as the difference between outcomes with and without the program.

HIV/AIDS prevention has other external effects. Many HIV/AIDS preventive interventions have additional positive effects that might be overlooked if cost-effectiveness calculations considered only the impact on HIV/AIDS. For example, condom programs also prevent other STDs, which are spread in the same ways as HIV and generate similar negative externalities. HIV education in the schools and encouragement of condom use among adolescents who engage in sex will complement efforts to postpone sexual activity, reduce teenage pregnancy and abortion, and as a result improve school completion rates among girls. Harm reduction programs among injecting drug users generate demand for drug rehabilitation programs and reduce transmission of hepatitis B and C and other blood-borne diseases. Although these benefits are difficult to quantify, they should not be overlooked.

Involving other actors can improve cost-effectiveness. While governments must support prevention efforts among those who are most likely to contract and spread HIV, governments are not necessarily the most effective agent to design and implement such programs. Enlisting and subsidizing NGOs to help in the design and implementation of interventions can greatly improve the cost-effectiveness of public programs, particularly if the NGOs are staffed by or representative of people at high risk of contracting and spreading HIV. The cost-effectiveness of all interventions can also be improved through other government measures, such as relaxing legal prohibitions and minimizing the stigma that people in high-risk subpopulations often face, so that NGOs can work more effectively. Public-private cooperation in confronting AIDS is discussed at greater length in chapter 5.

Which Interventions Are Cost-Effective?

Only a few HIV/AIDS interventions have been rigorously evaluated with respect to their impact on the incidence and prevalence of HIV; among those evaluated, interventions targeted to those who practice high-risk behavior tend to be more effective (Aral and Peterman 1996; Choi and Coates 1994; National Research Council 1996; Oakley, Fullerton, and Holland 1995). Appendix A of this report presents the results of 22 of the more rigorous evaluations of HIV/AIDS interventions that have taken place in developing countries. Unfortunately, information on the costs of these interventions is usually unavailable; their cost-effectiveness has rarely been evaluated.

The impact of alternative prevention strategies in four epidemics. The effectiveness of alternative interventions will be strongly influenced by the nature of the intervention itself and by the heterogeneity of the behavior that is fueling the epidemic. To illustrate this point, Van Vliet and others (1997) have simulated the impact of increased condom use and increased treatment of curable STDs (chlamydia, gonorrhea, and syphilis), on a heterosexual HIV epidemic in the four hypothetical populations described in chapter 2 using the STDSIM simulation package.[16] The simulations show the impact of increased condom use and STD treatment in various groups in each of the four populations fifteen years after the start of the epidemic. (As seen in chapter 2, in each population earlier intervention among those most likely to contract and spread HIV would be more effective than the later behavior change discussed here.)

The simulations show the impact of increased condom use in three groups of people with different rates of partner change—female sex workers, men with casual or commercial partners, and women in stable relationships. These groups are commonly the focus, respectively, of outreach programs to sex workers, socially marketed condom programs, and reproductive health services. In these simulations, the term " sex worker" refers to women with the highest rate of partner change—10 new partners per week, or more than 500 partners per year. In the real world, of course, some women who have very high rates of partner change do not regard themselves as "sex workers" and they may contact their male partners in a variety of settings. There are other men and women in these imaginary populations with large numbers of partners, but fewer than 500 per year. The impact of different interventions in all of these groups can be simulated, individually and simultaneously. However, for expositional purposes, we present simulated interventions with only three groups.

In the baseline scenario, before any intervention, we assume that only 20 percent of sex workers and 5 percent of men having sex with casual or commercial partners use condoms consistently, that is, in every act of intercourse. We also assume that none of the women in steady relationships are using condoms.[17] The simulations show the impact of instantaneously raising consistent condom use among sex workers to 90 percent, and among the other two groups to 20 percent.[18] These levels were selected because the authors believed they are realistically achievable in some countries. In other countries, it may be possible to exceed the levels of condom use simulated here. Among those who use condoms, the failure rate through breakage or misuse is assumed to be 5 percent.

The simulations also show the impact of increased STD treatment on HIV prevalence. The baseline scenario assumes that 25 percent of all STD cases that produce symptoms are effectively treated and that there is no specific screening or treatment program for sex workers. The simulations show the impact of increasing the share of symptomatic STDs treated to 75 percent in the general population and, in a separate scenario, the impact of implementing a monthly screening and treatment program that covers 90 percent of sex workers. In the latter intervention, it is assumed that 5 percent of sex workers are not cured. The assumptions behind the baseline scenario and the five condom and STD interventions are summarized in table 3.4.

The simulated impact of increased condom use and STD treatment on the HIV prevalence of adults in the four hypothetical populations is

Table 3.4 Summary of the Assumptions before and after Interventions, STDSIM Modeling

(percent)

Assumption	Baseline	After intervention
Using condoms consistently		
Sex workers	20	90
Men with casual or commercial partners	5	20
Women ages 15–50 in stable relationships	0	20
Symptomatic STDs treated	25	75
Sex workers with monthly screening and		
treatment for STDs	0	90

Source: Background paper, Van Vliet and others 1997.

shown in figure 3.3. Despite the different underlying patterns of sexual behavior, the impact of specific interventions shows some striking consistencies across populations:

■ Achieving 90 percent condom use among sex workers results in a dramatic drop in HIV prevalence in all three populations where there is commercial sex (a, b, d), even though sex workers comprise only a very small share of each population (0.25 percent or fewer women). STD screening and treatment for sex workers is far less powerful.

■ Increased STD treatment among the general population is less effective than raising condom use among those with many partners. This is not surprising, since people with high-risk behavior generate a disproportionate number of STD cases, and condoms prevent transmission of both HIV and STDs. STD treatment among the general population and condom use by women in stable relationships have the largest impact in the populations with concurrent casual sex (b, c).

■ The impact of greater condom use among women in stable relationships is very small, and in the serial monogamy population (d) it has almost no impact on the epidemic. In the population where the epidemic is driven by commercial sex (a), condom use by monogamous women slightly accelerates a decline in HIV prevalence, while in the other two populations it merely slows the growth of a still-expanding epidemic.

Figure 3.3 The Impact of Changes in Condom Use and STD Treatment in Four Populations with Different Patterns of Sexual Behavior

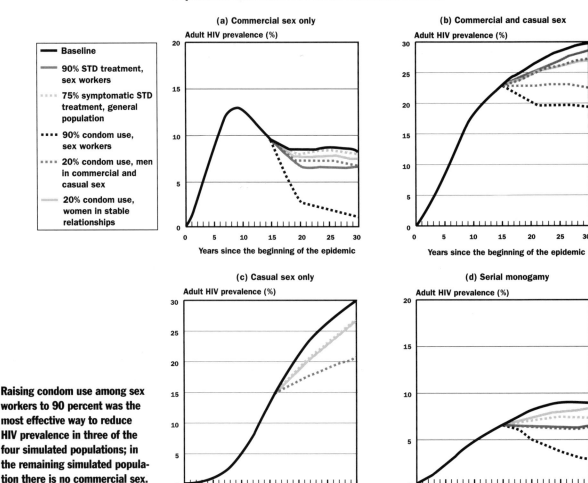

Source: *Background paper,* Van Vliet and others 1997.

Raising condom use among sex workers to **90 percent** was the most effective way to reduce HIV prevalence in three of the four simulated populations; in the remaining simulated population there is no commercial sex.

Looking at the ranking of interventions within specific populations, we see that:

- In the epidemic fueled by commercial sex alone (a), all of the interventions produce an absolute decline in HIV prevalence; in the epidemic fueled solely by casual sex (c), none of the simulated interventions is sufficient to cause an absolute decline.

- In the population with concurrent commercial and casual sex (b), prevalence declines in response to 90 percent condom use among sex workers; 20 percent condom use by men in casual sex keeps prevalence from rising. Other interventions in this population merely slow the expanding epidemic.
- In the population practicing serial monogamy (d), increased condom use by sex workers is the only intervention that results in an absolute decline in HIV prevalence.

In reality, no intervention is an "either/or" proposition. There is always spillover between interventions intended for people with different rates of partner change, so any particular intervention is likely to change behavior in more than one group of people but to differing degrees. Further, combined interventions to multiple groups will have greater impact than single interventions; for example, working only with sex workers to raise condom use will be less effective than simultaneously working with both sex workers and their clients. However, these simulations show that the greatest share of the impact will be achieved through the interventions that succeed in changing the behavior of those with the highest rates of partner change.

The broad conclusion that can be drawn from these simulations is that, although the overall pattern of sexual behavior in the population does affect the impact of interventions, prevention of infection among those with the highest rates of partner change has a large effect irrespective of the underlying patterns of sexual behavior in the population. Assuming that it is not a great deal more expensive to increase rates of condom use among those with high rates of partner change—sex workers or others—focusing condom subsidies and promotion efforts on changing their behavior is likely to be highly cost-effective.

Studies of the cost-effectiveness of HIV interventions in developing countries are rare and not transferable. Fewer than half a dozen studies have documented the costs *and* effects of preventive interventions in developing countries (Beal, Bontinck, and Fransen 1992; Gilson and others 1996; Moses and others 1991). An overview of the results of several cost-effectiveness studies in developing countries is in appendix B of this report. Most evaluation studies measure impact by changes in intermediate behaviors that are believed to affect risk—such as the increase in condom use, or knowledge about HIV prevention, or the number of people receiving sterile syringes. The number of HIV infections averted is then extrapolated on the basis of assumptions about the relationship between the behavior and HIV incidence. However, the lack of good data on sexual behavior and on the relation between sexual behavior and incidence makes these benefits very difficult to assess. Virtually no studies, except those based on simulations, have measured the impact of interventions on secondary cases where they are thought to exist (Over and Piot 1996; Stover and Way 1995; *background paper*, Van Vliet and others 1997). None to our knowledge have taken into account the external benefits of interventions or the issue of complementarities between interventions.

While cost-effectiveness studies can be very useful in deciding among alternative interventions in a given setting and stage of the epidemic, their conclusions are usually not easily transferable to other settings (*background paper*, Mills and Watts 1996). For example, an evaluation of the effectiveness of enhanced treatment of symptomatic STDs in reducing HIV incidence in rural Mwanza Region of Tanzania found that the intervention lowered HIV incidence by 42 percent at a cost of roughly $10 per person treated, or $234 per primary HIV infection averted (Gilson and others 1996; Richard Hayes, personal communication).[19] However, treatment costs clearly could be much higher in a middle-income country, and effectiveness may have been quite different in an area with lower HIV prevalence than the Tanzanian study site (4 percent of adults were infected).[20] Moreover, without estimates of the costs and impact of alternative interventions in the same area, we cannot say whether a particular intervention is more or less cost-effective than other interventions in reducing HIV transmission. Ideally, we would like to know the costs and effects of alternative interventions implemented in the same setting, but this has rarely been done (box 3.9).

Interventions focused on those most likely to contract and spread HIV should be more cost-effective from the public perspective, because preventing infection in a person with risky behavior prevents many secondary infections among individuals with whom they mix—some of whom practice high-risk behavior and some of whom practice lower-risk behavior. In fact, the extent to which ongoing programs affect those who practice high-risk behavior is often unknown. For example, condom social marketing programs improve the access of the poor to condoms, but it is still not known to what extent these condoms are used by those in the highest-risk groups. Information on the level, distribution, and type of high-risk behavior; the number of people involved; and their charac-

Box 3.9 The Cost-Effectiveness of Prevention among Those with the Highest Risk

HOW MANY HIV INFECTIONS COULD BE AVERTED by spending an additional $1 million per year on prevention in groups with different levels of risk of HIV infection? The answers generated in a recent study in the United States demonstrate the high cost-effectiveness of focusing prevention expenditure on those most likely to contract and spread HIV, as well as the additional benefits from early intervention (Kahn 1996).

The study defined four risk groups in terms of the level of steady-state HIV prevalence they would attain without intervention—high risk (50 percent prevalence); medium risk (15 percent); low risk (1 percent); and very low risk (0.1 percent). Steady-state HIV prevalence is defined as the point at which the number of new infections exactly matches the number of people exiting the group through death or eliminating the risk factor (for example, stopping injecting drug use). Examples of these groups in the United States include young homosexual men in San Francisco (high risk), injecting drug users in San Francisco (medium risk), and women attending STD clinics in California (low risk). Very low risk includes most of the general population, including women of childbearing age in 41 of the country's 50 states.

The results of the exercise depend on the assumptions with respect to the costs and impact of different interventions. The author initially assumed that prevention costs for one individual in any given group would be $200 per year, then examined the sensitivity of the results to this assumption. This $200-per-year figure was based on a survey of the annual cost of various prevention interventions for high- and low-risk groups in the United States.[1] In terms of the impact of interventions, the author assumed that interventions lower risk by 10 percent, which he believes is a conservative estimate.

Box table 3.9 shows the number of HIV infections prevented, given these assumptions, by spending $1 million on each of the several groups; either with late intervention (after steady-state prevalence has been reached) or with early intervention (before

(Box continues on the following page.)

Box Table 3.9 HIV Infections Averted by $1 Million Annual Spending on Prevention, U.S. Estimates

Risk group	Baseline HIV prevalence (%)	HIV infections averted	
		5-year horizon	20-year horizon
High-risk			
Steady state	50	164	681
Pre–steady state	10	93	837
Medium risk			
Steady state	15	58	348
Pre–steady state	3	14	112
Low risk			
Steady state	1	4	26.2
Pre–steady state	0.2	0.8	5.4
Very low risk			
Steady state	0.1	0.4	2.6

Source: Kahn 1996.

Box 3.9 *(continued)*

steady-state prevalence has been reached). The impact of each intervention is shown for a five-year time horizon and for a 20-year time horizon. One million dollars prevents the most cases if focused on the highest-risk group in the steady state. However, the benefits of intervening early with this group only become evident in the 20-year time horizon.

These figures understate the impact of prevention in high-risk groups, however, because secondary infections prevented in the partners or children of people in high-risk groups are not included. The number of infections averted in low-risk groups will not be much affected by this omission, but among high-risk groups the total of averted infections might be several times higher, depending on the group and the degree of sexual mixing with lower-risk groups.

The result—that prevention in high-risk groups is most cost-effective—is robust to large changes in the assumptions about the effectiveness of the intervention. If programs reduce risky behavior by 50 percent instead of 10 percent, the number of infections averted in the high-risk steady state rises to 830 for the five-year simulation and 3,750 for the 20-year simulation, while the number of infections

averted for the low-risk steady state rises to only 18 and 93, respectively. Even if prevention is substantially more successful at changing behavior in the low-risk groups, the higher effectiveness of prevention in high-risk groups remains.

While interventions in high-risk groups are more effective, they are also potentially more costly. However, the study estimated that interventions in the low-risk groups (steady state) would have to be roughly one-fortieth to one two-hundredth (1/40–1/200) the cost of an intervention in high-risk groups (steady state) to prevent the same number of infections averted by intervening in high-risk groups in the steady state. In other words, in the high-risk steady state, interventions in the low-risk population would have to cost $1 to $5 per person per year, compared with $200 per person per year in the high-risk population, to prevent an equivalent number of HIV infections for a $1 million outlay.

---

[1]The programs and costs per person per year included: annual testing and counseling ($40–110); bleach distribution and outreach ($60); 3-session group counseling for IDUs ($75); needle exchange ($40–800); peer workshops for high-risk gay men ($250); 5-session counseling for low-risk women ($269); 12-session counseling for medium-risk gay men ($470).

teristics is a basic public good and will enhance efforts to improve cost-effectiveness by helping to improve the targeting of programs.

Cost-effectiveness and the accessibility of target populations. Although it is highly desirable to focus public interventions on those who are most likely to contract and spread HIV, identifying and reaching these individuals can be difficult, especially where legal sanctions and social stigma cause these people to want to avoid being discovered. The costs of reaching those most likely to contract and spread the virus can have a significant impact on the cost-effectiveness of interventions.

Figure 3.4 shows a stylized classification of groups of people according to the extent to which they practice high-risk behavior and their pre-

Figure 3.4 Classification of Groups by Riskiness of Their Behavior and Their Accessibility

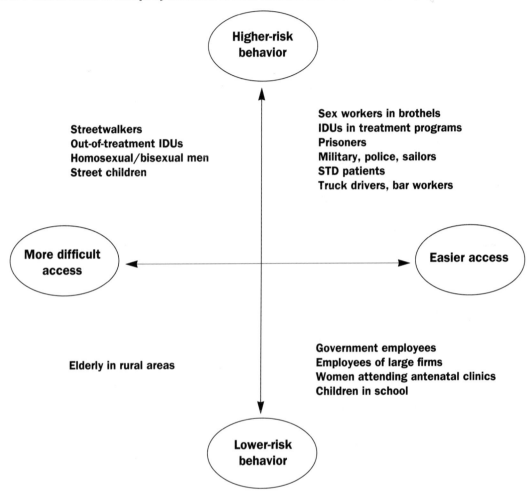

IDU Injecting drug user
Source: Adapted from Adler and others 1996, figure 8. Used by permission.

sumed accessibility. Of course, the extent to which members of these identifiable groups practice risky behavior varies considerably across settings and with the effectiveness of prior prevention efforts. Thus, such a figure would need to be modified according to the situation in a specific country, on the basis of the results of HIV and behavioral monitoring systems.

In the upper-right quadrant are groups with higher-risk behavior that are relatively easy for government agencies and collaborating prevention

Prevention programs should focus on people most likely to contract and spread HIV; some of these individuals are easy to identify and reach, others less so.

partners to reach. The benefits of behavior change in these individuals are relatively high, especially given the large number of secondary infections generated, while the costs of locating them will be relatively low, enhancing cost-effectiveness. In the upper left quadrant are groups who also practice high-risk behavior but who are less easy to reach. In these instances the benefits of behavior change will again be great but the costs of locating and working with these individuals may be high, reducing the net benefits. In the lower-right quadrant are people who, on average, are presumed to practice lower-risk behavior but to whom access is easy. The benefits of intervening in these groups may not be great, but the costs of reaching them may be very low. Inexpensive interventions for these groups may still be cost-effective relative to some alternatives (box 3.10). The lower left quadrant includes people who are very unlikely to contract and spread HIV and are very difficult and costly to reach; of the four types of groups, this is the lowest priority for public-sector HIV prevention efforts. Note that the accessibility of most groups can be improved through government actions to reduce stigma, decriminalize behavior, and educate the public on the nontransmissability of HIV by casual contact and the benefits of working with these groups.

Of course, the "groups" identified in figure 3.4 are not homogeneous with respect to their risky behavior. Since individuals with high-risk behavior cannot be easily identified, programs need to focus interventions on people with characteristics that are highly correlated with risky behavior. However, some sex workers consistently use condoms, and some government employees who have many partners do not. Intervening to change the behavior of people with specific characteristics like age, sex, occupation, or geographic area is not a perfect way to reach those with high-risk behavior. Some members of these groups will be exposed to interventions even though they practice low-risk behavior. Moreover, others with high-risk behaviors who don't belong to any of these groups will be missed. The lack of perfect criteria for focusing interventions on those with the highest-risk behaviors is one source of leakage of program resources. This reduces the cost-effectiveness of interventions if resources go to people with lower-risk behavior. On the other hand, leakage may improve cost-effectiveness if resources go to people with even riskier behavior than those targeted. Surveys of sexual behavior, such as those conducted by WHO/GPA and DHS, can help to overcome these problems by establishing the characteristics and geographic location of those who have unprotected sex and high rates of partner change. Unless programs

Box 3.10 Educating Adolescents on HIV/AIDS: A Sound Investment

IN COUNTRIES WHERE SEXUAL ACTIVITY BEGINS at an early age and young people have high rates of partner change, promoting safer behavior among adolescents is clearly important to slowing the spread of HIV. There are many possible interventions to address risky behavior among adolescents, both in and out of school. However, even in societies where sexual activity does not generally begin until after young people have completed their schooling, reproductive health education in the school system—which includes information on the benefits of postponing sexual activity as well as how to prevent pregnancy, STDs, and HIV for those who do not abstain—is a potentially powerful intervention. Besides preventing HIV among students who might otherwise adopt risky behavior, these programs have many other benefits. They prevent STDs and associated infertility, and they prevent unwanted pregnancy, which may lead to abortion or to girls' dropping out of school. More broadly, reproductive health education that includes education on preventing HIV may help to alter the social norms of the next generation of adults in ways that encourage safer behavior.

Such programs are sometimes unpopular with parents who worry that information on reproductive health, STDs, and contraception might cause their children to become sexually active at an earlier age. Research has shown that this is not the case. Reviews of school-based programs have found that participating youth have not begun sexual activity earlier (Gluck and Rosenthal 1995, Kirby and others 1994, UNAIDS 1997). Moreover, a review of school-based programs in the United States found that programs that included sexual health education and AIDS prevention not only delayed the start of sexual activity, but reduced the number of partners and raised contraceptive use among those who became sexually active (Gluck and Rosenthal 1995).

Given the other broad social benefits and the relatively low cost of adding HIV/AIDS education to existing programs, HIV/AIDS education is likely to be a good investment in preventing HIV. The overwhelming majority of AIDS program managers who responded to the *AIDS in the World II* survey felt that, as of 1993, reproductive health education needed expansion in their countries (Mann and Tarantola 1996). This was true for countries at all stages of the epidemic.

are carefully designed and implemented, targeting prevention programs to groups with specific characteristics may stigmatize members and lead to discrimination, making future prevention efforts more difficult and less effective.

Government Priorities, Resource Constraints, and the Stage of the Epidemic

Considerations discussed in the preceding sections suggest a broad prevention strategy that assigns priority to activities based on the principles of epidemiology, public economics, and cost-effectiveness. Whatever the stage of the epidemic, this strategy calls for a strong emphasis on prevention activities starting with those at highest risk of contracting and spreading HIV and covering as many others as available resources allow. As the epidemic expands, containing it will increasingly require efforts to prevent infection among people with relatively lower levels of risk, which will raise the cost of prevention activities. If the epidemic is to be contained, this expansion of activities must not weaken the fundamental commitment to work with those most likely to contract and spread the virus. This section presents a minimum set of activities to improve the efficiency and efficacy of national prevention programs, and suggests a framework for deciding the order in which to expand activities if additional resources are available.

Provision of *public goods* or ensuring their provision by regulation is an important role for government at all stages of an HIV/AIDS epidemic. Governments should invest in the information-collecting infrastructure they need to monitor the epidemic and to identify where high-risk behaviors are practiced and how to reach those at highest risk. In a nascent epidemic, understanding levels of HIV and STD infection in the subpopulations most likely to contract and spread HIV, the underlying patterns of sexual behavior, and the nature of the links to lower-risk subpopulations are critical pieces of information for assessing the probability of a more extensive epidemic. As the epidemic progresses, governments need to monitor, in addition, the spread of HIV to populations at lower risk and to support evaluation of the costs and effectiveness of alternative interventions.

The need to *reduce the negative externalities* of high-risk behavior, as well as the insights of epidemiology, argue for heavily subsidizing safer behavior among those most likely to contract and spread HIV in countries at all stages of the HIV/AIDS epidemic. It is not necessary to wait for HIV to take hold to make widespread knowledge, 100 percent condom use, and quick recognition and treatment of other STDs the norm among those more likely to contract and spread HIV, such as sex workers, bar maids, long-distance truck drivers, the military and police, miners or factory workers living away from home, and homosexual and bisexual men with multiple partners. The case for rapid action is even stronger in the case of injecting drug users because of the potential for very rapid increases in HIV prevalence in this subpopulation, and the subsequent spread of HIV to others, including their sex partners (some of whom may be sex workers) and their children. In a nascent HIV epidemic, such highly focused actions may be sufficient to dramatically slow the spread of the virus. In countries with concentrated and generalized epidemics, preventing HIV among those with the highest chances of contracting and spreading the virus is still essential to slowing the epidemic. However, behavior change among those with less-risky behavior who may nonetheless inadvertently spread the virus will be necessary to reverse the course of the epidemic.

With respect to ensuring *equity*, in countries with nascent epidemics, government can protect the poor best by acting early and decisively to prevent an epidemic. In countries with generalized epidemics, the risk of infection has risen for everyone and poverty should not restrict access to preventive services. Government can ensure that the poor have access to the knowledge, skills, and means to prevent HIV.

The *cost-effectiveness* of interventions aimed at people with different levels of risky behavior will also change as HIV spreads from those who practice more risky behavior to those at relatively lower risk. Interventions for those most likely to contract and spread HIV are still likely to be highly cost-effective. In countries where HIV has spread widely in the general population, the cost-effectiveness of preventive interventions for lower-risk populations, such as subsidies for STD treatment, provision of safe blood, and reproductive health and AIDS education in schools, improves. However, these programs do not generate many externalities: their benefits accrue mainly to the person who uses them. While such interventions will prevent HIV infections and save lives, they will not do so as efficiently as interventions focused on those more likely to contract and spread HIV and they will often not be sufficient to reverse the course of the epidemic. Furthermore, the costs of providing these services to the entire population at low risk of spreading HIV are potentially very large. In light of the substantial individual benefits of these services, people who are not poor will usually be willing and able to pay for them. In countries with scarce financial resources, the priority should be first on guaranteeing equity in access to these services by the poor.

Not all developing countries face equal resource constraints in pursuing this strategy. In the lowest-income countries, prevention should start among those whose behavior generates or is likely to generate the highest reproductive rate for HIV; subsidized prevention to others with lower HIV reproductive rates that are still greater than 1 can be extended as resources permit. Middle-income countries may have the resources, even at the earliest stages of the epidemic, to finance interventions for a much larger share of those for whom the HIV reproductive rates would be greater than 1. They may also have the resources to subsidize to a greater extent services for the poor and to extend prevention subsidies to subpopulations that are unlikely to spread HIV to others.

At an operational level it is impossible to determine the actual or potential reproductive rate of HIV for any group of individuals. However, using information about the average number of partners, condom use, and injecting behavior from surveys and epidemiological surveillance, subpopulations in a given country can be ranked from those with the highest-risk behavior (those most likely to contract and spread HIV) to those with the lowest-risk behavior. Figure 3.5 shows a stylized view of the ranking of several subpopulations in a hypothetical population according to the extent of risky behavior at a specific point in time, and how the scope of prevention efforts would expand to include groups with increasingly less-risky behavior depending on the availability of resources. Once the highest-priority subpopulations have been effectively reached, programs can be expanded to cover those with progressively less-risky behavior, provided that sufficient resources are available. Indeed, if sustained behavior change is achieved in the highest-priority groups, the relative priority assigned to other groups will increase. Of course, no ranking of this sort can apply to all countries, or even to a single country over time. To overcome problems in locating the people most likely to contract and spread HIV at any given point in time, it is essential that policymakers and program managers finance collection of the necessary information for cost-effective use of the scarce resources available for HIV prevention.

In concluding, we return to the important epidemiological point that countries at the nascent stage of the epidemic have a unique opportunity to act early, to make a few key investments, and largely prevent an HIV epidemic. Not all countries with low levels of infection will necessarily go on to experience an HIV epidemic, even without government action. However, our inadequate understanding of the distribution of different behaviors in the population and the links between different subpopulations make it difficult to predict which among the countries will be so lucky and which will not. Furthermore, even in countries where high-risk behavior is relatively rare, patterns of sexual and injecting behavior can change with economic and social conditions. Interventions at the nascent stage are the most effective and will likely involve far less total cost than if implemented after HIV has saturated subpopulations with high-risk behavior. Further, because the number of people in these sub-

Figure 3.5 Resource Availability and Program Coverage

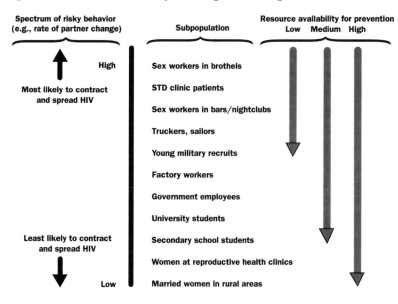

Note: This is a hypothetical example only and is not meant to reflect the situation in any particular country.
Source: Authors.

populations is small relative to the entire population, the absolute costs of prevention will be relatively low.

These recommendations are not meant to limit the scope of government involvement if there are ample resources and public will to undertake even more. Rather, our intention is to point out the minimum set of activities that all governments should be engaged in to improve the efficiency and equity of prevention programs and a rational order in which to expand these activities if HIV spreads or more resources become available.

The National Response

NEARLY ALL DEVELOPING COUNTRIES HAVE RESPONDED IN some way to the challenge of HIV/AIDS, often with the active assistance of donor countries and multilateral institutions. But developing countries' AIDS prevention efforts are diverse and generally undocumented, making it difficult to evaluate the extent to which high-priority policies are already being implemented.

AIDS in the World II includes a survey of managers of national AIDS control programs in 118 countries, covering such issues as political commitment, organization, coordination, management, preventive and treatment responses, program evaluation, and human rights (Mann and Tarantola 1996).[21] However, to date there has been no systematic evaluation of developing countries' responses to the epidemic, and in particular of the priority and effectiveness of different activities. The overwhelming impression left by the *AIDS in the World II* survey and by many national AIDS control plans is that many countries have launched a collection of AIDS prevention activities without clear-cut priorities; indeed, many programs do *not* focus on preventing infections among people most likely to contract and spread the virus. The reasons for this are only partly due to a lack of understanding of the importance of these priorities; domestic political considerations and the preferences of international donors for particular programs are probably also responsible. Nevertheless, in such an environment, limited resources are likely to be stretched very thin, and the cost-effectiveness of public expenditure on prevention is likely to be low.

How well have these prevention efforts corresponded to the priorities recommended in this chapter? Available information is scarce, but it suggests that policy performance could be improved in at least three areas: provision of the information necessary to combat the epidemic and plan sound programs (public goods); ensuring prevention of HIV among those most likely to contract and spread it (reducing negative externalities); and making sure that poor people have access to the means to protect themselves (equity).

Expanding the Information Set

The limited evidence available suggests that as much as one-quarter of all developing countries have yet to initiate systematic monitoring of HIV prevalence. A background study for *AIDS in the World II* classified countries into four groups according to whether HIV sentinel surveillance sites were planned (but not yet operating), limited, many, or extensive (Sato 1996). The distribution of 123 developing countries in these four groups is shown in table 3.5; data for individual countries are

Table 3.5 Distribution of Countries, by the Number of Sentinel Surveillance Sites and the Stage of the Epidemic, January 1995

Stage of epidemic	Planned sites	Limited sites	Many sites	Extensive sites	No information on sites	Total (%)	Number of countries
Nascent	21	59	3	3	14	100	29
Concentrated	7	48	36	5	5	100	42
Generalized	0	0	52	43	0	100	21
Stage unknown	16	58	3	0	23	100	31
Total (%)	11	46	23	10	11	100	
Number of countries	14	56	28	12	13		123

Source: Tabulations based on data in table 2 of the statistical appendix.

in table 2 of the statistical appendix to this report. The good news is that more than three-quarters of these countries reported having at least limited sites for HIV sentinel surveillance as of January 1995. Countries at the generalized stage of the epidemic were most likely to have many or extensive sites, while those at the concentrated and nascent stages were most likely to have limited sites. However, one in five developing countries at the nascent stage of the epidemic reported no sentinel surveillance sites, and for another 14 percent the extent of sentinel surveillance was unknown. If we add all the countries reporting only planned sites to those countries for which information about sites was inadequate to determine their extent, we find that 27 countries—more than one-fifth of countries at all stages of the epidemic—were not reporting any operating HIV sentinel surveillance sites in January 1995.

Even in those countries that have some form of sentinel surveillance, information on HIV prevalence among those most likely to contract and spread the virus is often lacking. Our classification of countries by stage of the epidemic requires information on HIV prevalence in at least one subpopulation presumed to practice riskier-than-average behavior and, if HIV prevalence exceeds 5 percent of one of these groups, prevalence among women attending antenatal clinics. Although delays in the receipt of reports probably account for some missing data, available information was insufficient to classify 31 developing countries by stage of the epidemic, usually because of a lack of information on those presumed on average to practice riskier behavior—sex workers, injecting drug users, men who have sex with men, the military, and STD patients.[22] Among the 123 developing countries we attempted to classify, 43 countries (35 percent) had no information on HIV prevalence in any group with presumed high-risk behavior during the last 5 years. While it is more convenient to monitor HIV trends among blood donors and women attending antenatal clinics, regular monitoring of HIV prevalence early in the epidemic among those who practice high-risk behavior is much more important. Because of the potential for explosive growth of HIV among injecting drug users, HIV prevalence in this group should be monitored at least once a year and preferably more often (AIDSCAP and others 1996, Chin 1990).

In addition to ensuring more and better monitoring of HIV prevalence, governments urgently need information on patterns of sexual behavior, condom use, and drug-injecting behavior. As we saw in chapter 2, the heterogeneity of behavior and the extent of mixing between people with high- and low-risk behavior determine the baseline shape of the AIDS epidemic. Countries at all stages of the epidemic need information on the prevalence and distribution of risky behaviors among representative samples of men and women to understand the likely path of the epidemic and how the epidemic can be minimized. However, this information remains scarce. Fewer than 20 developing countries are known to have undertaken sexual behavior surveys, such as those sponsored by GPA or DHS, that could provide such information.

Finally, evaluations of HIV prevention programs often fail to measure the costs of interventions or their impact, reporting instead on indicators of process and implementation (Mann and Tarantola 1996). Information on costs and impact is important not only for assessing the most efficient allocation of resources, but also for demonstrating the effectiveness of interventions to change behavior among those most likely to contract and spread HIV and the spillover benefits to the low-risk population. Developing countries need better information on the costs and effects of pilot interventions, which interventions affect whose behavior, and at what cost.

Preventing Infection Among Those Most Likely to Contract and Spread HIV

The *AIDS in the World II* survey shows that most countries have at least a few interventions that focus on people most likely to contract and spread HIV, but most of these programs achieve only limited coverage (Mann and Tarantola 1996). To arrest the epidemic, the coverage of preventive measures directed toward these groups needs to be expanded considerably.

For example, recent DHS surveys in seven African countries—all of which have been very hard-hit by the AIDS epidemic—reveal that only 40 to 70 percent of men and women with a recent nonregular partner named condoms as a means of preventing HIV transmission (figure 3.6). In countries such as Tanzania and Uganda, where nearly everyone knows someone who has died of AIDS, such low awareness of the benefits of condom use is shocking. Given this low level of knowledge, it is not surprising that consistent condom use is also low. In Malawi, for example, a recent survey found that only 30 percent of people who had nonregular partners consistently used condoms (Lowenthal and others 1995). In Côte d'Ivoire, only 5 percent of those in "high-risk relationships," including relationships in which one partner was infected, reported using

Figure 3.6 Percentage of People with a Recent Nonregular Partner Who Are Aware That Condoms Prevent HIV Transmission

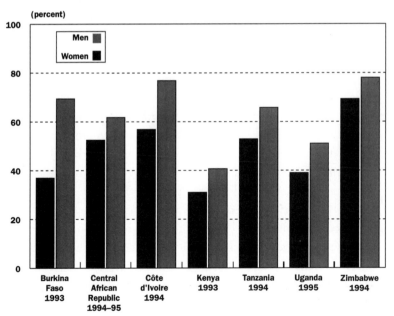

Even in countries with widespread epidemics, many women and men with a recent, nonregular sexual partners are not aware that condoms prevent the spread of HIV.

Note: A nonregular partner means a casual partner or a sex worker. The reference period for these countries varied.

Source: DHS data.

condoms during every act of intercourse (Coleman and others 1996). Condom use in Uganda has increased substantially, especially among the young, but is still far short of adequate coverage (Asiimwe-Okiror and others 1997, Stoneburner and Carballo 1997). In contrast, Thailand has been extraordinarily successful in raising condom use, particularly among those at higher risk of contracting and spreading HIV; and there is clear evidence of declines in HIV prevalence among some population groups (box 3.11).

Although many people practicing high-risk behavior are difficult to reach for prevention, some people whose circumstances may put them at higher risk of HIV infection are part of "captive" populations that can be readily identified: the military, police, and prisoners (box 3.12). Because governments usually have much easier access to these organized groups than to injecting drug users, sex workers, or others with many casual partners, it should be possible for governments to reach nearly all members of each of these groups with preventive interventions. Have governments done this?

The Civil-Military Alliance to Combat HIV and AIDS conducted a survey of prevention activities among the militaries of 50 countries, about half of which were low- and middle-income countries (Yeager and Hendrix 1997).[23] Although individual country results are not available (countries were assured that responses would be confidential), the aggregate data clearly reveal that prevention programs in the military often fall short of complete coverage. For example, 80 percent of the responding countries indicated they had policies supporting condom use by the military, but only 55 percent indicated they had "written plans to operationalize these policies." The most specific condom promotion policies were found in African militaries, which have been very hard hit by HIV. It is striking that about 20 percent of the responding militaries do not distribute condoms at all, while most of the others offer condoms free of charge but only if requested by a soldier.

Low coverage of other groups of people presumed to practice riskier-than-average behavior can be seen in a small survey of UNAIDS Country Programme Advisers conducted for this report. The survey asked the advisers who work in 32 developing countries to identify groups of people with high-risk behavior and to comment on the extent to which programs focused on prevention in these groups; the approximate percentage of each group covered; and the extent of government encouragement and finance of these programs. Although the results reflect the assess-

Box 3.11 Thailand's Response

BETWEEN LATE 1987 AND MID-1988, HIV PREVA-
lence among injecting drug users in Bangkok
exploded from 0 percent to more than 30 percent.
In response, local and national governments
launched an extensive risk reduction program
among injecting drug users that included training
on how to disinfect injecting equipment and education on HIV prevention. By 1989 a survey in Bangkok found that 59 percent of users had stopped
sharing needles, while others had reduced sharing or
were using sterile equipment. Nationwide, HIV
prevalence in injecting drug users stabilized at 35 to
40 percent (Brown and others 1994).

Meanwhile, the sexual transmission of HIV was
increasing. The first round of national sentinel surveillance in mid-1989 detected 44 percent HIV
prevalence among brothel-based sex workers in the
northern city of Chiang Mai; brothel-based sex
workers elsewhere in the country had HIV prevalence rates of 1 to 5 percent. In addition, a national
behavioral survey in 1990 found that 22 percent of
Thai men ages 15 to 49 had visited sex workers the
previous year. By then, Thai NGOs and the government had begun efforts to improve public awareness
of HIV and to promote condom use; knowledge of
how HIV is spread and how it can be prevented was
nearly universal, and condom use was increasing
nationwide.

Efforts were stepped up dramatically in 1991,
when the government of Prime Minister Anand
Panyarachun mounted an extensive and intensive
national prevention effort with a greatly expanded
budget. Government ministries, NGOs, businesses,
and communities began working together to promote condom use, reduce risky behavior, change
norms regarding commercial sex, improve STD
treatment, and provide care and support for those
affected by HIV. By 1996 the government was providing more than $80 million annually for HIV/
AIDS prevention and care.

An essential component of this response is the
"100% Condom Program" which aims to enforce
consistent condom use in all commercial sex estab-
lishments. Condoms are distributed free to brothels
and massage parlors, and sex workers and their clients
are required to use them. Local coalitions of government officials, health officers, and police check compliance by tracing the contacts of men seeking treatment at government STD clinics. Brothels that fail to
comply may be closed. Extensive efforts to reach the
clients of sex workers have been crucial to the success
of the campaign. Through mass media campaigns,
education and skills building in workplaces and
schools, and peer education efforts, condom use during commercial sex rapidly became the norm among
Thai men who purchased sexual services.

Results have been encouraging. Condom use in
brothels increased from 14 percent of sex acts in
1989 to more than 90 percent by 1992 (box figure
3.11a). The number of new STD cases among men
treated at government clinics dropped from nearly
200,000 annually in 1989 to about 20,000 in 1995.
Most strikingly, HIV prevalence among young male
conscripts entering the Royal Thai Army has
declined from a peak of 4 percent in mid-1993 to
1.9 percent by late 1996 (box figure 3.11b). The
Thai response is a compelling example of the principles presented in this chapter. Epidemiological and
behavioral data needed to design effective programs
were collected and broadly disseminated. Acknowledging and working with the commercial sex industry, rather than against it, the Thai program developed ways to change the behavior of sex workers and
their clients, while simultaneously promoting
changes in social norms. With commercial sex transmission slowed, increasing efforts are now being
devoted to addressing the social and developmental
determinants of risk through programs such as con-

(Box continues on the following page.)

Box 3.11 *(continued)*

tinued schooling and work opportunities for young rural women to keep them from entering sex work.

Of course, Thailand's response would have been even more effective—and the current epidemic smaller—had the extensive prevention efforts been launched earlier. Obstacles encountered in launching the program and the ways in which these were ultimately overcome are discussed in chapter 5.

Box Figure 3.11a Rising Condom Use by Sex Workers and Declining STDs in Thailand, 1988–95

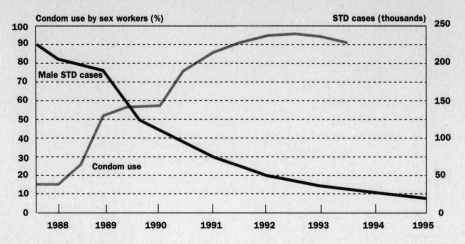

Source: Rojanapithayakorn and Hanenberg 1996.

Box Figure 3.11b Declining HIV Prevalence among Young Thai Army Conscripts, 1989–96

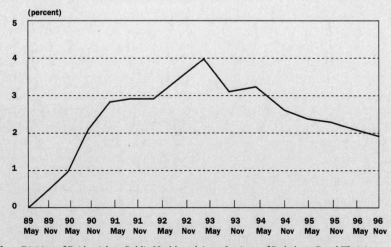

Source: Data from Division of Epidemiology Public Health and Army Institute of Pathology, Royal Thai Army.

Box 3.12 STDs and HIV in the Military

BOTH THEIR DEMOGRAPHIC CHARACTERISTICS and their occupation put members of the military at high risk of acquiring STDs and HIV and of passing them on to others (Miller and Yeager 1995). Military recruits are generally young, sexually active men, and often unmarried. They may be easily influenced by peer pressure, especially when stationed away from home. In time of war, the risks of contracting HIV and STDs may seem low relative to the risk of death in combat. For these reasons, military personnel often have STD and HIV infection rates higher than the general population (box figure 3.12).

Sexually transmitted diseases are likely to be particularly rife in military units far from home. Over a five-year period in the 1830s, for example, 32 to 45 percent of British soldiers stationed in India were hospitalized for sexually transmitted diseases, compared with only 2 to 3 percent of Indian soldiers (Farwell 1989). While the Indian soldiers were often married and lived with their wives and families, very few British soldiers were allowed to marry and all were far from home, where social norms might have tempered their sexual habits. In the early 1890s, the mean rate of admission for STDs among British troops on home soil was half the rate among British troops in India. During the 1960s, STD rates among U.S. Army troops stationed in the continental United States (30 per 1,000 troops per year) were one-ninth to one-fifteenth the rates among troops stationed in Vietnam (262 per 1,000), the Republic of Korea (344 per 1,000) and Thailand (453 per 1,000) (Greenberg 1972).

The military is one group—a potentially large one—in which government can act decisively to prevent STD and HIV transmission through information, condom programs, and STD treatment. Monitoring interventions and their impact in the military is also easier than for other subpopulations.

Box Figure 3.12 HIV Prevalence in the Military

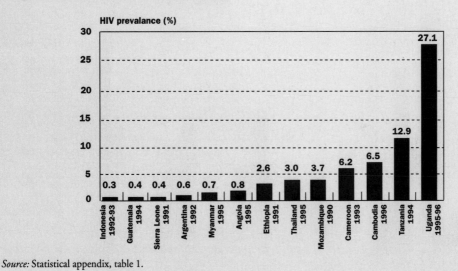

Source: Statistical appendix, table 1.

ment of the advisers rather than the national AIDS program managers and cannot be generalized beyond the countries surveyed, they nonetheless reveal tremendous scope for improving outreach to those who are most likely to contract and spread HIV. [24] Respondents named adolescents as the subpopulation most likely to be the recipient of an intervention, financed either by government or the private sector (figure 3.7). All countries had at least one program focused on youth, even though in many of the countries it is not clear to what extent adolescents engage in risky behavior. About nine out of ten countries reported a public or private program focused on sex workers, while about seven out of ten countries had a program focused on injecting drug users; a slightly lower proportion of countries had programs that focus on the military and on men who have sex with men. However, respondents estimated that these programs on average covered only about one-half of the relevant group with high-risk behavior. Coverage was highest for adolescents and the military, and lowest for men who have sex with men and injecting drug users.

In most of the countries assessed by UNAIDS Country Programme Advisers, prevention programs did not reach the majority of people most likely to contract and spread HIV.

The UNAIDS Country Programme Advisers also reported that governments were least likely to finance and most likely to impede prevention programs targeted to men who have sex with men and injecting drug users (figure 3.8). Although six out of ten governments funded

Figure 3.7 Coverage of Subpopulations with High-Risk Behavior, Estimates of UNAIDS Country Programme Advisers in 32 Countries

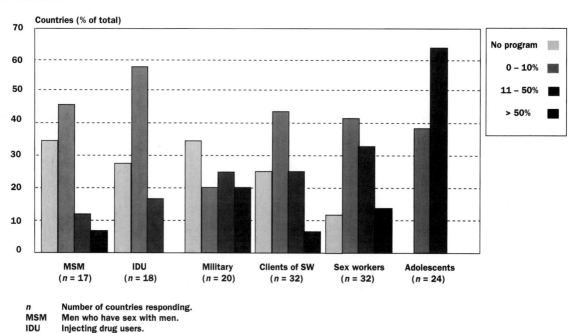

n Number of countries responding.
MSM Men who have sex with men.
IDU Injecting drug users.
SW Sex workers.

Source: Authors' calculations, based on Country Programme Adviser survey results.

Figure 3.8 Government Support for Prevention Targeted to Groups with High-Risk Behavior, Estimates of UNAIDS Country Programme Advisers in 32 Countries

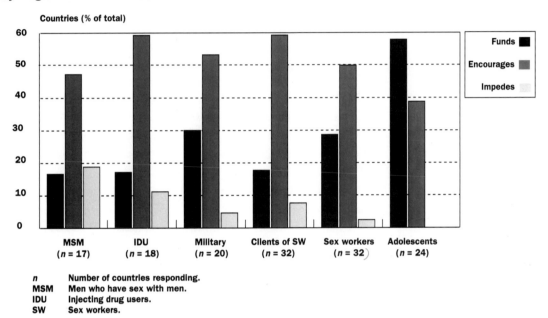

Countries (% of total)

	Funds ■
	Encourages ▤
	Impedes ▢

| MSM | IDU | Military | Clients of SW | Sex workers | Adolescents |
| (n = 17) | (n = 18) | (n = 20) | (n = 32) | (n = 32) | (n = 24) |

n	Number of countries responding.
MSM	Men who have sex with men.
IDU	Injecting drug users.
SW	Sex workers.

Source: Authors' calculations, based on Country Programme Adviser survey results.

prevention programs for adolescents, only about one-third did so for the military and for sex workers. Two advisers indicated that the government in their country promoted prevention for the general population of heterosexuals, but neither encouraged nor discouraged programs for those most likely to contract and spread the virus.

To summarize, while some prevention programs have attempted to encourage safer behavior among those most likely to contract and spread HIV, coverage is generally low. The fact that governments may have logistic and political difficulties in reaching groups such as sex workers and injecting drug users, while understandable, does not detract from the urgent need to assure the fullest possible coverage of these groups. Often these obstacles can be overcome through government funding and support of NGOs. Moreover, in many countries coverage is low even among "captive" populations, such as the military; where this is the case, governments have an opportunity to inexpensively reach these groups with information and other prevention interventions. Effective interventions with broad coverage of those with high-risk behavior will go a long

Most of the countries assessed did not fund prevention programs focused on those most likely to contract and spread HIV; they were most likely to impede programs targeted to injecting drug users and men who have sex with men.

way toward preventing infection among others engaging in high-risk activities and among the lower-risk population.

Improving the Equity of Prevention Programs: Expanding Condom Use

The effectiveness of government programs in ensuring access to prevention for the poor has rarely been evaluated. However, improving equity of access to condoms is one of the major objectives of condom social marketing programs and free government distribution of condoms. Have they improved equity?

Condom availability and use in general have expanded considerably, partly as a spontaneous response to HIV and partly as a result of social marketing and other public, private, and donor-sponsored programs. As of 1996, 60 developing countries had functioning condom social marketing programs, although not all were on a national scale. This was twice the number in 1991.[25] Many of these programs are supported by international donors through three major contractors—DKT International, Population Services International (PSI), and Social Marketing for Change (SOMARC); others, for example, in Botswana, India, South Africa, and some Latin American countries, are also subsidized by national governments. In some countries, such as Indonesia, condom brands launched through social marketing have been taken up by for-profit distributors. Social marketing aside, nearly three-quarters of the 70 countries that responded to condom distribution questions in the *AIDS in the World II* survey provide condoms through a national AIDS control program (Mann and Tarantola 1996). The likelihood of having a condom social marketing program is more strongly related to the increased spread of HIV/AIDS than is government condom distribution (table 3.6). This is partly because government condom distribution includes distribution through government family planning clinics and health services. Finally, in many countries, such as Brazil, Thailand, and Vietnam, unsubsidized commercial sales have risen.

However, the extent to which these programs disproportionately help poor people to obtain condoms is not clear. As we have seen earlier in this chapter, in most countries, people with higher incomes and education are more likely to use condoms. Providing subsidized condoms to low-risk individuals who would have purchased them at market prices would neither improve equity nor reduce the epidemic. Likewise, while

Table 3.6 Condom Social Marketing and Government Condom Distribution Programs, by Stage of the Epidemic

Stage of the epidemic	Percentage of countries with	
	CSM programs, 1996	Condom distribution by NACP, 1992
Nascent	31	71
Concentrated	67	79
Generalized	90	100
Unknown	13	58
Total (%)	49	77
Number of countries	123	70

CSM Condom social marketing
NACP National AIDS control program.
Source: Statistical appendix, table 2.

condom use has risen in both the subsidized and commercial market, it is still not clear the extent to which subsidized programs have squeezed out private sales. This is likely to be an important issue in older condom social marketing programs, after their initial effect on popularizing condoms and generating greater demand has worn off.

A second way in which these programs promote equity is by encouraging condom use among those most likely to contract and spread HIV, forestalling or slowing the epidemic before it reaches the poor. Unfortunately, relatively little is known about the extent to which condom social marketing programs are used by those with the highest rates of partner change—which is the key to their effectiveness in slowing the epidemic. Surveys of sexually active adults confirm that people are far more likely to use a condom for sex with a casual or extramarital partner than a steady partner or a spouse (Agha 1997, Coleman and others 1996, Lowenthal and others 1995, Tchupo and others 1996). But they do not show whether these programs reach people with the *highest* rates of partner change. Do these programs lower the costs of condom use sufficiently to bring about high-use rates among sex workers, soldiers, truck drivers, and other people with many partners? By selling through nontraditional outlets like bars and hotels, condom social marketing programs are probably much more likely to reach people with risky behavior than are conventional programs that distribute condoms through health clinics. If the majority of individuals with the highest rates of partner change *are* reached through these programs, there may be substantial cost-saving if the programs involve a lower subsidy per condom than free distribution. Furthermore, such programs may avoid the political controversy and possible stigmatization that may arise with more targeted programs.

Additional research on the sexual behavior and the economic status of those who use subsidized condoms, and the extent to which those with the highest rates of partner change use condoms from these programs, will help greatly to demonstrate and improve their cost-effectiveness.[26]

That said, many countries still lack vigorous condom programs that specifically prevent HIV and other STDs. Many condom social marketing programs, for example, in Bangladesh, Colombia, Costa Rica, Pakistan, and Sri Lanka, as well as recently launched Chinese programs in Yunnan Province and Shanghai, are oriented primarily toward family planning, with little if any marketing for STD and HIV prevention (DKT International 1997; Kang 1995; "Signs of Change . . ." 1996; "Sri Lankan Condom Sales . . ." 1996). Even in some African countries with concentrated or generalized HIV/AIDS epidemics—for example, Mali, Niger, and Senegal—family planning and reproductive health are the main themes of the programs. Depending on the country, such themes may be less controversial than HIV and STD prevention. However, they may also fail to reach those with the highest rates of partner change. For example, sex workers and young, sexually active men do not frequent health or family planning clinics. Furthermore, women who need condoms for STD prevention may be reluctant to obtain them from community health or family planning clinics, even if free of charge, because of inconvenience, an unreliable supply, or a desire for anonymity. These problems can be overcome if condoms are promoted specifically for HIV and STD prevention and if they can be readily and cheaply obtained in nontraditional outlets readily accessible to people in situations that tend to be conducive to casual and commercial sex. Such locations include pharmacies, kiosks in red-light districts, bars, nightclubs, hotels, truck stops, and military bases (box 3.13). In Peru, socially marketed condoms are sold in three-quarters of the pharmacies and in strategically placed vending machines (Futures Group International 1995a).

One way for governments to stimulate demand for condoms for disease prevention is to end restrictions on condom advertising. But even when there are no legal barriers, open promotion of condom use can be extremely controversial if it is seen as encouraging promiscuity. Messages

> ## Box 3.13 Preventing HIV on the Road to Ho Chi Minh City
>
> VIETNAMESE TRUCK DRIVERS HAVE A SAYING: Never, ever, ever hit a child along the road because, well, he might be yours. It's no secret that with all those days and nights on the road and all that time away from home, truckers seek out amorous company.
>
> Ho Duc Cu is matter-of-fact about the issue. It is near sunset, and he is sitting at a noodle shop at the Goods Transportation Company truck stop on the outskirts of Hanoi, drinking tea from a ceramic bowl and getting ready to haul 10 tons of tractor equipment south to Ho Chi Minh City (formerly Saigon).
>
> "It's a three-and-a-half day drive from here to Saigon," Cu says. "For a lot of drivers, that means two to three women along the way."
>
> The Vietnamese government has widely publicized the risks of HIV/AIDS, so most truck drivers understand what the disease is and how to avoid it. However, it was only after DKT International, a U.S.-based condom social marketing company began promoting the sales of *Trust* and *OK* condoms in Vietnam that truckers had ready access to reliable protection.
>
> Cu finishes his tea and walks over to his Russian-made truck. Inside the driver-side door is a pouch containing a handful of *OK* condoms. "I'm gone from my wife and kids 26 days out of every month," Cu says with a faint grin. He adds that not only are *OK* condoms dependable, but you can buy them at locations along most roads throughout Vietnam.
>
> *Source:* DKT International (n.d.). Used by permission.

must convey useful information and at the same time be directed to the appropriate populations so as to avoid offending influential leaders and segments of the public. Religious leaders in particular may have strong negative reactions to condom promotion if they are not informed about the benefits of condom use or if they are confronted with messages they find offensive. In Uganda, religious sensitivities led to an unofficial ban on promoting condoms on television and radio from 1991 to 1995 (Buwembo 1995). In the Philippines, with a nascent epidemic, opposition by the Catholic Church to artificial contraception extends to condoms for prevention of HIV and STDs (SOMARC 1996). In Niger, conservative religious groups defaced billboards advertising the SOMARC-sponsored social marketing of condoms (Futures Group International 1995b).

Sponsors of social marketing of condoms and other birth control methods have nevertheless found ways of generating support, even among critics. For example, they work closely with religious leaders, potential critics, and local spokespeople before launching condom promotion campaigns to explain the many advantages of condom use (prevention of HIV and other STDs; infertility arising from STDs; unwanted pregnancy, abortion, and withdrawal of pregnant teenagers from secondary school; and the promotion of child spacing, which reduces child and maternal mortality). They test promotional messages with their intended audience and with potential critics to avoid giving offense and maintain a low profile until a basis for success is firmly established. Condoms marketed under names like *Trust, Protector, OK,* and *Couples' Choice* encourage the view that condom use is safe, modern, and socially responsible, irrespective of whether they are used for family planning or disease prevention. More explicit messages about condoms will sometimes be more readily accepted among the people who need condoms the most.

■ ■ ■

This chapter has provided evidence that people will adopt safer behavior, particularly people at high risk of contracting and spreading HIV, and that governments have many ways, direct and indirect, to influence individual behavior. It has identified prevention activities in which governments have a unique role, since private individuals will not finance them sufficiently, and it has outlined important considerations in determining the cost-effectiveness of public spending on HIV/AIDS prevention. The chapter highlighted two areas in which most governments can greatly improve the effectiveness of their efforts to prevent HIV, given sufficient political commitment. The first is to increase the amount and quality of information collected concerning the nature and extent of risky sexual and injecting behavior in the population, trends in the incidence and prevalence of HIV, and the costs and effects of alternative preventive interventions in the local context. The second is to use this information to ensure that prevention programs result in safer behavior among the subpopulations that are most likely to contract and spread HIV and to ensure access to prevention among the poor. Neither of these issues is easy to resolve; however, both are easier to tackle than the very difficult decisions that are thrust upon governments in countries with widespread epidemics. These are the topics of our next chapter.

Notes

1. Other useful references include Adler and others (1996); European Commission (1997); Dallabetta, Laga, and Lamptey (1996); Gerard and others (1995); Lamptey and Piot (1990); and Nicoll and others (1996).

2. See, for example, the work of Becker (1981) and reviews of the literature by Birdsall (1988) and Strauss and Thomas (1995).

3. "Costs" here are not limited to the monetary costs of treatment or prevention. Costs of becoming infected include suffering and premature death and the stigma and discrimination sometimes suffered by people with AIDS and their families. The costs of engaging in safer behavior include, for example, any social stigma associated with purchasing condoms or obtaining treatment for STDs, as well as the time, inconvenience, embarrassment, or monetary costs of obtaining them.

4. Information programs encouraging safer behavior should not be expected to have much of an effect on the behavior of the low-risk general population, since these individuals may correctly conclude that they face relatively little risk. This explains the lack of relation found in many studies between knowledge of the risks of HIV (which in some hard-hit countries approaches 100 percent) and behavior change in the general population (Sepulveda 1992, for example).

5. Since 1996 the import tax and sales tax have been reintroduced. Nearly one-third of the cost of running the condom social marketing program goes to pay the sales and import taxes (*background paper*, Pyne 1997).

6. An estimated one-third of the 750,000 heroin users in the United States, for example, are considered to be occasional users who are not addicted (National Research Council 1989). However, the addictiveness of drugs depends on their purity. In Yunnan Province, China, located adjacent to the Golden Triangle of opium production in Southeast Asia, injected heroin is more than 80 percent pure, most likely making it far more addictive than in the United States and more difficult to stop (McCoy and others 1997).

7. A 620-place drug rehabilitation center in Kunming, the capital of Yunnan Province in China, features a three-month program, primarily for injecting heroin users, that encourages their complete rehabilitation and support from family members (McCoy and others 1997). Among the patients are those who were arrested and many who voluntarily enroll. The price charged to patients is $120 for those who are mandated treatment and $220 for those who voluntarily enroll. Families are reportedly willing to pay this price, which includes all therapy, drugs, room, and board, finding it cheaper than supporting the drug habit of their relatives. However, among patients followed after treatment, 80 percent return to injecting within two years.

8. The same general arguments—high costs and high relapse rates—apply to programs that rely on methadone, a synthetic drug that, taken orally, removes the craving for heroin without inducing euphoria. Moreover, because methadone is only effective against heroin addiction, it does not substitute for other injected drugs.

9. This change in behavior occurred at a time when Thailand had neither needle exchange nor methadone treatment programs.

10. However, the programs are much less successful at promoting condom use than at modifying risky injecting behavior (Normand, Vlahov, and Moses 1995). Once drug injectors are infected, preventing the spread of HIV to others through sex is extremely difficult; thus, early implementation of harm reduction strategies among injecting drug users is critical to preventing the spread of HIV.

11. The evaluations were undertaken in Australia, Canada, the Netherlands, Sweden, the United Kingdom, and the United States.

12. Both the study by Serwadda and others (1992) in Rakai and Barongo and others (1992) in Mwanza note that education is no longer significant in multivariate regressions. However, other intermediate behavioral variables are typically included as explanatory variables in these studies, masking the effect of education (which may be a determinant of all of them) and leading to bias in the esti-

mates. The educational differences in rural Mwanza are statistically significant for women and men, and they remain significant in multivariate regressions (although whether endogenous regressors were included is unknown).

13. This result from the GPA sexual behavior surveys is net of the impact of age and occupation. Several other studies had similar findings. For example, greater schooling is associated with a higher probability of casual partnerships among men in DHS data from Burkina Faso, the Central African Republic, Côte d'Ivoire, and Uganda (*background paper,* Filmer 1997). In rural areas of Kenya, Tanzania, and Zimbabwe, educated women are more likely to engage in casual sex than are uneducated women, but in urban areas the relation is reversed. In Côte d'Ivoire, men and women from wealthier households (with a car and good housing) were more likely to have casual partners.

14. The absolute levels of condom use in figure 3.2 are not comparable across countries, since the reference period for the DHS question about casual partners and condom use was as short as one month (in the Central African Republic and Zimbabwe) and as long as one year (in Haiti and Tanzania).

15. Prevalence among those with 0–6 years of schooling was 1.46 percent, among those with 7–9 years was 1.06 percent, and among those with more than 9 years of schooling, 0.65 percent. These prevalence rates were measured per 100 person-years of observation (Carr and others 1994). Since the WHO/GPA data on sexual behavior from about the same time period showed that men with higher income and education were more likely to have commercial or casual partners, it is likely that condom use was already on the rise among Thai men before the brunt of the HIV/AIDS epidemic hit.

16. The underlying sexual behaviors in these four populations are summarized here as: (a) commercial sex; (b) commercial and casual sex; (c) casual sex only; and (d) serial monogamy. The first three populations allow some concurrent partnerships, but the last does not. Also, both (b) and (d) have commercial and casual sex. For more detail, refer back to chapter 2.

17. Women in stable relationships are monogamous in populations (a) and (d), but are not necessarily monog-

amous in populations (b) and (c). However, their rate of partner change is very low. Although condoms are used for contraception in many developing countries, they are usually not the preferred method of family planning for married couples; DHS surveys conducted in the 1990s found that condom use among married couples ranged from 0 to 3 percent (Curtis and Neitzel 1996).

18. In the population with an epidemic fueled by commercial sex, the increase in condom use among men with casual and commercial partners from 5 to 20 percent represents 20 percent consistent condom use with sex workers only. For the simulations of condom use, it is assumed that if either partner wants to use a condom, a condom will be used.

19. The cost-effectiveness might have been substantially higher had the authors included estimates of the number of secondary infections averted.

20. Mills and others (1993) found that STD treatment costs for similar interventions in Mozambique and South Africa also amounted to roughly $10 per episode of STD treated.

21. The survey of the managers of national AIDS control programs in 187 countries was conducted between December 1993 and June 1994. Of these, 118 responses were received, for a response rate of 75 percent. However, the quality of the responses varied from "complete and detailed" (about one-quarter of the responses) to "sparse and general" (half of the responses, which received individual follow-up). For more information on the survey methodology, see Mann and Tarantola (1996), box 30.1, pp. 315–17.

22. For these countries there were no data whatsoever on groups presumed to have high rates of partner change, data were from very small samples (fewer than 100 people), or data were too old (from 1990 or earlier).

23. Questionnaires were sent to 120 countries; the response rate was 42 percent. The low response rate and the high participation of industrial countries means that the results cited here are not representative of developing countries but are nevertheless true for the 50 countries that participated in the survey. Countries that responded included 15 in Africa, 8 in Latin America, 6 in Asia and

the Pacific, 12 members of NATO, and 9 European countries not in NATO.

24. Only the 43 countries with UNAIDS Country Programme Advisers were surveyed. Responses were received from 26 advisers representing 32 countries, for a response rate of 70 and 74 percent, respectively. Among the 32 countries, 15 were from Africa, 7 from Asia, and 5 each were from Eastern Europe and Latin America and the Caribbean. The countries are: Barbados, Belarus, Benin, Bulgaria, Burkina Faso, Cambodia, China, Congo DR (formerly Zaire), Côte d'Ivoire, Cuba, Dominican Republic, Eritrea, Ethiopia, Ghana, Haiti, Indonesia, Kazakhstan, Kenya, Lao PDR, Moldova, Mozambique, Pakistan, Philippines, Rwanda, Senegal, South Africa, Togo, Uganda, Ukraine, Venezuela, Vietnam, and Zambia.

25. Condom social marketing began in eleven countries in 1996: Albania, Chad, China (Yunnan Province and Shanghai), Republic of Congo, Guinea-Bissau, Lesotho, Madagascar, Myanmar, the Russian Federation, Senegal, and Uzbekistan.

26. The costs of condom social marketing programs are better documented than is their impact on HIV transmission or the extent to which they are used by the poor. The cost per condom sold over a five- to six- year period through 1995 in eighteen Sub-Saharan African social marketing programs was $0.19 (1995 dollars), including the cost of the commodity and overhead (Guy Stallworthy, PSI, personal communication). The net costs ranged from $0.08 to $0.20, depending in part on whether the project was new, which raised costs. Cost recovery to the program is only about $0.01 per condom. A review of CSM programs in ten countries (Bolivia, Congo DR [formerly Zaire], Côte d'Ivoire, the Dominican Republic, Ecuador, Ghana, Indonesia, Mexico, Morocco, and Zimbabwe) found that net costs ranged from $0.02 to $0.30 per condom sold, including the value of donated condoms (Mills and others 1993).

CHAPTER 4

Coping with the Impact of AIDS

WHILE SOME COUNTRIES STILL HAVE THE opportunity to avert a full-scale AIDS epidemic by acting early to change the behavior of those at highest risk, others already have large numbers of infected people across many groups in the population. Chapter 1 presented evidence of the terrible impact of HIV/AIDS on individual welfare, in terms of human suffering and losses in life expectancy. What can be done to mitigate the impact of the AIDS epidemic on people and society? There are many impacts of the AIDS epidemic that cannot be quantified—for example, the emotional pain experienced by infected individuals and their families and the psychological damage wrought on surviving family members. These impacts are very important, but how to respond to them is beyond our expertise and best left to others. This chapter considers the economic aspects of three types of impacts—on infected individuals, on the health sector generally, and on surviving household members—and the ways in which government policies can help people to cope, given the many other pressing demands for scarce public resources.[1]

The first part of the chapter shows that there are affordable, effective, and humane ways for governments in low-income countries to help ease the suffering of individuals infected with HIV. However, both governments and individuals in the poorest countries should be wary of funding expensive treatments with uncertain benefits. The second part of the chapter suggests how governments can cope with the increased demand for and scarce supply of health care brought on by the AIDS epidemic in ways that are effective and compassionate, as well as fair and affordable. The third part proposes a strategy for developing countries to address the needs of poor families hit by the AIDS epidemic in the context of other poverty programs. The chapter concludes with a summary of the policy recommendations for governments attempting to cope with the impact of HIV/AIDS on health care and poverty.

Health Care for the Person with AIDS

WHAT IS THE HEALTH IMPACT OF HIV/AIDS ON AN infected individual over the course of the disease? Are there effective, affordable treatments for people with AIDS in low-income countries? To answer these questions, this part of the chapter reviews the many illnesses that often afflict people with HIV/AIDS, the available treatments, and their cost. It distinguishes between three types of care: relief of symptoms, such as headache, pain, diarrhea, and shortness of breath, which is sometimes called palliative care; prevention and treatment of opportunistic illnesses (OIs); and antiretroviral (ARV) treatments, which attempt to combat HIV itself. Next it presents the amounts that developing countries are actually spending to care for people with HIV/AIDS. While this amount is often large relative to a country's GNP per capita, it is usually too little to buy all the drugs needed to treat opportunistic illnesses, much less to pay for antiretroviral therapy. The section closes with a review of programs to assist with the home care of people with HIV/AIDS.

The discussion finds that although treatment of HIV itself is difficult and extremely expensive, some of the symptoms and opportunistic illnesses typically suffered by people with AIDS can be treated simply and at low cost. Some infectious diseases associated with HIV, especially tuberculosis, are somewhat more expensive to treat, but because they are infectious there are sound reasons for governments to subsidize treatment of any infected individual who would not otherwise get treated, regardless of the individual's HIV status.

Palliative Care and Treatment of Opportunistic Illnesses

The pattern of opportunistic illnesses differs from country to country, depending on which diseases are prevalent, and the quality and amount of treatment available. The natural history of HIV illness and several of the most important opportunistic illnesses are defined in box 1.2. Figure 4.1 presents the proportion of AIDS patients who suffer from each of three OIs—tuberculosis, cryptococcosis, and *Pneumocystis carinii pneumonia* (PCP)—in six developing countries and the United States. Tuberculosis is most common in the three poorest countries, the Congo DR (formerly Zaire), India, and Côte d'Ivoire, becoming less common as per capita income rises. At the other end of the income gradient, PCP is most common in the United States, and is also common in the middle-income developing countries, Brazil, Mexico, and Thailand, but is rarely reported in the three lower-income countries. Cryptococcosis, a generic name for a group of fungal diseases that includes cryptococcal meningitis, shows no consistent pattern by income level, but infects at least 5 percent of people with HIV in all six countries. Among these three diseases, and indeed among all OIs, tuberculosis spreads most readily from people with HIV to others. As we discussed in chapter 1, tuberculosis greatly exacerbates the health impact of HIV in many developing countries, particularly in Africa and India, where it is the most common opportunistic infection.

Figure 4.1 Percentage of AIDS Patients with Three Opportunistic Infections, Seven Countries

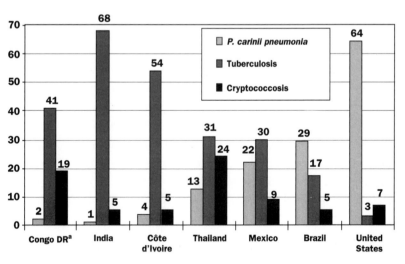

The pattern of opportunistic illnesses differs across countries, depending on which diseases are prevalent and the quality and amount of treatment available.

Note: Since only three of the 20 or more OIs are included, and since a patient may suffer from many OIs before death, percentages for a given country need not total 100 percent.

a. Formerly Zaire.

Source: Background paper, Perriëns 1996; Kaplan and others 1996.

Because of the variation in symptoms and in opportunistic illnesses, the cost and number of health care episodes for an HIV-infected person vary widely. Table 4.1 presents rough estimates for the average costs of pharmaceutical and inpatient palliative care of symptoms, prevention of tuberculosis and PCP, and curative care of the more common opportunistic illnesses. Estimated lifetime cost per patient for this care ranges from $300 to $1,000, depending on which drugs are used and the cost per day of inpatient care.

How effective are these treatments? In the early stages of HIV illness, palliative treatment can inexpensively relieve some of the pain, discomfort, and incontinence that otherwise rob people of the ability to enjoy life and contribute to their family and their community. Without symptomatic treatment, dehydration that results from diarrhea and nausea can kill in a few days. Fever and headache can be disabling for days or weeks. As shown in the top panel of table 4.1, drugs for palliative care are quite cheap. Hence, all but the poorest HIV-infected patients and their families are likely to be willing and able to buy these drugs, provided they are available. The sad truth is that these drugs are often not available, an issue we discuss below.

Moving down the table, we see that the opportunistic illnesses that commonly arise early in the course of AIDS can also be treated quite inexpensively. Treatment for thrush, toxoplasmosis, and pneumonia/septicemia can buy one to four years of life at an additional drug cost of $30 to $150—all but the very poor would probably be willing and able to pay for these treatments.

Rarer opportunistic illnesses like the fungal diseases tend to occur later in the course of the HIV infection and are more difficult and expensive to treat. For example, in the United States the average life expectancy after diagnosis with cryptococcal meningitis, the most common of the cryptococcosis diseases, is 320 days, while in the Congo DR, perhaps because of later diagnosis, this drops to 180 days, even with expensive state-of-the-art drugs (*background paper,* Perriëns 1996). Since a patient in the Congo DR might survive 30 days without treatment, such drugs would extend life by about 150 days for about $870. In Thailand earlier diagnosis would result in treatment extending life by perhaps 330 days for $1,740. Many patients in these two countries might decide against buying these drugs, even if they have the money to do so.

In the final stage of AIDS, the immune system is so weak that a variety of infections spread throughout the body, leading to death. At this point, morphine to assuage extreme pain and the sensation of suffoca-

Table 4.1 Annual Cost per Patient of Palliative Care and Treatment of Opportunistic Illnesses, Sub-Saharan Africa and Thailand

(1996 dollars)

Symptom or illness	Diagnosed episodes per 100 patient-years		Cost per episode[a]		Average cost per patient-year	
	Sub-Saharan Africa	Thailand[b]	Sub-Saharan Africa	Thailand	Sub-Saharan Africa	Thailand
Palliative care[c,e]						
Diarrhea	63		13.00		8.19	
Scaling skin rash	15		1.50		0.23	
Itching skin rash	52		2.00		1.04	
Cough	120		1.40		1.68	
Fever	105		0.60		0.63	
Headache	52		0.25		0.13	
Pain, mild	52		1.12		0.58	
Pain, severe	17		14.00		2.38	
Nausea	75		1.75		1.31	
Shortness of breath	43		6.50		2.80	
Subtotal	*594*	*594*			*18.96*	*18.96*
Treatment of inexpensive OIs						
Tuberculosis[d]	47.5	40	37.00	261.88	17.58	104.75
P. carinii pneumonia	3	20	8.00	207.76	0.24	41.55
Toxoplasmosis	0	2	8.00	207.76	—	4.16
Oral thrush	77	77	2.00	2.48	1.54	1.91
Esophageal thrush	14	14	10.00	4.96	1.40	0.69
Pneumonia/septicemia	20	20	60.00	25.38	12.00	5.08
Subtotal	*161.5*	*173*			*32.76*	*158.14*
Treatment of expensive OIs						
Cryptococcosis	5	25	870.70	1,741.40	43.54	435.35
Herpes simplex virus	5	18	140.00	46.80	7.00	8.42
Penicilliosis	0	9	1,852.50	697.40	—	62.77
Other OIs including: Cytomegalo virus Mycobacterium avium/complex	19.5	19	717.88	717.88	139.99	136.40
Subtotal	*29.5*	*71*			*190.52*	*642.94*
Inpatient days[f]	3,000	3,000	7.25	22.44	217.50	673.34
Outpatient visits[f]	1,200	1,200	2.50	13.60	30.00	163.20
Grand totals per case						
Palliative plus inexpensive OIs					*299.22*	*1,013.65*
Palliative plus all OIs					*489.74*	*1,656.59*

a. Costs per episode are estimated in Perriëns (*background paper,* 1996).

b. Frequency of various symptoms and OIs for Thailand are from Perriëns (*background paper,* 1996), or Kaplan and others (1996) or, where both give a value, an average of the two.

c. The frequency and treatment costs of symptoms listed under palliative care are assumed to be the same in Sub-Saharan Africa and Thailand.

d. In Zambia, experience suggests that preventing tuberculosis in a group of HIV-infected patients costs approximately the same as treating those in the group who get the disease (Foster, Godrey-Faussett, and Porter 1997).

e. Since palliative care involves generic drugs, these costs should be similar in all countries able to buy essential drugs in bulk through international tenders.

f. Patients in Thailand pay about 30 percent of these costs for drugs, inpatient days, and outpatient visits. The cost per patient-day in Sub-Saharan Africa is taken from Chela and others (1994).

tion provides relief to the dying patient, and this in turn helps to ease the distress of the patient's family. If purchased in bulk at international generic prices, enough morphine to ease the last two weeks of life would cost less than $4. But because of international controls on morphine distribution, this essential drug is rarely legally available in poor countries at any price.

The foregoing discussion has shown that many of the symptoms and opportunistic illnesses that occur in the early stages of AIDS can be effectively treated at low cost. Unfortunately, the low-cost generic forms of the needed drugs are often unavailable; even when they are available, people often lack information about their efficacy. Thus, many people pay much more than the $10 to $20 cited in the table for palliative treatments, while achieving no additional benefit. Governments can address these problems by facilitating the availability of generic drugs needed for palliative care and common opportunistic illnesses. For example, countries with a concentrated or generalized epidemic could add these medications to their list of "essential drugs," which are widely distributed. Governments can also help patients to make informed decisions by ensuring access to reliable information about the efficacy of various treatment options, both pharmaceuticals and traditional remedies. The degree of government subsidy for treatment will depend on the country's overall health financing policy. We discuss this issue later in the chapter.

Antiretroviral Therapy Is Expensive, Uncertain

The treatments discussed above ease suffering and prolong life but ultimately fail to save the patient's life because none attacks the underlying cause of illness—the continued spread of HIV within the body and the consequent decline of the immune system's ability to recognize and repel biological threats. A few drugs have reduced the levels of HIV in the patient's blood below the ability of laboratory tests to detect it. Unfortunately, these drugs are expensive and complex to administer, their long-term benefits are uncertain, and their efficacy varies greatly from one individual to another.

The first drug that showed evidence of inhibiting the spread of the virus in an infected patient was Zidovidine (AZT, or ZDV). When AZT was introduced in the late 1980s, the cost of a year's dosage was about $10,000 in industrial countries. By 1997 the cost of a year's dose had fallen to about $2,738 in industrial countries, while Thailand and a few other developing countries had negotiated bulk purchases for as little as $657 per patient per year. However, except for prevention of mother-to-child transmission, AZT rarely provided dramatic benefits, adding perhaps six months of healthy life for the average patient (Prescott 1997; Perriëns and others 1997)

A more effective therapy involving the use of three antiretrovirals was announced in June 1996. A year later, the U.S. government issued draft guidelines recommending early, aggressive treatment of HIV-infected individuals with triple-drug therapy (Brown 1997). However, it was clear that more time would be needed before the new therapies could be fully assessed. Some individuals taking the medicines in clinical trials have dramatically improved their health and no longer have detectable levels of viral RNA activity. Yet even among these patients the virus may only be hiding and could re-emerge. And other patients show little or no reduction in viral levels, while still others cannot tolerate the drugs. As of mid-1997 no studies had yet been completed estimating the average percentage of patients who could benefit from triple-drug therapy or the characteristics of patients most likely to respond favorably—or to relapse.

Does triple-drug therapy offer reasonable hope for treating the disease in developing countries? Even if the therapy is shown to be generally effective, three substantial problems will remain: the cost of the drugs themselves, the costs and difficulty of the monitoring needed for the therapy to be effective, and problems with patient compliance. Although all of these problems also exist in industrial countries, they are likely to be especially severe in developing country medical settings.

Table 4.2 shows the cost of the drugs and necessary monitoring in Thailand, one of the few developing countries where the therapy is available, and the United Kingdom or United States, and hints as well at the great complexity of regimen. Because most of the drug costs and all of the monitoring costs are lower in Thailand than in the two industrial countries, overall costs are a minimum of about $8,000 per year in Thailand, compared with a minimum of about $12,000 per year in the United Kingdom and United States. These costs are likely to decline over time, perhaps substantially. But even if costs fell to *one-hundredth* of current costs, or about $80 dollars per person per year, they would still be several times the total annual per capita expenditure on health in many low-income countries. Moreover, effective antiretroviral therapy requires a highly trained, specialized physician working in a well-equipped clinic with experience performing a wide range of sophisti-

Table 4.2 Annual Cost of Antiretroviral Therapy, Thailand, and the United Kingdom or the United States

(dollars)

| Inputs | Daily dose (mg) | Daily or unit cost | | Annual cost | |
| | | Thailand | U.K. or U.S. | Thailand | U.K. or U.S. |
Drugs					
Nucleoside RT inhibitors					
Zidovudine (AZT)	500	1.80	7.50	657	2,738
Didanosine (ddI)	400	5.80	5.75	2,117	2,099
Zalcitabine (ddC)	2.25	5.40	6.81	1,971	2,486
Stavudine (d4T)	80	—	7.95	—	2,900
Lamivudine (3TC)	300	—	7.37	—	2,690
Protease inhibitors					
Saquinavir (SQV)	1,800	19.08		6,870	
Ritonavir (RTV)	1,200	21.95		8,010	
Indinavir (IDV)	2,400	11.84		4,320	
Monitoring	*Times per year*				
Blood count	12.00	2.00	21.00	24	252
Blood chemistry	4.00	12.00	35.00	48	140
CD4 count	4.00	30.00	157.00	120	628
RNA viral load	3.50	50.00	100.00	175	350
Additional outpatient visits	12.00	13.60	100.00	163	1,200
Total for triple-drug therapy [a]					
AZT, ddI, and IDV				9,595	19,803
AZT, ddI, and RTV				13,285	23,493

— Data not available or not applicable.

a. Triple-drug therapy consists of two of first group of drugs plus one of second group plus monitoring. Drugs are given daily. Which three drugs should be combined is a matter of current research and probably varies by patient.

Source: Background paper, Perriëns 1996; Prescott and others 1997; and Moore and Bartlett 1996.

cated tests and procedures, all of which are in critically short supply in most developing countries.

In the event that cost and infrastructure problems could somehow be overcome, patient compliance would continue to pose serious difficulties. Patients undertaking triple-drug therapy must swallow up to 20 pills a day according to a complex schedule related to sleep and meal times. Failure to follow the schedule increases the chance that the virus will become resistant or that the patient will be too sickened by the drugs to continue treatment. Even well-educated patients with good clinical support have difficulty adhering to this demanding regimen; moreover, patients in the early stages of HIV infection are sometimes not willing to take drugs that make them nauseous when they otherwise feel healthy. In clinical trials in industrial countries, for example, as few as 26 percent of patients complied with the instructions (Stewart 1997). Problems with patient compliance are likely to be worse in low-income countries due to lower education levels and the many other problems that poor people in developing countries face.

Even with all these difficulties and uncertainties, many patients in developing countries will ask their physicians for triple-drug therapy, just as patients have attempted to obtain AZT. Governments will in turn face pressure to buy these drugs and to subsidize the necessary clinical services. When very few people have AIDS, total costs will also be small relative to other government expenditures. But as the epidemic progresses, the number of AIDS cases and the cost of the subsidy will escalate rapidly, drawing resources from other pressing social needs. At some point it will become evident that such a subsidy is unaffordable and also unfair to the many people who for a variety of reasons want government help but do not have HIV.

Individual Treatment Costs for AIDS Are High, Even in Poor Countries

We have seen that medical responses to HIV/AIDS range from a few pennies to thousands of dollars. How much a country actually spends to treat a case of AIDS depends on many factors besides the differing cost of health care inputs. The most important of these is the amount of treatment that the HIV-infected person, his or her family, and any third party payers such as insurance companies or the government are willing and able to buy, and how much the government subsidizes health care and AIDS treatment. Figure 1-8 showed that across countries this amount is strongly correlated with per capita income. An in-depth study of AIDS expenditure in four countries and São Paulo State, Brazil, confirms this general pattern; the average total (public and private) AIDS expenditure varies from 0.6 times per capita GDP in Tanzania to 3.0 times per capita GDP in São Paulo; the average is a ratio of about 1.5 (*background paper,* Shepard and others 1996).

Alternatives to Expensive Inpatient Care

Where the AIDS epidemic is severe, health policymakers inside and outside government have sought ways to provide compassionate care at low cost. Three alternatives to expensive inpatient care are outpatient

AIDS clinics, hospice care (residential low-technology care for the terminally ill), and home-based care.

One innovative program to deliver high-quality treatment of symptoms and opportunistic illnesses without the expense of hospitalization was an outpatient clinic started in 1989 in São Paulo, Brazil. Such clinics are especially well suited to serve urban HIV-positive and AIDS patients who are able to leave their homes. Later in the course of the disease, when the patient is less mobile, the hospice or nursing home provides a lower-cost substitute for inpatient care in a sophisticated referral hospital. However, since such facilities are rarely available in developing countries, the main alternative to the hospital is care at home.

What sort of home-based care is most effective? An analysis of the cost of eight home-based care programs in Zambia found that community-initiated programs were more effective and much less expensive than hospital-initiated programs (Chela and others 1994; Martin, Van Praag, and Msiska 1996). Assuming that the average patient with AIDS would survive six months with either type of care, the benefits of the care must be measured in reductions of hospitalization cost; reduced travel time to the hospital for the patient and the patient's caretakers; increased patient satisfaction and comfort; and ancillary benefits to the community, such as improved understanding of the ways to prevent AIDS and decreased stigma toward HIV-positive people. Since the study found that patients who received home-based care reduced their hospitalization before death by only two days, the expenditure on the hospital-initiated home-based care programs of about $312 (6 months x 2 visits per month x $26 per visit) was much more than the $14.50 saving in hospital charges (2 days x $7.25 per day). On the other hand, the costs for six months of community-initiated home-based care averaged just $26, less than one-tenth the cost of the hospital-initiated program, and could almost be justified on the basis of reduced hospital use alone.

The tenfold cost difference between hospital- and community-initiated home care programs was due to the much larger expenditure on transport and staff time for the hospital-based programs. For example, on a typical day a team of trained hospital-based nurses could visit only four to eight patients, about a quarter of whom were away from home when the team arrived. As a result the hospital-initiated teams spent on average about two hours on the road in order to spend only fifteen minutes with the patient. In contrast, the community-initiated teams walked only a few minutes and spent an average of two hours with the patient.

If the low cost of the community-initiated home-based care program in Zambia can be generalized to other settings, it is possible that such care would be financed by the patients, their families, and their communities. Indeed, the community-initiated Zambian programs function well because of strong volunteer support from the local communities. Since the benefits of the program include the public ones of improved knowledge about HIV prevention and reduced stigma, there may be a government role in financing such programs, at least until their private benefits to patients' families are sufficiently well understood for these families and communities to support such programs on their own. Where policies exist to facilitate access to health care for the poor, they should be extended to include community-based home care programs using the same eligibility criteria.

Difficult Health Policy Choices in a Severe AIDS Epidemic

THE PREVIOUS SECTION DESCRIBED THE IMPACT OF AIDS ON the individual HIV-infected person and demonstrated that limited treatment of symptoms and opportunistic illnesses, especially when performed partly by community-initiated home care programs, can provide compassionate care at relatively low cost. In this section the need to keep costs low becomes more apparent as we widen the focus from the individual HIV-infected person to the health care needs of all people in a country. To better understand the difficult tradeoffs involved, we first estimate the magnitude of the impact of AIDS on the health sector, and then discuss how government policies can mitigate this impact.

How HIV/AIDS Will Affect the Health Sector

AIDS will affect the health sector in two ways: by increasing demand and by reducing the supply of a given quality of care at a given price. As a result, some HIV-negative people who would have obtained treatment had there been no epidemic will be unable to do so, and total national expenditure on health care will rise, both in absolute terms and as a proportion of national product.[2]

Increased demand for care. Most people who develop AIDS are prime-age adults. Without AIDS, this 15-to-50 age group accounts for only 10 to 20 percent of all deaths in a developing country, but these deaths typically generate a disproportionate share of total health care demand (Over, Ellis, Huber, and Solon 1992; Sauerborn, Berman, and Nougtara 1996). Moreover, since several studies suggest that adults with AIDS use more health care prior to death than those who die of other causes, or even of other prolonged illnesses, the percentage increase in the demand for care by adults is likely to exceed the percentage increase in their mortality due to AIDS. As a result of these two factors, in a country where prime-age adults utilized one-quarter of all health care before AIDS, a given percentage increase in their demand for health care will increase *total* demand by *at least* one-quarter of that percentage. For example, a 40 percent increase in the mortality rate of prime-age adults will increase total demand by at least 10 percent, even though total mortality has increased by only 4 percent to 8 percent.[3] If AIDS patients use expensive antiretroviral therapies, the increase in demand will be much greater.

How much the demand for care increases in the aggregate depends on the increase in the prime-age adult death rate, which in turn depends on the level of HIV prevalence and the median time from infection to death (table 4.3). A stable prevalence rate of 5 percent among prime-age adults eventually increases their annual mortality by about five deaths per 1,000 adults if the median time from infection to death is ten years, or by about ten deaths if the median time is only five years.[4] A prevalence rate of 30 percent, such as is observed in Lusaka, Zambia, will increase the number of deaths per 1,000 adults by 30 to 60, depending on the median time to death. In Sub-Saharan Africa, where mortality rates in this age group were as high as five per 1,000 before the epidemic, even a 5 percent infection rate will double or triple the adult death rate. In a middle-income developing country with adult mortality of one per 1,000, the same endemic level of HIV infection will increase prime-age adult mortality five- or tenfold.

Given these parameters, how much will the epidemic increase the demand for care? *In a country where adults consume one-quarter of health care prior to the AIDS epidemic, HIV prevalence is constant at 5 percent of adults, the median time to death is ten years, and the baseline mortality rate among prime-age adults is 5 per 1,000, the epidemic will cause a 26 percent increase in the demand for health care at every price.*[5] If the prevalence rate

Table 4.3 Deaths per Thousand Adults Caused by a Constant Rate of HIV Infection

HIV prevalence rate (percent)	Median time from infection to death	
	10 years	5 years
0	0	0
5	5.3	11.1
10	10.5	22.2
15	15.8	33.3
20	21.1	44.4
30	31.6	66.7
50	52.6	111.1
100	105.3	222.2

Note: The death rates in columns 2 and 3 are calculated by multiplying the prevalence rate from column 1 by $20/(2M-1)$, where M is the median time from infection to death. This formula assumes a steady-state epidemic in which incidence is constant and a proportion $1/(2M)$ of those infected in a given year die in each of $2M$ subsequent years. In the absence of HIV, the baseline mortality rate per thousand adults age 15 to 50 ranges from 0.8 in industrial countries to as high as 5 in some parts of Sub-Saharan Africa.

is higher, the median time to death shorter, or the baseline adult mortality rate smaller, the percentage increase in demand will be correspondingly greater.

A final important factor that may increase demand is insurance. This may take the form of private insurance, a government-run insurance program, or, more typically, health care financed through general taxation. Because a portion of health care costs is often covered by one or more of these types of insurance, the price paid by the patient is usually a fraction of the cost of providing the care. Since insurance enables patients to purchase more care than they would otherwise, it increases the demand for care arising from any given level of illness, thus magnifying the price shock of an AIDS epidemic. For example, if the proportion of cost of providing care paid by patients (i.e., the coinsurance rate) is 25 percent, patients will reduce their utilization in response to increased cost by only a quarter as much as they would if they had to pay the full increase.

Reduced supply of health care. In addition to increasing the demand for care, the AIDS epidemic will reduce the supply available at a given price, in three ways. The magnitude of these effects, discussed below, will generally be larger in the poorest countries with the largest epidemics.

The first and largest effect is the increased cost of maintaining a given level of safety for medical procedures. Even without HIV, hospitals and clinics in poor countries may pose a risk to health. Needles and other instruments are not always sterilized, rooms are often overcrowded and poorly ventilated, and care providers may lack rubber gloves and sometimes even soap. Without modern blood banks, a transfusion might infect the recipient with hepatitis B. In such situations, infections of all types spread rapidly; some, including such common illnesses as pneumonia, may kill. Before HIV, however, infections picked up in a clinic or hospital were rarely fatal to persons not already in a seriously weakened state.[6]

Because the AIDS epidemic has greatly increased the risk to patients of existing medical procedures, simply maintaining the level of safety that existed before HIV requires additional hygiene and blood screening, both of which increase the cost of care. In middle- to high-income countries, where blood screening and sterilization of injecting equipment are already the norm, the impact of AIDS is confined to the incremental costs of adding an HIV test to existing tests and using rubber gloves and face masks in situations where they were previously not used. In poor countries, where blood screening and needle sterilization were lacking before the epidemic, the resources needed to maintain the quality of care in the face of the AIDS epidemic can be substantial. For example, the annual recurrent budget of the Ugandan Blood Transfusion Service, which was established in response to the epidemic and meets the demands of the entire Ugandan national health care system for clean blood, is estimated to be about $1.2 million, including capital and recurrent costs. This amounts to about 2 percent of national public health expenditures or about 1 percent of total national health expenditures (European Commission 1995a). Despite the potentially high costs of blood screening, HIV has greatly increased the justification for a government role in ensuring a safe blood supply. However, there is no convincing rationale for government to subsidize the entire cost of running such a service indefinitely (see box 4.1). Blood screening and improved collection procedures will protect blood donors and recipients. However, since average donors and recipients do not engage in unprotected sex with a large number of partners, a person infected while giving or receiving blood is not likely to pass the infection to many others. Thus, in developing countries where the cost of establishing a safe blood supply is high, blood screening will not be among the more cost-effective approaches to preventing an epidemic based on sexual transmission (see box 4.2).

To be sure, blood screening and better hygiene will help to prevent the spread of other infectious diseases besides AIDS. Such measures will also reduce the occupational risk of AIDS and other diseases that health care workers face, and therefore reduce the amount of additional compensation needed to offset their occupational risk—an issue we discuss below. A careful accounting of the net cost of protecting patients from HIV by screening blood would need to take into consideration these additional benefits, for which data are lacking. However, it seems likely that even if these benefits are taken into account, the remaining cost of screening blood and improving hygiene to protect patients from HIV/AIDS would substantially increase the unit cost of medical care.

The second factor reducing the supply of medical care at a given price is the increased attrition of health care workers who become infected with HIV. Like all adults, health care workers may become infected with HIV as a result of sexual contact or use of unsterile injecting equipment. They also face an additional risk of becoming infected in the course of their work; however, this risk is generally much smaller than the risk from sexual contact. Thus whether the AIDS mortality rate among health care

Box 4.1 The Government Role in Ensuring Clean Blood

THE HIV/AIDS EPIDEMIC HAS DRAMATICALLY increased the importance of clean blood. Where the most serious common infection that a transfusion recipient previously had to fear from unscreened blood was hepatitis B, which is rarely fatal and communicated in only about 2.5 percent of unscreened transfusions, recipients in some countries now face a one-fourth chance of HIV infection (Emmanuel, WHO, as cited in Fransen, personal communication). As a result of the HIV/AIDS epidemic, the transfusion required for a surgical procedure or childbirth that might have been relatively routine in a developing country ten years ago now requires the guarantee of clean blood to be equally safe.

What should be the government's role in the provision of safe blood? Setting aside poverty, which is addressed in the text, five justifications can be identified for the public to subsidize or otherwise play a role in the provision of blood: (1) to prevent HIV infections in blood recipients; (2) to prevent infections in the sexual partners of blood recipients; (3) to avoid the sudden onset within a community of the health risk from unscreened blood; (4) to provide the economies of scale that apply to a blood bank service; and (5) to avoid the difficulty that a citizen would have in judging the quality of a blood bank.

While a high-quality blood bank will obviously be quite effective in preventing the transfusion of infected blood, and thereby in preventing the hospital from infecting transfusion recipients, this fact does not, by itself, imply that the government should play a role in supplying the clean blood. Setting aside for the moment considerations (2) through (5), the provision of clean blood is comparable in importance to the provision of clean needles, clean bandages, and clean hands of the nurses who change those bandages. Any arguments for government financing of decent quality of care, including basic cleanliness in the hospital, also apply to clean blood. If one accepts the argument that hospital care is a basic need, which should be heavily sub-

sidized by the government, then the same argument would apply to clean blood. If, however, one believes that there is no obvious reason to favor curative health care over other necessities, such as clothes, housing, and clean water, then clean blood should receive as little subsidy from the government as other curative health care services.

Yet even those who believe that most curative care deserves little subsidy admit that the treatment of infectious diseases confers positive externalities and thus should be subsidized. This brings us to an evaluation of the second consideration. Assuming that transfusion recipients recover from the medical procedure and then become sexually active, preventing their infection may prevent them from infecting others. How large are these positive externalities? For one country, Uganda, box 4.2 shows that a highly effective program prevented 517 secondary infections in 1994 at a cost of $1,684 each. While this cost is much less than the lifetime treatment cost of an HIV-infected person in an industrial country, it is more than any reasonable estimate of the cost of preventing secondary infections in Uganda. Thus, the prevention of secondary infections does not appear to be sufficient to justify government subsidy of the entire cost of the program, although it could justify a partial subsidy.

Considerations (3) and (4) appeal to the same economic arguments often used to justify government infrastructure investments. The sudden increase in risk from blood transfusion is a shock to the health care system, too rapid for individuals and private institutions to make new blood-screening arrangements quickly. As the insurer of last resort against catastrophic changes in the environment, the government has a role in assisting society in adjusting to the new higher cost and complexity of health care in the presence of AIDS. Furthermore, as demonstrated by box figure 4.1, a blood transfusion service entails substantial economies of scale. Since a

(Box continues on the following page.)

Box 4.1 *(continued)*

single transfusion service can serve all local needs without exhausting its economies of scale, it would be a natural monopoly without the fear of competition to ensure quality service at the best price. It would be forced to charge prices above marginal cost in order to cover its costs and might charge prices well above average costs in order to maximize its profits. Just as for electric utilities and other natural monopolies, there is a well-established justification for government intervention to regulate, if not own and operate, them in such circumstances. However, they do not justify a 100 percent subsidy for blood.

Consideration (5) involves the inability of the public to judge the quality of a blood bank. This argument is not particular to blood transfusion services, since patients have an equally difficult time judging the quality of their physicians.[1] Yet patients can choose among many different physicians, but,

because of the economies of scale, are unlikely to have a choice of blood banks. The government and the public should not assume that any monopoly, whether it produces electricity or blood bank services, and whether it is "for-profit" or "nonprofit," will indefinitely perform in the public's best interest. In this situation there is an argument for the establishment of a regulatory board to whom the blood transfusion service is responsible.[2] The board should consist of representatives of the medical establishment, government, and patients and should produce an annual report on the quality of the blood bank service, which should then be widely disseminated in the press.

In sum, the appropriate role of government in financing blood supplies depends first on one's view of the degree of financing the government should provide to curative health services. The argument for curative services extends directly to the provision of blood. The number of secondary infections averted through blood screening is unlikely to be a powerful argument for government subsidies. Even so, there is a strong argument for the government to launch and nurture a blood bank service as a subsidized "infant industry," before subjecting it to the rigors of the financing arrangements provided for the rest of the health care system. Finally, because economies of scale will tend to make the blood bank a monopoly in most communities, blood bank services should be subject to strict regulatory review.

Box Figure 4.1 The Cost per Unit of Blood Transfused in Uganda

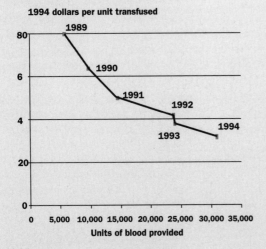

1994 dollars per unit transfused

Source: European Commission 1995a, p. 94. Nominal amounts converted to current dollars at 1.2 dollars per ECU and then to 1994 dollars using the U.S. consumer price index.

---

[1] Information is asymmetrically distributed between the firm producing and selling blood and the hospitals, physicians, or patients who consume it.

[2] While patients should be charged the same percentage of the marginal cost of a unit of blood that they are charged for other curative care of noninfectious diseases, it does not follow that donors should be paid for blood. The observations by Richard Titmuss (1972) regarding the benefits of recruiting voluntary donors have been found to apply in many different national settings.

Box 4.2 Cost of Preventing Secondary HIV Infections through Blood Screening in Uganda

HOW COST-EFFECTIVE IS BLOOD SCREENING IN preventing secondary HIV infections? One answer to this question can be seen in the results of the Uganda Blood Transfusion Service (UBTS) for 1993. Having established its ability to supply Kampala with clean blood in 1991, by 1993 the UBTS was reaching out to cover the entire country. That year the service transfused 20,156 patients throughout the country at an average cost of approximately $38 per unit of blood, and an average of 1.2 units per patient, for a total budget of approximately $929,900. Box table 4.2 breaks out the HIV prevention benefits of the service, showing that its use averted HIV infection in an estimated 1,863 surviving transfusion recipients.

But to measure the positive externalities of the program, and thus the rationale for government subsidies, we need to look beyond these primary infections to consider secondary infections. Children who are infected by transfusion are unlikely to live long enough to infect others, but some of the adults may be sufficiently young and sexually active to engage in risky sexual behavior later in their lives. Since many of these people are quite sick, the evaluation study estimated that each of these adults would have only a 50 percent chance of infecting one other person with HIV (European Commission 1995). Thus the total number of secondary infections averted would be 415.[1] If the entire justification of the blood supply service is prevention of these secondary infections, the cost-effectiveness of the service is $929,900 divided by 415, or $2,240 per such infection averted. If Uganda had had a sustainable blood supply system, the cost of preventing these 415 infections would have been only $319,894, or $771 each. This much smaller amount is still substantially larger than the cost of preventing secondary infections in other ways (see box 2.6).

---

[1] The authors point out that the counseling provided to blood donors may have averted additional primary infections (European Commission 1995). Any secondary infections averted through this route should be added to the 415 to compute the total positive externalities of the program.

Box Table 4.2 Effectiveness of Blood Transfusion at Averting HIV Infection, Uganda, 1993

Effects of blood transfusions	Benefits		
	Children	Adults	Total
Patients transfused	11,515	8,641	20,156
Patients expected to die without transfusion	5,758	3,898	9,656
Patients who died despite transfusion	3,801	2,592	6,393
Number of deaths prevented	1,957	1,296	3,253
Number of primary HIV infections prevented	1,033	830	1,863
Number of secondary HIV infections prevented	0	415	415

Source: Based on the results achieved by the Ugandan Blood Transfusion Service as reported in Beal, Bontinck, and Fransen (1992); European Commission (1995a); and Fransen (1997, personal communication).

workers is higher or lower than among the general population depends mostly on the effects of income, education, and social status on sexual behavior. Two studies of HIV prevalence among health care workers from Africa suggest that doctors and nurses are at least as likely to become infected as other people (Mann and others 1986, Buvé and others 1994). If this is true elsewhere, a country with stable 5 percent HIV prevalence can expect that each year between .5 and 1 percent of its health care providers will die from AIDS; a country with 30 percent prevalence would lose 3 to 7 percent of its health care workers to the epidemic. This attrition from AIDS deaths may substantially increase the cost of health care. For example, if labor costs are half of total health care costs, and training or recruiting a replacement worker requires a one-time expenditure equal to the worker's annual salary, then a 7 percent increase in attrition will increase total costs in the health sector by 3.5 percent.

The third way in which AIDS reduces the supply of health care is through the additional risk it imposes on health care workers. Even though most HIV-infected health care workers acquire their infection through sexual contact, in a society with a large proportion of HIV-positive patients, health care work will be more dangerous than if there were no HIV. Some students who would have become doctors and nurses will therefore choose alternative occupations, unless they are compensated with higher pay for the increased risk. A recent survey of medical and nursing students in the United States found that AIDS had indeed reduced the attractiveness of specialties in which contact with HIV-positive patients was more likely (Bernstein, Rabkin, and Wolland 1990; Mazzullo and others 1990). This problem is likely to be most severe in hard-hit developing countries, where HIV prevalence is much higher and rubber gloves and other protective equipment are often in short supply. In Zambia, for example, some nurses have demanded special payments to compensate for increased occupational risk due to HIV (Buvé and others 1994).

The magnitude of increased costs of medical staff has not been estimated. As noted above, improved precautions in hospitals and clinics may reduce these costs. But because people respond to perceived risk rather than actual risk, such improvements may have little impact on the demand for increased compensation. Thus, it seems clear that health care workers' perception of risk will increase the cost of care.

The total impact of these three effects—increased cost of preventing infection in medical facilities, attrition of health care workers due to HIV, and additional pay that health care workers demand to compensate them for increased risk—will depend most importantly on HIV prevalence and whether modern blood banks and hygiene were already in place. *In a country that has 5 percent HIV prevalence among prime-age adults and lacked blood banks and blood screening before the epidemic, a conservative guess is that the cost of providing care of a given quantity and quality will rise by about 10 percent.*

Scarce care, higher expenditures. Taken together, increased demand and reduced supply have two related impacts: first, health care becomes scarcer and thus more expensive; second, national health care expenditure rises. The size of the increases in health care prices and national health care expenditure depends partly on the price-responsiveness, or "elasticity," of the demand for and supply of care. For most goods, higher prices reduce demand, as consumers switch to substitutes or forgo an intended purchase altogether. This same principle holds true for health care, but the price-responsiveness or elasticity of demand for adult health care is usually small, since there are no close substitutes, and people who are sick and who have the ability to pay will often pay whatever is needed to get well. For the purposes of our simulation, we assume that a price increase of 8 percent would decrease utilization by only about 8 percent, for an elasticity of 0.8.[7]

Higher prices also generally increase supply. Here, too, however, the nature of the health sector affects the supply response. In the very short run, perhaps a month, the supply of care is unlikely to change much. Over the long run, the supply of physicians and inputs to health care can expand as much as necessary. Over the medium run, five years or so, we would expect the supply of care to respond somewhat to increased demand and the resulting higher price. One response observed in Canada, Egypt, India, Indonesia, and the Philippines is that physicians who work in the public sector rearrange their schedules to offer more health care privately, after their obligations to the government have been met. The elasticity of this response has been estimated at about 0.5, meaning that every 10 percent increase in the price of care elicits a 5 percent increase in supply (Chawla 1993, 1997; Bolduc, Fortin, and Fournier 1996).

We have argued in the previous two subsections that a constant 5 percent seroprevalence rate would eventually increase the demand for care by about one-quarter and the cost of care of a given quality by 10 percent. Drawing on the assumptions in this subsection about the elasticities of the demand and supply responses, and assuming that patients pay half the cost of health care, box 4.3 shows that total national health expenditure, and also the government's share of expenditure, would both increase by about 43 percent. The increase would be less in a country like India,

Box 4.3 Estimating the Impact of AIDS on the Health Sector

HOW MUCH WILL THE EFFECTIVE PRICE OF CARE increase as a result of AIDS? Box figure 4.3 demonstrates how the approximate size of these increases can be estimated for a hypothetical country with elasticities of demand and supply for health care of 0.8 and 0.5 and a government policy to subsidize half of the cost of care. The two solid lines show the amount of health care that is demanded and supplied at each price prior to an HIV epidemic. (The demand curve is drawn with an elasticity of only 0.4 in order to incorporate the effect of the government

Box Figure 4.3 The Impact of a 5 Percent Infection Rate on the Quantity and Price of Health Care

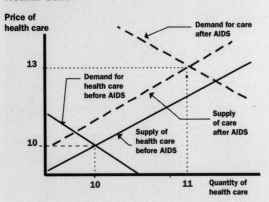

Note: The demand and supply curves are constructed so that the price elasticities at the point (10, 10) are 0.8 and 0.5, respectively. The impact of AIDS is illustrated by shifting the demand curve to the right by 25 percent at every price and the supply curve up by 10 percent at every quantity. See the box text for an explanation of these assumptions.

subsidy on consumers.) The figure is constructed so that the market equilibrium occurs at a price of 10 currency units per unit of health care, at which a total of 10 units of care are delivered. Total health care expenditure in this hypothetical country is thus 10 times 10 or 100 currency units in the absence of the AIDS epidemic.

Now assume there is an HIV/AIDS epidemic that levels off at a constant seroprevalence of 5 percent of the adult population. The arguments in the chapter suggest that the amount of health care demanded at every price is likely to increase by 25 percent, while the cost of purchasing any given amount of care of a given quality will increase by 10 percent. These two impacts of the AIDS epidemic are illustrated by a rightward shift of the demand curve by 25 percent (to the dashed, downward-sloping line) and an upward shift of the supply curve by 10 percent (to the dashed, upward-sloping line). The impacts on the equilibrium price and quantity can be read from the figure. The price of a unit of health care will increase about 30 percent, and the amount of care provided will increase about 10 percent. Total national expenditure, the price per unit of care times the number of units, will increase 43 percent to 143 currency units (since 13 x 11 = 143).

We have seen that a third-party payment, such as insurance or a government subsidy for treatment, makes people less sensitive to changes in cost of health care. By reducing the price elasticity of demand, such third-party payments make the demand curve steeper, both before and after the introduction of AIDS.

where only about one-fifth of the cost of care is paid by the government, and substantially more in countries like those of Latin America and Eastern Europe, where three-quarters or more of the cost are subsidized.

Does the available empirical evidence support these conclusions? Although there are significant data problems, the short answer is yes.

Measuring the scarcity of medical care through changes in the price of care of given quality is problematical because of the difficulties in measuring quality. This is especially true in developing countries, where a general lack of data is compounded in the health sector by government subsidies and nonprice forms of rationing. In such cases, the *effective* price of care may rise even though nominal prices remain constant (see box 4.4). Furthermore, because of the lag between infection and death, the time between the attainment of a given HIV prevalence rate and the full impact of that rate on the demand and supply of health care can be ten to 20 years. For these reasons, we cannot accurately assess changes in scarcity of health care in developing countries by observing changes in nominal price. Nonetheless, we can get some sense of the extent to which HIV/AIDS increases the effective price of health care by considering whether the epidemic makes it more difficult to obtain care. Studies of hospital admissions data strongly suggest that this is the case.

Table 4.4 shows the percentage of beds occupied by HIV-positive patients in six referral hospitals in developing countries with large epidemics. The hospitals are the top health care institutions in each country, providing the best care available outside of a few expensive private clinics. Because these hospitals are at the apex of their health care pyramids, we would expect that AIDS patients account for a significant proportion of their patients. Even so, the percentage of beds occupied by HIV-positive patients is striking, ranging from 39 percent in Nairobi, Kenya, to 70 percent in Bujumbura, Burundi.

Box 4.4 The Effective Price of Care

SOME READERS MAY OBJECT THAT THE PRICE PAID BY patients need not increase in countries where the government guarantees free care. However, as we have seen, even with HIV prevalence rates of 5 percent or less, the demand for medical care is likely to increase faster than the government's ability to supply it. When this happens, means of rationing medical care other than price come into play. People in countries where health care is officially "free" are familiar with these mechanisms. Some systems rely on waiting time. In others, a patient dissatisfied with inferior care in a public facility can pay for better care during a doctor's private office hours. In still other cases, side payments to a nurse or other gatekeeper are necessary in order to get access to "free" care. The effective price of health care to the consumer is the value of all of the consumer's sacrifices, in time and in money, needed to obtain care of a given quality. The AIDS epidemic increases the effective price, even if care is supposedly "free."

Table 4.4 Evidence of Possible Crowding Out of HIV-Negative by HIV-Positive Patients, Six Countries, circa 1995

City	Hospital	Percentage of beds occupied by HIV-positive patients
Chiang Mai, Thailand	Provincial	50
Kinshasa, Congo DR[a]	Mama Yemo	50
Kigali, Rwanda	Central	60
Bujumbura, Burundi	Prince Regent	70
Nairobi, Kenya	Kenyatta National Hospital	39[b]
Kampala, Uganda	Rubaga Hospital	56

a. Formerly Zaire.

b. Since Floyd and Gilks found the average length of stay to be identical across HIV-positive and -negative patients, the ratio of HIV-positive to total admissions is a useful estimate of the proportion of beds occupied by HIV-positive patients. Thus this entry is calculated from figure 4.2 as 9.6/24.9.

Sources: First four hospitals, van Praag 1996; Kenyatta Hospital, Floyd and Gilks 1996; Rubaga Hospital, Tembo and others 1994.

If the hospitals were operating well below capacity before the epidemic, they might have accommodated the HIV-positive patients without reducing care for HIV-negative clients. Although no data on occupancy prior to the epidemic are available for these specific hospitals, bed occupancy rates in such hospitals typically were well above 50 percent even before AIDS.[8]

The best evidence that AIDS is making it more difficult for people not infected with the virus to get medical treatment comes from an in-depth study of Kenyatta National Hospital (KNH), the premier teaching hospital in Nairobi, Kenya. The KNH study compared all patients admitted during a sample 22 days in 1988 and 1989 with all patients admitted during a sample 15 days in 1992 (Floyd and Gilks 1996). Panel A of figure 4.2 shows that while the average number of patients admitted per day increased from 23 to 25, the number of HIV-positive patients more than doubled, while the number of HIV-negative admissions shrank by 18 percent. Since the number of HIV-negative people in the hospital's "catchment area" could not have shrunk by this much, this evidence suggests that the AIDS epidemic did in fact result in some HIV-negative patients being dissuaded or barred from admission to the hospital.

There are no data on what happened to the HIV-negative patients who were not admitted. But hospital records show that the mortality rates for those who were admitted increased between the two periods,

Figure 4.2 Impact of AIDS on Utilization of and Mortality at Kenyatta National Hospital, Nairobi, 1988/89 and 1992

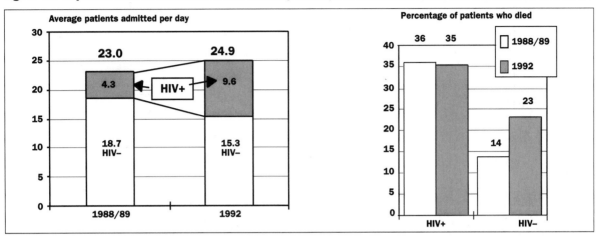

Panel A. Utilization increased for HIV-positive patients but fell for HIV-negative patients between 1988 and 1992.
Source: Floyd and Gilks 1996

Panel B. In-hospital mortality remained constant for HIV-positive patients, but rose by 66 percent for HIV-negative patients between 1988 and 1992.

from 14 to 23 percent (panel B of figure 4.2). The mortality rate for the HIV-positive patients did not increase, and other indicators of the quality of care remained constant. Thus, the most likely explanation for the increased mortality rate among the HIV-negative patients is that the rationing scheme used to allocate increasingly scarce beds had the effect of changing the mix of HIV-negative patients toward those with more severe illnesses. Whether the rationing was imposed by hospital staff or was a response by prospective patients to their perception of a higher effective price of care (box 4.4), it is likely to have excluded some patients whose lives the hospital could have saved.

Since the HIV-infected make up an increasingly large fraction of the sick people in a severely affected country, it is appropriate that they occupy an increasing share of hospital beds and consume an increasing share of health care resources. The pressure of this increased demand for care will naturally be felt by all citizens, whether or not they are HIV-infected. However, the extent of the shift in health care resources away from the HIV-negative can be exaggerated, as indeed it may have been in Kenyatta National Hospital, if the government provides special subsidies for people with HIV.[9] We discuss this issue, and the broader issue of how the level of government health care subsidies affects the demand for care and health care expenditure, in the next section.

Data from Kenyatta National Hospital on admissions and mortality suggest that the increased need to care for people with HIV has squeezed out some HIV-negative people who would otherwise have received care.

Policies To Mitigate the Impact on the Health Sector

Scarcer and more expensive care and increased total health expenditure present society with difficult choices. Because a large share of the increased expenditure is typically financed through tax revenues, governments and their constituencies will confront tradeoffs along at least three dimensions:

- treating AIDS versus preventing HIV infection
- treating AIDS versus treating other illnesses
- spending for health versus spending for other objectives.

The need to confront these difficult choices can be reduced somewhat if a government is willing and able to increase tax revenues. But few countries will be able to avoid the choices entirely, especially developing countries facing a severe epidemic. Unable to pay for everything, most governments will subsidize some goods and services more than others, thereby disproportionately benefiting certain groups of citizens.

As the number of AIDS cases increases, governments are likely to face mounting pressure for two responses that on first consideration seem rational and humane. One is to pay a larger share of health care costs; the other is to provide special subsidies for the treatment of HIV/AIDS. Unfortunately, these responses can have unintended consequences. For reasons discussed below, governments that wish to minimize the impact of HIV on the health sector should try to avoid both courses of action. However, this does not mean that governments should do nothing to help alleviate the suffering caused by HIV/AIDS. The section concludes with a list of compassionate and affordable measures that governments can and should undertake to mitigate the health sector impact of an HIV/AIDS epidemic.

No increase in the overall subsidy to health care. One obvious and politically appealing response to the HIV/AIDS epidemic is to increase the government share of health care costs and thus the overall subsidy for health care. Such a course of action may be especially attractive early in the epidemic, when few people are sick with AIDS. There is an argument for it on economic grounds as well: it would fill the gap created by the failure of the private market to offer health care insurance in poor countries. However, increasing the subsidy to curative care increases the demand for a limited supply. As a result, both effective price and total expenditure will rise by a greater proportion than the increased subsidy alone, or the increased demand arising from the epidemic alone, or even the sum of the two, would suggest. As more and more people become sick with AIDS, this effect becomes evident in escalating health care expenditures; in a severe epidemic, the burden on the government budget is likely to become unsustainable.

To understand how changes in the level of government subsidies affect the impact of the epidemic on the health care sector, we first look at the extent to which governments already subsidize care. Then, taking India as an example, we project the impact of an expanding epidemic at the current subsidy level and an increased subsidy level. As we shall see, increasing the overall subsidy to care can greatly exacerbate the impact of the epidemic on the health sector.

Most governments subsidize a large share of health care expenditures. The balance includes payments by private insurers and all "out-of-pocket" payments at private or government-subsidized facilities, whether traditional or modern. The average overall subsidy to health care varies widely but generally rises with GDP. As can be seen in figure 4.3, the poorest countries, with average per capita income of about $600, typi-

Figure 4.3 Public Share of Health Expenditure in Selected Countries, by Income Group, Various Years, 1990–97

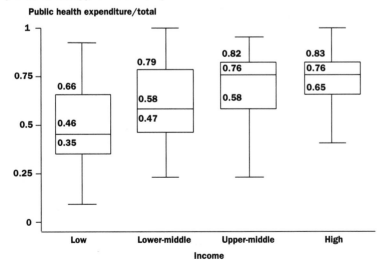

The public share of health care expenditures tends to be higher in countries with higher incomes.

Note: The middle line in each box shows the median proportion subsidized; the top and bottom of the boxes are the 75th and 25th percentiles, and the "whiskers" give the minimum and maximum subsidy rates observed.

Source: World Bank data.

Figure 4.4 Simulated Impact of a Severe AIDS Epidemic on Health Expenditure, India, 1990–2010

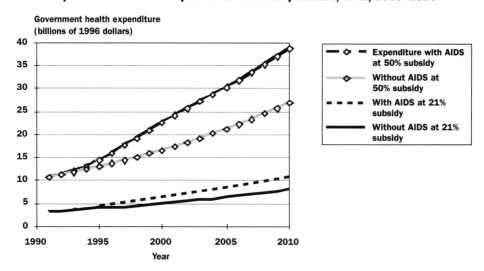

Note: The projections follow box 4.3 in assuming that the elasticity of demand for healthcare is 0.8 and of supply 0.5. If the demand elasticity in India were smaller or the supply elasticity greater than these assumptions, all the expenditure impacts would be correspondingly smaller.

Source: Ellis, Alam, and Gupta 1997; authors' calculations.

If India maintains its current level of health care subsidies, a severe AIDS epidemic would increase government health care expenditure by about $2 billion per year by 2010. If subsidies are increased to the 50% level, the same size epidemic would increase annual government health expenditure by an additional $30 billion.

cally subsidize less than half of the cost of health care, while upper-income countries subsidize about three-quarters of the cost.

In India in 1990 the government subsidized about 21 percent of total health care expenditures, a small share even compared with other low-income countries. The bottom line of figure 4.4 projects government health expenditure if India were to have no AIDS epidemic and continued to spend 6 percent of a constantly growing GDP on health care, of which the government continued to finance 21 percent. In this baseline scenario, India's government health expenditures grow from $3.2 billion in 1991 to $8 billion in 2010. The second line from the bottom shows the increase in government health spending if India's current steep rise in HIV prevalence continues until 2000, then levels off at a stable 5 percent. This is about the growth in prevalence seen in countries such as Zambia and Botswana, where focused prevention was not implemented early in the epidemic. The result in India would be to increase the government's expenditure on health in the year 2010 by about one-third, from $8 billion to $10.5 billion.

What if India in 1990 had increased health care subsidies to about 50 percent, the level seen in many Latin American countries? The top pair of projections in figure 4.4 shows the impact of the higher subsidy on expenditure. Even without an AIDS epidemic, expenditure more than triples to $11 billion in 1991 due to the more than doubling of the government's share of existing expenditure combined with the demand stimulus caused by the greater subsidy. Subsequent growth of health expenditure proportional to GDP brings health expenditure to $27 billion in 2010 (third line from the bottom). Now again suppose a serious AIDS epidemic that reaches a stable 5 percent HIV prevalence rate in 2000. The fourth line from the bottom of figure 4.4 gives the projected result: health expenditures in 2010 would reach $39 billion. Thus, not only has the increased subsidy tripled health care spending, as might have been expected, but it has also increased the vulnerability of the budget to the AIDS epidemic, adding $12 billion (43 percent of $27 billion) rather than just $2.5 billion (31 percent of $8 billion) to government health care expenditures.

The large expenditure shocks that will result from the AIDS epidemic will create new pressures on health budgets, especially in countries that enter the AIDS epidemic with higher subsidy rates. For example, although Mexico's infection rate was estimated to be only 0.4 percent in 1994 and it subsidized only 49 percent of the cost of AIDS treatment, compared with 76 percent for other sicknesses, AIDS was already consuming 1.2 percent of its health budget. In contrast, Tanzania has kept the subsidy rate for AIDS treatment down to 28 percent in line with the subsidy it provides to other illness categories. As a result, despite a prevalence rate of 5 percent, more than ten times higher than Mexico's, the AIDS share of total government health care expenditure is only 3.5 percent, just three times larger in Tanzania than in Mexico.[10]

Although a discussion of the design of health financing systems is beyond the scope of this book, the evidence suggests that countries in the nascent or concentrated stages of the epidemic, like India, should carefully consider not only the immediate budgetary consequences of any expanded commitment to fund curative care, but also the multiplication of these consequences that would occur if the AIDS epidemic spreads. A prudent course would be to consider any expansion of government-financed health care subsidies or insurance only in conjunction with vigorous prevention programs that enable people most likely to contract and spread HIV to protect themselves and others.

Equal subsidy rates regardless of HIV status. A second common health sector response to the HIV/AIDS epidemic is to offer a different subsidy rate depending on whether or not the person receiving care is infected with HIV. Especially in the countries in the nascent stage of the epidemic, HIV-infected people all too frequently experience discrimination, including restricted access to or higher effective prices for health care. As the epidemic advances, however, governments are often pressed to provide special subsidies for the treatment of HIV/AIDS. This section points out the government's role in limiting discrimination against the HIV-infected in health care settings and then considers the consequences of preferential subsidies for HIV treatment.

AIDS treatment subsidies vary greatly from country to country. Figure 4.5 presents data on the percentage of AIDS-related and total 1994 health care expenditure funded by the government. In three of the five countries, the subsidy rates for AIDS treatment are significantly different from that for total health care expenditure. For example, although Mexico subsidized a generous 49 percent of the cost of AIDS treatment, this was much less than the 76 percent share of total health care expenditure. Brazil and Thailand subsidized AIDS care at a *higher* rate than all types of care, while Tanzania and Côte d'Ivoire subsidized AIDS treatment and total health care expenditure at roughly the same rate.

A bias against those with HIV/AIDS can take many forms, ranging from a singling out of AIDS-specific drug therapies for exclusion from public funding, to outright refusal of service. There are many anecdotes

Figure 4.5 Percentage of AIDS-Related and Total Treatment Expenditures Financed by the National Government, Four Selected Countries and São Paulo State, Brazil, 1994

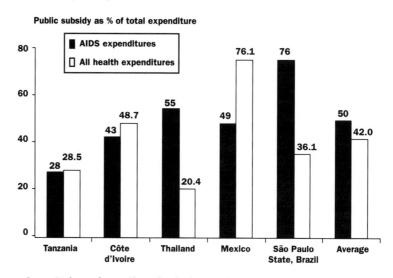

Governments often provide different levels of health care subsidies depending on whether or not the patient is infected with HIV.

Source: Background paper, Shepard and others 1996.

about discrimination against the HIV-infected in health care settings. In some hospitals, the HIV-infected were placed in special AIDS wards, which were subsequently shunned by fearful health care workers. In others, the HIV-infected were required to pay extra costs for rubber gloves or a private room. In still other cases, the HIV-infected have been denied treatment for common illnesses, perhaps because doctors and nurses mistakenly believed that nothing could be done to help a person with HIV/AIDS. Such discrimination is unfair, unprofessional, and unethical. Moreover, it displays ignorance of the many ways, discussed above, in which inexpensive treatments for symptoms and opportunistic illnesses can prolong and improve the lives of people with HIV/AIDS. Government has an important role to play in training medical personnel in order to eradicate all vestiges of discrimination against HIV-infected patients.

Yet it is equally unfair, and also inefficient, for government to subsidize a higher proportion of the costs of care for patients with HIV than for other patients. Aside from the issue of poverty, to be addressed in the next section of this chapter, there are three ways to justify government subsidies for curative health care: (1) as an incentive for those with an infectious disease to seek a cure and avoid infecting others, (2) as health care insurance with universal coverage and mandatory participation through general taxes, or (3) as government support for a "merit good" or "basic need." No treatment has yet been shown to reduce the infectivity of sexual contact with an HIV-infected person (see box 4.5). AZT treatment of HIV-infected pregnant women has been shown to reduce transmission at birth, but is still too costly an approach to preventing secondary infections in the poorest countries (see box 4.6). With the prominent exception of TB, the treatment of which should be subsidized in all countries, most opportunistic illnesses that afflict the HIV-infected are infectious only to other equally sick HIV-infected people. Thus the argument for treating the diseases on the grounds that they are infectious is weak. If government subsidy is considered as an insurance payment, efficiency criteria argue for a higher coinsurance (i.e., lower subsidy) rate for any condition in which the patient is likely to be highly price-responsive.[11] The first section of this chapter established that the drugs and medical services to treat AIDS can amount to a great deal of money, although some of the most expensive of these treatments purchase the patient little additional life span and decrease, rather than improve, the quality of life. Thus on efficiency grounds, where the objective is to limit the responsiveness of expenditure to insurance, AIDS

Box 4.5 Is Antiretroviral Therapy an Effective Way To Prevent Sexual Transmission?

UNTIL RECENTLY, ANTIRETROVIRAL (ARV) THERAPY could not be considered as a possible way to prevent sexual transmission because the available drugs for treating HIV/AIDS had little impact on infectivity. The 1997 discovery that protease inhibitors and triple-drug therapy suppress HIV below the level of the most sensitive blood tests to detect it has raised hopes that these drugs might prevent the spread of HIV, in addition to greatly extending the life of the patient. Even if this proves true, however, policymakers deciding whether to provide public subsidies will need to consider that the $10,000 to $20,000 cost of treating a single patient would prevent many more cases if spent on focused prevention in high-risk groups. Furthermore, we saw in chapter 1 that even without the expense of antiretroviral therapy, current expenditure for treating an AIDS patient would buy a year of primary school for ten students in most developing countries. In the poorest countries, the much higher cost of antiretroviral therapy would buy a year of primary school for 400 students. For this reason, even if the cost of antiretroviral therapy is shown to reduce the infectivity of sexual contacts, and even if the cost falls substantially, decisionmakers will still want to consider very carefully before initiating such subsidies.

patients should face somewhat lower subsidies, not higher ones. The final possibility, that AIDS treatment is a basic need, is difficult to justify in poor countries where the opportunity cost of treating one adult for AIDS may be measles vaccines for 100 to 200 children or, as shown in figure 1.8, ten student-years of primary school. Thus none of these economic arguments justifies higher subsidy rates for AIDS.

What policy recommendations can be drawn from these two observations? The prudent, efficient, and equitable course is to place the financing of health care for HIV/AIDS on the same footing as other diseases. The treatment of particularly infectious illnesses striking HIV-infected people, including TB and STDs, should be subsidized relatively generously because of the secondary infections treatment will prevent. Other health care problems of the HIV-infected should be subsidized at the same rate that applies to other adult health problems that are equally infectious. Assuming that Brazil subsidizes about one-third of other health care costs (as figure 4.5 indicates it does in São Paulo) and that the infectious proportion of illness episodes is similar among the HIV-infected and uninfected populations, this policy would lead Brazil to reduce its subsidy to antiretroviral therapy from 100 percent to one-third. Similarly Thailand would reduce its subsidy of antiretroviral therapy from 100 percent to about 20 percent. Mexico, on the other hand, would increase its subsidy to AIDS patients to approximately the same rate it offers other patients.

In mid-1997, none of these countries appeared to be following this recommendation precisely. Brazil and Mexico were continuing their former policies, with a tilt of subsidies toward AIDS treatment in Brazil and away from it in Mexico. Having spent $108 million on antiretroviral medication in 1996, Brazil was projecting an expenditure four times that large for 1997 (Chequar 1997). Thailand had recently embarked on an experiment that held out the possibility of an equal percentage subsidy for the treatment of AIDS and other diseases, on average, if not for the individual patient. In 1996, the Thai Ministry of Public Health found that, given the rising patient load, its policy of a 100 percent subsidy for antiretrovirals and drugs for the opportunistic illnesses would soon consume considerably more than the entire budget allocated to the National AIDS Program (Prescott and others 1996). As a result, the government revised its policy to provide free antiretroviral therapy only to HIV-positive pregnant women, where it might prevent mother-to-child transmission, and to participants in nationally approved clinical trials, where patients receive the support they need to maximize compliance (Kunanusont 1997). This policy makes sense for antiretroviral therapy, on the assumption that low levels of compliance outside clinical trials would have little therapeutic effect on patients and might cause negative externalities in the form of drug-resistant strains of HIV. Furthermore, participants in clinical trials produce a positive externality in the form of the knowledge that can be used to benefit many other patients, and therefore should receive a higher subsidy than other patients. Thailand's decision to subsidize AZT for prevention of mother-to-child transmission can be justified as a "merit good," which might be affordable in a middle-income country (see box 4.6).

Affordable, humane responses to the epidemic. We have argued that governments should avoid two types of health care responses to the epidemic: increasing the overall subsidy to all types of treatment, and providing disproportionately large subsidies to treatment of HIV/AIDS. There are nonetheless several ways in which governments can intervene to mitigate the health impact of HIV/AIDS on infected individuals and their families and on the overall health sector. Each of these interventions is justified on public economic grounds, either because it has large positive externalities, or because it improves the efficiency or the equity of the health care market in other ways.

Box 4.6 Preventing Mother-to-Child Transmission

OF THE MANY TRAGEDIES CAUSED BY THE HIV/AIDS epidemic, perhaps none is more disturbing than that of children who contract the virus from their mothers at birth or through breastfeeding. Methods exist for preventing mother-to-child transmission; sadly, most of the methods so far developed are difficult to implement in the very poor countries where most mother-to-child transmission occurs.

About one-half to two-thirds of mother-to-infant transmission is believed to occur at the time of birth (Reggy, Simonds, and Rogers 1997). The risk of HIV transmission from mother to newborns can be reduced by two-thirds, from 25 percent to about 8 percent, by administering zidovudine (AZT) to the mother before and during birth, and to the non-breastfed newborn for six weeks after birth (Connor and others 1994). The total drug and related medical costs for the AZT regimen currently recommended by the U.S. Centers for Disease Control and Prevention (CDC) for reducing mother-to-child transmission amounts to $1,045 per case treated in the United States (Mauskopf and others 1996). In Thailand, where some inputs are less costly, the total cost is about half as great (Prescott and others 1996). Even so, this is roughly 50 times the average per capita health expenditure of low-income countries in Sub-Saharan Africa, where about two-thirds of mother-to-child transmissions take place. And at roughly $3,000 per HIV infection averted, this approach to prevention does not compare favorably with other approaches discussed in chapter 3 and would be affordable only in middle- or upper-income countries.

Several research efforts are under way to find a lower-cost means of reducing mother-to-child transmission. One involves trying to identify the most efficacious part of the AZT regime, in order to reduce the total amount of AZT needed. Trials are also under way in industrial and developing countries to investigate various other medical approaches to reducing transmission (Biggar and others 1996, DeMuylder and Amy 1993). However, it is unclear whether any of these strategies, if found effective, would be affordable or technically feasible in many developing country settings.

Newborns of HIV-positive mothers who escape infection at birth may nonetheless be infected later through breastfeeding. As a result, public health officials have had to weigh the advantages of breastfeeding for child health against the possibility of HIV transmission. In areas where the primary causes of infant deaths are malnutrition and infectious diseases, UNAIDS recommends that women continue to breastfeed their children. If a woman is known to be HIV-positive, she should be provided with the means to make an informed choice about infant feeding methods. In areas where there are safe alternatives for infant feeding, however, children will be at less risk of illness and death if not breastfed (UNAIDS 1996a). While it may be possible to simply reduce the duration of breastfeeding, it is not known what impact this might have on reducing transmission, since there is no consensus on when the risk of transmission is highest within the breastfeeding period (*background paper*, Saba and Perriëns 1996).

■ *Provide information about the efficacy of treatments.* Because people with HIV/AIDS are often desperate for treatment and cannot easily research what works, they are especially vulnerable to quackery. Governments can serve the interests of everyone by promptly investigating unproven treatments and providing credible information about their validity. So long as this is done through existing media channels—for example by issuing press releases and arranging media interviews with credible experts—it can be done quite inexpensively.

■ *Subsidize the treatment of infectious opportunistic illnesses and STDs.* Subsidized treatment is especially appropriate for tuberculosis, one of the most common opportunistic illnesses to infect AIDS patients, since curing a single case can avert many secondary infections. Treatment of gonorrhea, syphilis, and the other classic STDs should be subsidized, not only because they are highly contagious, but also because they exacerbate HIV transmission, as discussed in chapter 3. Because few people are susceptible to them, treating toxoplasmosis, cryptococcosis, or one of the other infectious opportunistic illnesses that develop only in people with severely disabled immune systems prevents few secondary cases and thus should be subsidized at a lower rate, closer to the subsidy rate for chronic, noninfectious disease. Whether a subsidy to antiretroviral treatment of HIV itself is justifiable as a way to prevent secondary HIV infections will depend on the efficacy of the treatment and on its cost relative to the cost of other HIV prevention measures. In mid-1997, such treatments were far too expensive and uncertain to warrant subsidies on these grounds (see box 4.5).

■ *Subsidize the start-up costs for blood safety and AIDS care.* The AIDS epidemic has increased the willingness of individuals to pay for certain types of services, such as screening of blood for transfusions and care for the terminally ill. Where these services are lacking, government help with the start-up costs is justified, just as governments subsidize other large indivisible investments, such as an electric utility or a water system, so long as the users then pay for the services they receive. Thus, governments in poor countries should establish blood banks but should not indefinitely provide free blood. Similarly, government should help establish AIDS treatment facilities, especially community-based home care programs, but should not permanently subsidize the care they provide.

■ *Provide special assistance to the poor.* Most countries already make special provision for medical care to the poor. As the AIDS epidemic increases the demand for care, governments may wish to focus such assistance even more on those who can least afford it. Sliding fee scales and other measures to make care available to the poor should apply to people with HIV/AIDS just as they do to people with other illnesses. This principle of providing assistance to those who need it most, regardless of their HIV/AIDS status, is discussed more fully in the next section on ways to mitigate the impact of HIV on poverty.

AIDS and Poverty: Who Needs Help?

IN ADDITION TO ITS DEVASTATING IMPACT ON INFECTED INDI-viduals, HIV hurts all those who are linked to them by bonds of kinship, economic dependence, or affection. The grief suffered by survivors, and the possible lasting psychological damage, especially to young children who lose a parent, are potentially the most damaging consequences of the epidemic. They are, however, difficult to measure, probably unreachable by public policy, and therefore beyond the scope of this book. In addition, survivors often suffer economically. This harm from a prime-age adult death constitutes the most important economic impact of an HIV/AIDS epidemic and is the topic of the remainder of this chapter. It can be measured by the impact of adult death on such social indicators as orphanhood, child nutrition, schooling, and poverty. By worsening these measures and widening the gap between the poor and others, HIV can exacerbate poverty in poor countries and delay attainment of national economic development goals. We look first at how HIV/AIDS affects poverty, then at the implications of these findings for poverty policy in a severe AIDS epidemic.

How HIV/AIDS Affects Poverty

It is sometimes said that "AIDS is a disease of poverty." In what sense might this be true—or false? First, are the poor more likely to become infected with HIV than others? Second, what proportion of people infected with HIV are poor? Answers to these questions are important, for they will influence both the focusing of prevention measures and attempts to mitigate the impact of the AIDS infections that do occur. In considering the impact of AIDS on poverty, we first examine the available evidence to answer these two basic questions; then we ask how the impact of an AIDS death compares with other shocks that households suffer, and how households of different income levels cope.

HIV infects the rich and the poor. In developing countries, the relationship between income and HIV infection rates has been best documented in eastern and central Africa.[12] Whether the patterns observed in this part of Africa will also emerge elsewhere remains to be seen. Several factors have exacerbated the epidemic in hard-hit areas of Africa: most people who had HIV at the time of the studies had become infected years earlier, when little was known about HIV prevention; moreover, the area is traversed by major transport routes and has suffered from war. Yet each of these factors is also evident in other developing regions to varying degrees: knowledge about HIV prevention is still often scant, and other regions also have major transport routes and wars. Thus, until other data are available, the experience in eastern and central Africa may offer worthwhile clues about how infection rates in other regions are likely to differ across income groups as the epidemic progresses.

As we learned in chapter 3, early in the epidemic in Sub-Saharan Africa men and women who travel more, and men who had higher incomes, were more likely than others to contract the virus. There are reasons to believe that this may hold true elsewhere. Studies show that sex is similar to other pleasurable pastimes: the number of partners per year rises with income. Also, a person with a higher income is likely to attract more prospective partners, and will have more money than a person with lower income to compensate sexual partners or to support any offspring. These factors, combined with the fact that HIV, unlike other STDs, cannot be readily cured, has made HIV unique among widely prevalent infectious diseases in striking rich people in the same proportion, or larger proportions, than it strikes the poor. That HIV infects the rich as well as the poor is important to keep in mind when considering which households need help the most.

Of course, we would expect that more-educated people with higher incomes would be in a better position to learn about the epidemic and alter their behavior to avoid infection. Chapter 3 presents evidence that this is already occurring: in some countries, highly educated people have higher frequencies of condom use than the less well educated. Also, recent studies in developed countries have shown AIDS incidence to be highest among the very poor. If these trends are replicated worldwide, AIDS will become like other infectious diseases, in that the poor will be more likely to become infected than the nonpoor. Ultimately, AIDS may become most prevalent in the poorest urban slums of developing countries.

Already, most people with HIV/AIDS are poor. Although lack of data makes it impossible to calculate the precise proportions of poor and nonpoor who are infected, knowledge of income levels and infection rates across countries suggests that many more poor people are infected than nonpoor people. For example, according to an internationally adjusted standard of absolute poverty, Sub-Saharan Africa has about four times as many poor people as nonpoor people. Thus, even if poor people were infected at *just slightly more than one-quarter the rate* of the nonpoor, poor people would account for the majority of HIV infections in Africa. Since poor people in many parts of Sub-Saharan Africa have infection rates that are well above one-quarter the rates of the nonpoor, we know that, in Africa, at least, there are many more poor people than rich people with HIV. Although magnitudes are less striking, the same general principle will tend to apply in other developing regions.[13]

We have seen that AIDS is already a disease of poverty in the sense that it affects *more* poor people than nonpoor, and it may eventually become a disease of poverty in the sense of infecting a *higher proportion* of poor than nonpoor. If we assume that one of government's main responsibilities is to make it possible for people to escape from poverty, these findings lead us to new questions. What is the impact on a poor household when the mother, father, or another prime-age adult who is a member of the household dies from AIDS? How do poor households cope with AIDS deaths? We examine these questions in the next two subsections. Box 4.7 describes three sets of characteristics that determine the initial impact of an adult death and how well an afflicted household copes.

What is the direct impact of an AIDS death? The death of a prime-age adult is obviously a tragedy for any household. Survivors must contend not only with profound emotional loss, but also with medical and funeral expenses, plus the loss of income and services that a prime-age adult typically provides. How serious is the shock of an AIDS death to the economic welfare of the survivors? The direct impact of a death consists of the medical costs prior to death and the costs of the funeral. To assess the direct cost of a death from AIDS, we compare the medical and funeral costs of an AIDS death with those of a prime-age adult death from other causes. Finding that the difference is not large, we then consider how the death of a prime-age adult, regardless of cause, affects household consumption patterns.

Our analysis is based on findings from several household surveys described in box 4.8. In particular, we rely on the most extensive of these, a study done in Kagera, Tanzania, since detailed data from that study provide a basis for our subsequent analysis of how households cope with

> ## Box 4.7 Three Factors Determine the Household Impact of a Death
>
> THE OVERALL ECONOMIC IMPACT OF AN ADULT DEATH ON THE surviving household members varies according to three sets of characteristics:
>
> - those of the deceased individual, such as age, sex, income, and cause of death
> - those of the household, such as composition and assets
> - those of the community, such as attitudes toward helping needy households and the availability of resources.
>
> The first set of characteristics determines the basic impact of the death on the surviving household members; the second and third influence how well the afflicted household copes. Although disentangling the three is very difficult, it is nonetheless important when attempting to assess the household impact of an adult death to consider all three sets of factors.

AIDS deaths. Although the data are very limited, based on the available information, it is reasonable to expect that the impacts and coping responses described in this chapter will prove to be broadly consistent with future findings.

In the Kagera study, people diagnosed with AIDS were somewhat more likely to seek medical care than people who died from other causes, and they were more likely to incur out-of-pocket medical expenses.[14] Moreover, household medical expenditures tended to be much higher for AIDS than for other causes of death, as shown in figure 4.6. Strikingly, for all groups except men with AIDS, medical expenses were overshadowed by funeral expenses. On average, households spent nearly 50 percent more on funerals than they did for medical care. Moreover, funeral expenditures for AIDS deaths and non-AIDS deaths differed less than did medical expenditures. Thus, even though a significant proportion of funeral costs were covered by gifts from other households (about 45 percent on average), the difference in the household impact of an AIDS death and a non-AIDS death is smaller than the differences in medical costs alone would lead us to expect.[15]

Box 4.8 Studies of the Household Impact of Adult Death from AIDS and Other Causes

WITHIN THE PAST FEW YEARS, FOUR IN-DEPTH studies have examined the impact of adult death from AIDS on surviving household members. Compared with other studies, these four studies used more detailed survey instruments; applied these to larger, more representative samples of households; and followed the households over longer periods. The four studies were carried out in the following locations (the number of households surveyed is shown in parentheses):

- Chiang Mai, Thailand (300)
- Abidjan, Côte d'Ivoire (107)
- Rakai, Uganda (1,677)
- Kagera, Tanzania (759)

The studies used many similar parameters. All but the Thai study visited the sample households several times. All but the Côte d'Ivoire study included households that did not experience an AIDS sickness as well as those that did.[1] All but the Rakai study were done expressly to study social and economic impact and thus had extensive questionnaires about consumption and other social and economic measures of well-being.[2] All but the Côte d'Ivoire

study included deaths from causes other than AIDS as well as those from AIDS.

Two broad findings emerged from these studies and are discussed in the text. First, households use a variety of informal mechanisms to cope with misfortunes like an adult death in the household. Second, although these coping mechanisms cushion the impact of the shock, households are not entirely successful in protecting their well-being. In general, the poorer the household, the greater and more persistent the impact of a prime-age adult death from AIDS and similar shocks.

---

1. Although the Côte d'Ivoire study did not include an explicit control group, Béchu (*background paper*, 1996) is able to use differences across households in the severity of the AIDS cases combined with the sequence of six observations on each household to estimate the impact of fatal adult illness on consumption.

2. The Rakai study is part of a study of the effect of mass STD treatment on the incidence of AIDS. The household questionnaire focused on epidemiological issues and only asked a few questions related to economic well-being.

Sources: For Thailand, Pitayanon, Kongsin, and Janjaroen (1997) and Janjaroen (*background paper,* 1996); for Côte d'Ivoire, Béchu (*background paper,* 1996); for Uganda, Menon and others (*background paper,* 1996a); and for Tanzania, Over and others (forthcoming).

In Thailand, where per capita income is 10 times that in Tanzania, households in Chiang Mai province spent more than ten times as much on medical care prior to death as did the Tanzanian households (Pitayonon, Kongsin, and Janjaroen 1997). The households with an AIDS death spent $973 on average, which in contrast to Tanzania was only about 10 percent more than the $883 spent by the non-AIDS households. But, just as in Tanzania, the households spent much more on funerals than on medical care.[16]

The relative amounts spent for medical care and funerals will, of course, vary from country to country and even across communities within a district. Nevertheless, two broad observations are likely to apply in most situations: first, medical costs are only a portion of the cost of

Figure 4.6 Average Medical and Funeral Expenditures, by Gender and Cause of Death, Kagera, Tanzania, 1991–93

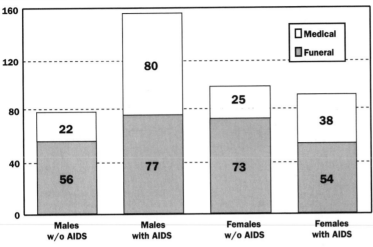

1996 dollars

Medical expenditures were higher for people who died of AIDS than people who died of other causes. But because funeral expenses are large, the overall difference in expenditure for AIDS deaths and deaths from other causes is smaller than the difference in medical expenditure alone would suggest.

Note: Throughout this report currency amounts have been converted from current Tanzanian shillings to 1996 U.S. dollars. The conversion procedure involves three steps: (1) convert current shillings to 1991 shillings using the project's price deflator; (2) convert 1991 shillings to 1991 dollar amounts at 289 shillings per dollar; and (3) inflate to 1996 dollars by multiplying by 1.15. Sample: deaths of 264 adult household members ages 15 to 50.

Source: Over and others, forthcoming.

a prime-age adult death; and second, nonmedical costs are likely to be similar, regardless of the cause of death. Where these observations hold true, the direct impact of an AIDS death will not be much different from that of a non-AIDS death, despite higher medical expenditures for AIDS. Thus, the high cost to households from AIDS will usually be due to the large number of deaths caused by the epidemic rather than by the fact that they are caused by AIDS. Given that the impact of a prime-age adult death is likely to be similar, regardless of cause, how does a prime-age adult death affect household consumption? Figure 4.7 shows household consumption during the previous twelve months for two groups of households in the first wave of the Kagera survey: those who had experienced a death during this period and those who had not. Households that suffered a death had lower overall expenditures and, as we would expect, devoted a larger share of the expenditure to medical and funeral costs. Also, these households spent one-third less on the "other nonfood" category (i.e., clothing, soap, and batteries). Finally, in households that suffered a death, food produced by the household was a larger share of

Figure 4.7 Consumption in Kagera, Tanzania, Households by Whether the Household Experienced a Death in the Past Year (Results from Wave 1 of Kagera Study)

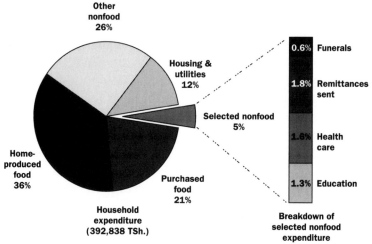

Households not experiencing an adult death in the past year

Note: Of the 6,395 TSh. spent on health care in the average sample household, TSh. 748, or 11%, was spent on the care of nonadult household members who had died.

Purchased food accounted for a smaller share of total consumption in households that suffered a death (bottom panel) than in households that did not (top panel). Households that suffered a death increased their consumption of home-produced food, but this only partially offset reduced consumption of purchased food.

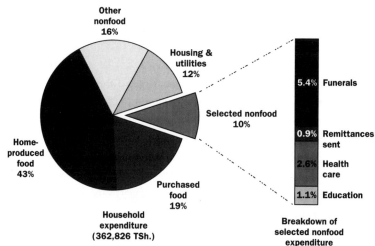

Households experiencing an adult death in the past year

Note: Of the 9,453 TSh. spent on health care in the average sample household, TSh. 6,069, or 64%, was spent on the care of household members who had died.

Source: Over and others, forthcoming.

Figure 4.8 The Time Pattern of the Impact of Adult Illness on Per Capita Household Expenditure, Côte d'Ivoire Study

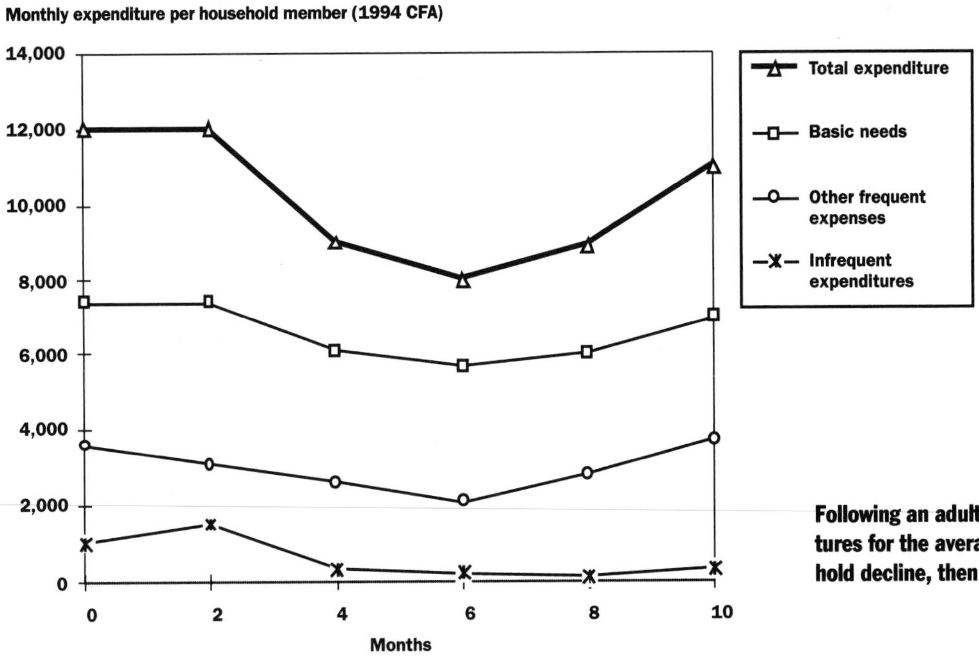

Monthly expenditure per household member (1994 CFA)

Following an adult death, expenditures for the average bereaved household decline, then partly recover.

Source: Background paper, Béchu 1996.

consumption than in households without a death, while purchased food was a smaller share of consumption.[17] These differences reflect the fact that the members of households that experienced a death cut back on the number of hours they worked for wages and thus had lower incomes with which to purchase food (Beegle 1996). They were only partially able to replace this lost income with additional production of food at home.

In the two studies that followed, detailed household consumption over time, the Kagera study and the one from Côte d'Ivoire, the time pattern of consumption demonstrates the resiliency of the average household to the impact of the death. Figure 4.8, drawn from the Côte d'Ivoire study, shows changes in expenditure per household member for three components of expenditures, and total expenditure during ten months after the households in the survey lost someone to AIDS.[18] Two patterns are immediately evident. First, total consumption dips and then partially recovers, still trending upward at the end of the survey. Second, basic needs, which include food, dip less than other categories of expenditure, then almost fully recover as families reduce other categories of

213

spending to minimize the impact on necessities (*background paper,* Béchu 1996). The household surveys from Chiang Mai and Rakai also suggest a partial recovery in per capita consumption but do not have sufficient data to confirm this pattern.

How households cope with the impact of adult death. The economic shock of a prime-age adult death, described above, would have been larger and more persistent, except that households use a variety of strategies to cope. Before the AIDS epidemic, prime-age adult deaths were much less common, so these mechanisms were used mostly to cope with other shocks. As a result, early assessments of the household impact of AIDS tended to overlook household coping and assume that AIDS would be catastrophic not only for the infected individual but also for the entire household. Press accounts that presented devastated households as typical contributed to the widespread belief that most AIDS-affected households in developing countries would collapse. To be sure, some households *are* destroyed by AIDS; this is especially true if both parents become ill or die while their children are still very young. However, such instances may be less typical than is generally assumed because of the typically long lag between HIV infection and death. Moreover, while premature death of a loved one is always tragic, leading to emotional pain and sometimes lasting psychological damage in survivors, survey data suggest that when it comes to coping with the *economic* impact of such a loss, households in general are surprisingly resilient.

The degree of household resilience to the economic impact of a prime-age adult death has important implications for society's response to a generalized epidemic. On the one hand, if nearly all AIDS-affected households collapsed, resources for mitigating the household impact of the epidemic would be stretched so thinly that governments and social welfare organizations would be overwhelmed. In such a situation, policymakers might easily conclude that those currently affected were beyond help and that the only reasonable response for the government would be to redouble prevention efforts. On the other hand, if many households were able to cope, governments and NGOs could focus the limited resources available for mitigating the impact of the epidemic on the households that needed help the most.

Understanding the variety of household coping mechanisms and how these will affect different groups of households is important. The mix of responses attempted by a specific household in response to a prime-age adult death depends on countless factors, some of which will vary across countries and communities. Since the available data on household impact comes mostly from Sub-Saharan Africa, the following discussion unavoidably reflects this bias. Policymakers in all countries faced with the possibility of a generalized epidemic will want to assess the extent to which these responses are evident in their own country. To varying degrees, however, three coping mechanisms observed in Africa—altering household composition, drawing down savings or selling assets, and utilizing assistance from other households—are all likely to be attempted whenever households confront the tragedy of a prime-age adult death. This section discusses each of these informal mechanisms in turn and then discusses formal assistance, such as that provided by governments and NGOs.

Before this analysis, it would be useful to consider whether income that had been devoted to health care and funeral expenses could be diverted after the death to other expenses. Some household coping certainly involves such responses. However, the potential should not be overestimated, since much of the cost of medical care and an even larger proportion of the cost of funerals is financed by transfers from outside the household. Since these transfers typically cease after the funeral, households must draw on additional coping strategies, described below.

Altering household composition. Households everywhere fulfill economic as well as social functions. In rural areas of developing countries, households are often the main production unit for subsistence farming and, in some instances, for cash crop farming as well. In such a situation, the economic shock of the death of a prime-age adult can be cushioned to some degree by altering household composition. Examples of such changes could include sending one or more dependent children to live with relatives, or inviting an unmarried aunt or uncle to join the household in exchange for assistance with farming and household tasks. Results from three of the four available household surveys—two from Africa and one from Chiang Mai, Thailand—show that the degree to which household composition is used to cushion the shock of the death varies according to the size and flexibility of local household structure.

Among the 759 households in the Kagera study interviewed once every six months for two years, 130 household members of all ages died, but roughly nine times as many people left the households alive over the same period and seven times as many joined the households. In addition, about 200 children were born to household members. As a result, the average size of all the households declined only slightly, from about 6.0 to 5.7 members.

During the six months between any two interviews, economically active adults left or joined about one-fifth of the households that did not have an adult death and about 40 percent of the households that did suffer a death. Since most households that suffered a death added at least one member, the average size of these households declined by less than one, from 6.4 to 5.7 members—so that the average household size after a death was the same as in households that did not suffer a death. Similarly, the dependency ratio rose only slightly in households with an adult death, from 1.2 to 1.4, slightly *less* than the 1.5 dependency ratio in households without an adult death.

A striking fact is that household size and dependency ratios changed very little, even though Kagera has high adult mortality from AIDS. The same phenomenon was observed in the survey in Rakai, Uganda, which, like Kagera, has a severe AIDS epidemic: 15 percent or more of adults in roadside communities are infected with HIV. This suggests that, even in a generalized AIDS epidemic, most African households that suffer an AIDS death will be able to adjust household size and dependency ratios in ways that make them similar to households that did not suffer a death.

The Chiang Mai survey reveals that, at least in this area of Asia, households are much smaller and less mutable than in Africa. The 108 households in the sample that had not experienced a death had 432 members, or exactly four per household. In contrast to the no-death households in Kagera, the Chiang Mai households experienced almost no change in membership, receiving among all of them only one new member and losing only 6 members over the reference period. The 216 households that experienced a death had an average of 4.1 members, of which they lost one each because of the death. Unlike the Kagera households, these Chiang Mai households remained a full person smaller (that is, with 3.1 persons per household) at the time of the interview, which was up to two years after the death (*background paper*, Janjaroen 1996).

There are two points of similarity between the household composition responses to death in the Kagera and Chiang Mai studies. First, the Chiang Mai households with deaths, like their Kagera counterparts, suffered an increased dependency ratio due to the deaths. Because of the smaller number of adults in the Thai households, the dependency ratio there almost doubled after the death. Second, in both countries households with a death were twice as likely to experience membership change as the households without deaths. However, the proportion of households that experienced a membership change and the rates of turnover were only about one-quarter as large in Chiang Mai as in Kagera.

When we later discuss possible policy responses to adult mortality, the possibility will be raised that households may respond opportunistically by moving people into a household that is benefiting from an assistance program. The evidence here suggests that, even if this turns out to be a problem in Africa, it is much less likely to be an issue in places like Chiang Mai where households are much smaller and apparently less able or willing to adjust their membership in response to outside stimuli.

Dissavings and the sale of assets. Drawing down savings and selling assets is an obvious potential mechanism for coping with prime-age adult death. Because assets may have been accumulated as part of a strategy to cushion unanticipated shocks, drawing upon them is one of the least painful ways of coping, much less painful than reducing food consumption, for example. Evidence from Kagera, Rakai, and Chiang Mai suggests that households do draw down savings or liquidate assets in response to a prime-age adult death.

The surveys in Kagera and Rakai both asked respondents about their ownership of three types of durable goods: a car or truck, a bicycle, and a radio. Less than 2 percent of the households owned a car or lorry, and changes in ownership did not show any clear pattern in relation to whether households suffered an adult death. However, ownership of bicycles and radios, which is much more widespread, did reveal a pattern. Table 4.5 shows how ownership of these assets changed over the course

Table 4.5 Asset Ownership in Households with and without an Adult Death

(percentage of total households)

Asset	Rakai District, Uganda		Kagera Region, Tanzania	
	Households w/o adult death	*Households with adult death*	*Households w/o adult death*	*Households with adult death*
Bicycle				
First visit	34	39	27	26
Last visit	41	35	29	28
Radio				
First visit	30	40	31	36
Last visit	37	36	35	35

Source: Authors' calculations; *background paper,* Menon and others 1996.

of the surveys, depending on whether or not the household suffered an adult death. In both surveys, radio ownership increased among households that had no death and decreased among households that had a death.[19] A similar pattern can be seen for bicycle ownership in Rakai, although not in Kagera. Thus the evidence from Rakai, partially supported by the Kagera data, suggests that some households that suffer an adult death may be selling durable goods as part of their coping strategy. The alternative is that these goods may have belonged to the deceased and been willed to someone outside the household. Even in this instance, however, the loss of the asset may help with coping efforts, if the recipient feels an increased obligation to assist the bereaved.

Additional evidence of households drawing down savings to cope with an adult death can be seen in Kagera data on membership in traditional rotating savings and credit associations (ROSCAs). In wave 1 of the survey, 51 percent of the 80 households that would experience an adult death during the eighteen-month survey period were members of a ROSCA; by the end of the survey, participation had dropped to 36 percent. Among households that did not experience a death during the survey period, ROSCA participation varied less, from 41 percent in wave 1 to 36 percent in wave 4.

Although the Chiang Mai survey does not permit comparisons of financial variables over time between households that experienced a death and those that did not, fully 41 percent of households with a death report having sold land, 57 percent report some dissavings, and 24 percent report borrowing from a cooperative or revolving fund (i.e., a ROSCA) to finance the adjustment to the death. Perhaps it is the greater wealth of the Thai households that permitted them to cushion the shock in these ways, rather than through readjustment of household composition.[20]

Assistance from other households. For all households confronted with an adult death, help from relatives and neighbors is a potentially important supplement to the household's own efforts. Policymakers considering how to best use the limited resources for mitigating the household impact will want to avoid displacing such private transfers. To do so, they will need information about assistance from households in the specific communities concerned. The discussion that follows is not intended to substitute for this information, but only to suggest the types and possible relative magnitude of such responses.

An important feature of the social organization of Kagera households, and indeed of most African communities, is interdependence in time of need. In Kagera, both bereaved and nonbereaved households were very likely to receive cash or in-kind assistance from other households. (About three-quarters of the nonbereaved households received such assistance, compared with 80 to 90 percent of the households that suffered a death.) But among households that received private transfers after a death, the median amount received during the half-year of the death ($53) was more than twice that received during the year before the death, as well as twice that received by households that did not suffer a death. [21]

New organizations established to help cope with the costs of AIDS death may be one explanation for this large difference. Focus group interviews in 20 of the sample villages found that besides traditional savings and mutual assistance associations, such as ROSCAs, residents of many villages had launched associations specifically to help families affected by an AIDS death. Most of these associations were launched and operated by women; many have regular meetings at which members make contributions in cash or in kind (Lwihula 1994).

Figure 4.9 shows the amounts of private transfers received by households according to whether or not they suffered an adult death. Note the dramatic response of private transfers between wave 1 and wave 4 for the households that experienced a death in that interval (i.e., during the "panel"). The figure also shows the much smaller amount of program transfers, an issue we turn to below.

Figure 4.9 Median Value of Assistance Received among Sample Households Receiving Transfers, by Source, Wave, and Occurrence of Adult Death, 1991–94

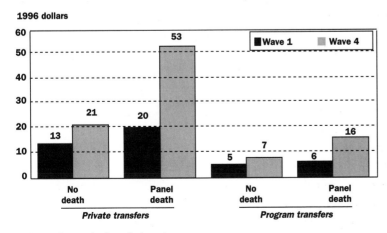

Households that suffered a death between the first and fourth "wave" of data collection received larger private and program transfers than households that did not suffer a death during this period. For these bereaved families, private transfers were much larger than program transfers.

Source: Over and others, forthcoming.

Assistance from government and NGOs. Regarding assistance from government and NGOs, two types of questions must be asked. First, which families receive such assistance and how much do they receive? Second, how much does it cost to provide this assistance? Cost considerations did not figure in our discussion of private assistance, since only private resources are involved. But governments and NGOs use public resources, whether generated by taxes or by voluntary contributions. Accordingly, we should ask whether these funds are used in the most effective way possible. In this discussion, assistance from government and NGOs is referred to as program transfers or formal assistance, to distinguish it from the private and informal assistance provided by households and village associations.

According to the Kagera survey, program transfers reached fewer households and provided a smaller amount of assistance than private transfers. In the last wave of the survey, one-fifth of the households that had not had an adult death in the past eighteen months had received assistance from an organization in the past six months; almost two-fifths of the households that had experienced a death received such assistance. The median value of program assistance received was small relative to total household expenditure, and to the amounts of private assistance received.

But while, on average, households that had experienced a death received larger amounts of assistance, as shown in figure 4.9, this was not always the case. Nor were program transfers always small relative to annual income. In one village, 50 percent of the households, including some that had never reported a death, received more than $110 in program transfers during the six months before wave 4.

To analyze the relative costs of various types of programs for assisting households affected by AIDS and other causes of adult death, the survey collected data from one governmental and eleven nongovernmental organizations operating in the Kagera Region. Figure 4.10 presents the average cost per year of operating each of the programs where cost data were available from at least two agencies. In considering the implied cost comparisons, it is important to remember that the services provided may be very different; for example, home care focuses on the sick household member, while educational support helps dependent children, who may not be sick, to attend school. Furthermore, even the programs averaged in a single category often contain disparate program elements and vary in quality.

Despite these caveats, the figure reveals that there can be very large differences in the cost per beneficiary of different types of programs. A

Figure 4.10 Average Cost per Year of Survivor Assistance in 1992 by Government and Nongovernmental Organizations, Kagera, Tanzania

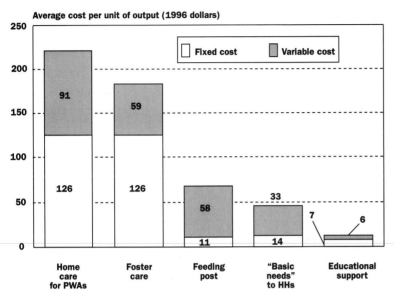

Note: PWA = person with AIDS, HH = household. Other costs are per child per year.
Source: Over and Koda, forthcoming.

Some types of assistance for households affected by AIDS consume much more resources than others.

particularly telling comparison is between the cost of supporting a child in a foster home, estimated at $107 per year, and the cost of supporting a child in an orphanage, which averages $1,063, or ten times larger (not shown on the figure). For children who cannot be placed in a foster home, it may be necessary to consider the alternative of an orphanage. However, policymakers and NGO providers should keep in mind that every child sent to an orphanage will consume the resources that could have been used to support ten children in foster homes.

The economic impact of AIDS is larger in poor households. We have seen that households that experience an adult death draw on their assets to cushion the shock of this catastrophe. It follows that households with lower levels of assets can be expected to have more difficulty coping with the death than households with more assets. In this section we examine the effect of a household's initial assets on its ability to cope with adult death. First we show how a household's assets affect the short-term impact of an adult death on per capita food consumption; then we consider the long-term harm to children, through worsening malnutrition and reduced school enrollments.

In examining this evidence, it is useful to keep in mind the key question that policymakers are likely to face in deciding how society in general and governments in particular, can mitigate the impact of a generalized epidemic: who needs help?

The impact on food consumption. The greater impact of a prime-age adult death on poorer households appears most starkly in changes in food expenditure and food consumption. Figure 4.11 shows the changes in per capita food expenditure and consumption (which includes both purchased food and home-produced food) for the poorer half of the Kagera households and the less-poor half of the Kagera households during the six months when the death occurred. For the better-off households, both measures of food intake increased. The picture is quite different for the poorest 50 percent of the households: food expenditure, which was already lower in these households than in the others, dropped by nearly a third. The resulting drop in per capita food consumption was cushioned by an increase in the consumption of home-produced food (not shown). Even so, per capita consumption in the poorer households

Figure 4.11 Short-Term Impact of the Death of an Adult Household Member on Food Expenditure and Consumption per Adult Equivalent Member, Kagera, Tanzania, 1991–93

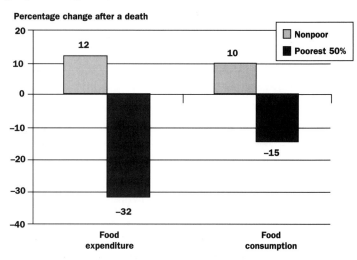

Among households in Kagera, Tanzania, that suffered a prime-age adult death, food expenditure and consumption declined for the poorer half of households but increased for the others.

Note: The poorest 50 percent of households are those with less than the median value of assets per member in wave 1 of the survey, which was about $415 per adult equivalent member in 1996 dollars. The sample is 64 households that experienced an adult death between the first and last waves of the Kagera survey.

Source: Over and others, forthcoming.

fell by 15 percent. Even if these households eventually return almost to the predeath level of per capita food consumption, as did the Ivoirian households, lack of adequate nutrition for a year or more can have a profound effect on the development of children. We turn to this topic.

The impact on child nutrition. Childhood malnutrition is potentially one of the most severe and lasting consequences of a prime-age adult death. The death of a parent or other adult may lower the nutritional status of surviving children by reducing household income and food expenditure, and by reducing adult attention to childrearing. Because childhood malnutrition can impede intellectual development and thus reduce a person's long-run productivity, improving childhood nutrition has long been an important development goal. Policymakers seeking to mitigate the impact of the AIDS epidemic will therefore be particularly concerned about minimizing the impact of the increasing number of prime-age adult deaths on childhood nutrition.

The impact of adult death on childhood nutrition is likely to vary according to many factors, not least of which is the nutritional status of children in the overall population. Little information is available on how adult death affects child nutrition. Moreover, the impact is likely to differ across countries and communities. The following discussion of the findings in Kagera illustrates some of the issues that policymakers will want to consider in attempting to mitigate the impact of the epidemic. In this discussion, the term "orphan" is used to indicate a child who has lost one or both parents.

We would expect that the drop in food consumption among the poorer bereaved families described above would result in an increase in malnutrition among children in these households, since these children are likely to be malnourished or at risk of malnutrition before the adult death. As figure 4.12 shows, among the poorer households in Kagera, stunting (very low height for age) among children under 5 is indeed substantially higher for orphans (51 percent) than for children whose parents are both alive (39 percent). What is surprising, however, is that the difference between orphans and nonorphans in the better-off households is even larger; indeed, orphans in the better-off households are stunted at almost the same rate as orphans in the poorer households.

This unexpected result raises difficult operational issues. If orphans in the poorer households were much more likely to be stunted than orphans in the less-poor households, as we might have expected, the policy prescription would be straightforward: to minimize childhood malnutri-

Figure 4.12 Stunting among Orphaned and Nonorphaned Children under 5, by Household Assets, Kagera, Tanzania

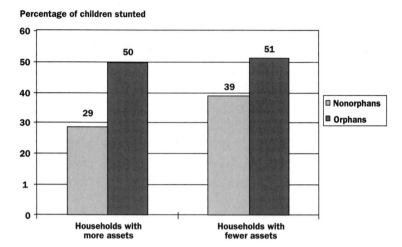

Percentage of children stunted

Source: Kagera data, authors' calculations.

In Kagera, Tanzania, half the children who had lost one or both parents were stunted, regardless of the level of household assets.

tion, focus nutritional assistance on the poor households that suffer a prime-age adult death. Instead we find that, at least in Kagera, half of the children who have lost one or both parents are stunted, *regardless of whether they live in a poorer household or a less-poor household.*

There are several possible explanations for this surprising observation. One is that stunting among both groups of orphans is due in part to pediatric AIDS and to other illnesses that a child may contract from an HIV-infected adult, such as tuberculosis, which would not necessarily be closely linked to household asset levels. Another possible explanation is that some stunted orphans in the households with more assets originally resided in poorer households, and their stunting is a legacy of that earlier poverty. Finally, the fact that stunting is about equal among both groups of orphans suggests that, for this population, stunting of 50 percent may be approaching an upward limit beyond which any additional deterioration in childhood nutrition results in increased child mortality rather then increased stunting. In all three cases, child nutrition could indeed be worse in the poorer bereaved families than in the less-poor bereaved families.

It may also be the case, however, that childhood nutrition does deteriorate sharply after a prime-age adult death, even in households with comparatively high levels of assets. This could happen, for example, if grief and psychological depression in the surviving parent interferes with childrearing, including obtaining food and providing meals. If this is true, young orphans stand a high probability of being malnourished, irrespective of the economic status of the household in which they are living.

One policy approach that could be appropriate in either case would be to focus nutritional assistance on young children who show evidence of being malnourished or who are likely to be at risk of becoming malnourished (by virtue of losing one or more parents).

There are several advantages to such an approach. First, because the percentage of children under 5 who have lost a parent will be small, even in a generalized epidemic, such a response is likely to be much less costly, and therefore more feasible, than the alternative of providing assistance to all households that suffer a prime-age adult death from AIDS.

Furthermore, because of the long illness that often precedes death from AIDS, it will often be possible to identify young children who will soon be orphans before the mother or father dies and to enroll them in programs to minimize the nutritional impact. In cases where the mother is HIV-positive, this supplemental feeding could perhaps simultaneously reduce the risk of mother-to-child transmission through breast milk.

In addition, programs that provide food directly to malnourished children and to orphans, rather than to households that include orphans, may avoid the problem of households fostering children primarily to obtain benefits intended for the orphans. While creating an incentive for households to foster children may be desirable in a severe AIDS epidemic, too large an incentive can increase the number of children shifted between households, to the detriment of their welfare. A better approach may be to include children in households where a death is anticipated in community-based nutrition monitoring and feeding programs on the model of the UNICEF-sponsored "village feeding posts."

Finally, including orphans in a program designed to address malnutrition more broadly is more equitable than focusing assistance on AIDS orphans alone. This is particularly true in very poor countries, where an alarmingly high proportion of all children are malnourished. In Kagera, for example, even in households that are less poor and have both parents alive, nearly one-third of children are stunted. In such a situation, providing assistance only to AIDS orphans would neglect a large number of children who are also very needy.

The impact on child schooling. Besides increasing childhood malnutrition, a prime-age adult death in a household is likely to reduce school enrollment. This lack of schooling, perhaps exacerbated by inadequate nutrition, will make it particularly difficult for child survivors of a prime-age adult death to escape poverty. The effects of a prime-age death that could decrease enrollment among children in the household include:

- reducing the ability of families to pay for schooling
- raising the demand for children's labor
- reducing the expected return to adults of investments in children's schooling.

We have already seen how changes in income and expenditure that occur before and after a death would tend to reduce the ability of families to pay school fees and other education costs. Children may also be withdrawn from school to work outside the home, help with chores and farming, or care for an ailing family member. In addition, where prime-age adult mortality is high, parents may be less willing to invest in their children's schooling, either because they fear that the children will not live long enough to realize the higher earnings schooling promises, or because the parents themselves do not expect to live long enough to benefit from their children's future earnings. Similarly, relatives who take in an orphan may be less willing than the parents would have been to invest in the child's schooling. For all these reasons, children who have lost one or both parents are likely to have lower enrollment rates than those whose parents are alive.

Data from the Demographic and Health Surveys on enrollment and orphanhood in nine countries generally support this premise. Figure 4.13 shows the predicted enrollment rates for children by orphanhood status from a regression that holds constant within each country the children's age, gender, urban residence, and quality of housing—a crude proxy for wealth. All of these areas are characterized by low incomes and all except northeast Brazil are in the midst of a generalized HIV/AIDS epidemic. Orphans have significantly lower enrollment rates in every area except Uganda and Zimbabwe; differences in enrollment rates of orphans and nonorphans are greatest in the five sets of bars on the left.

But while the data support the view that orphans are less likely to attend school than other children, they also clearly demonstrate that in most of these countries a very large proportion of children who are *not* orphans are also not attending school. This indicates that at least in these low-income areas orphanhood is not the major reason for nonenrollment of children; other demand- or supply-side problems in the education sector or the labor market are leading to low enrollments among

Figure 4.13 Enrollment Rate for Children Ages 7 to 14, by Orphan Status, Nine Countries

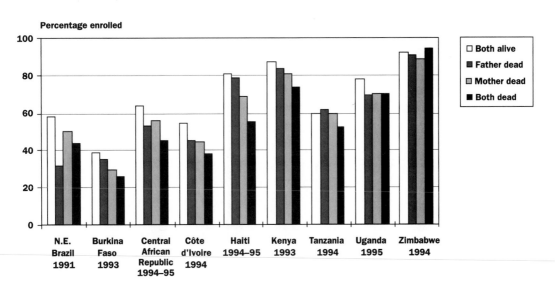

Source: DHS data, authors' calculations.

children irrespective of orphan status. As with nutrition, orphans appear to be specially disadvantaged in terms of education, but because enrollment levels are generally low in these low-income countries, special measures to boost enrollment among orphans would neglect the needs of the many nonenrolled children who are not orphans. Significant improvements in enrollments are therefore likely to require a systemic approach; this is beyond the scope of this book but is already among the educational goals in these countries.

Using the Kagera data, it is possible to consider how other factors may influence the relative enrollment levels of orphans and nonorphans. Using the same distinction between households according to assets, figure 4.14 shows that children in the poorer households are less likely to be enrolled than children in the less-poor households, regardless of orphanhood. Also, the difference in orphan and nonorphan enrollment rates is significant only among the poorer households. Most striking, however, are the low enrollment rates of children ages 7 to 10 *regardless of the level of household assets.* Thus, although orphans in the poorer households have the lowest enrollment rates, enrollment rates among all young children in Kagera are disturbingly low. Higher enrollments in the 11-to-14-age group are due mostly to the widespread practice of

Orphans often have lower school enrollment rates than nonorphans; but orphans are not the only children who do not enrol.

Figure 4.14 Enrollment Rates by Age, Orphanhood, and Household Assets, Kagera, Tanzania, 1991–93

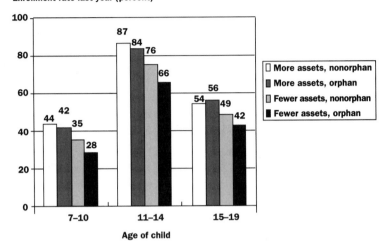

Enrollment rate last year (percent)

Legend:
- More assets, nonorphan
- More assets, orphan
- Fewer assets, nonorphan
- Fewer assets, orphan

Age of child

Source: Ainsworth and Koda 1993 and authors' calculations.

In Kagera, Tanzania, orphans in households with fewer assets were least likely to be enrolled in school; but even nonorphans in households with more assets and both parents alive had low enrollment rates.

delaying enrollment, which results in many older children attending primary school. These observations, combined with the fact that school fees are lower in Tanzania than in neighboring Kenya and Uganda (both of which have higher enrollment rates), suggest that for some reason households in Kagera are choosing not to enroll their young children in primary school. Whatever the reasons for nonenrollment, they are probably not solely financial, even among the poor.

While household asset levels had only a relatively minor impact on whether or not orphans were enrolled, the study also found significant differences according to whether the deceased household member was female or male. In households where a prime-age female had recently died, children had lower enrollment rates and were more likely to engage in activities in which women typically specialize—cooking, shopping, laundry, cleaning, and collecting water and firewood. For younger children, enrollment was often merely delayed, while older children were likely to drop out and not return to school. Even those children in households with a female death who did not drop out nonetheless spent fewer hours in school than children in other households. Because these effects were not observed in households where a prime-age male died, it appears that children are dropping out of school to perform tasks that had been done by the woman before her death.

Are these findings from Kagera likely to be true elsewhere? Baseline levels of enrollment vary enormously within and across regions, according to the financial costs of schooling, levels of household income, the opportunity costs of children's time, and the economic benefits of schooling. For low-income countries such as those shown in figure 4.13 it is reasonable to expect that the enrollment rates for orphans and nonorphans in the poorer households will be lower than for children in less-poor households, and that the differential enrollment rates between orphans and nonorphans in poor households are greater. Similarly, we would expect the death of a prime-age female to have a larger impact on enrollments than the death of a prime-age male in any community where women provide households with crucial services that, in the event of a woman's death, could be provided by children.

In middle-income countries, where enrollment rates among all children are much higher, orphans in low-income households may account for a larger share of nonenrollments than in poorer countries. In Thailand, for example, as of 1992, 93 percent of primary-school-age children were enrolled and the enrollment rate for secondary schooling was growing rapidly (Shaeffer 1995; Brown and Sittitrai 1995). Although we lack country-wide data on the extent to which orphans are underenrolled in Thailand, one small study found that 13 percent of the school-age children in families where someone was ill and dying of AIDS were withdrawn from school to help support the family (Pitayanon, Kongsin, and Janjaroen 1997).

Two broad policy conclusions can be drawn from this evidence. First, in areas where enrollments are very low, a systemic effort to improve overall enrollment will be more fair, and likely to yield larger benefits, than special programs focused only on orphans. Second, as enrollment rates improve, it becomes increasingly likely that poor orphans will have much lower enrollment rates than other children. Yet even in these situations, special programs for orphans may not be the fairest or most effective answer. So long as a significant proportion of poor children are not enrolled in school, interventions that aim to raise enrollments among the poor will address the schooling of the neediest children, including the neediest orphans.

Poverty Policy in a Severe AIDS Epidemic

The preceding analysis has highlighted several key factors of the impact of AIDS on poverty. First, HIV infects both the rich and the poor.

Although it already infects more poor people than rich people and will probably eventually infect a higher proportion of the poor, many HIV infections continue to occur among the nonpoor. Second, the short-term impact of a prime-age adult death is higher for AIDS than for other causes of death, primarily because of the long illness that often precedes an AIDS death. However, because other costs are incurred in all cases of adult death, the overall difference in the short-term impact of AIDS deaths and other deaths is not large. Third, rather than being destroyed by AIDS, households use a variety of mechanisms to cushion the short-term impact of an AIDS or other prime-age adult death. Fourth, we have seen that these coping mechanisms are much less effective in poor households, where more of the coping is at the expense of children's school enrollment and nutritional status. These latter effects are a permanent legacy of the AIDS epidemic that will hamper national efforts to achieve development goals for years to come.

These findings are based on a combination of theoretical analysis and empirical observation. Because the problems are relatively new and there are very few comparable data across countries, or indeed even across regions within a country, we have relied heavily on a single survey in Kagera, Tanzania, with additional observations from three other recent surveys in other countries. As experience with the AIDS epidemic increases and additional data become available, our understanding will certainly improve and some of the observations here will be challenged.

Despite these caveats, the broad findings described point out a key question that policymakers must consider in deciding how to mitigate the impact of AIDS on poverty: which households need help? The short answer, of course, is that the poorest households are most in need of assistance, and these are not necessarily the households hit by AIDS. In poor developing regions, many households that have not been affected by AIDS are likely to be very poor. Among the Kagera households that had not experienced an adult death, one-third of the children under 5 were stunted. Similarly, even in households where both parents were alive, 50 percent of the children under 11 were not enrolled in school.

The lack of a clear correlation between poverty and AIDS is strikingly evident in figure 4.15. The figure presents estimates of the percentage of the population in each district of Kagera living below the absolute poverty line of $124 dollars per person per year in 1991.[22] In parentheses below the name of each district is an indicator of the severity of the AIDS epidemic in that district during that year, the mortality rate of

Figure 4.15 Poverty in Kagera Region, by District and Adult Mortality Rate, 1991

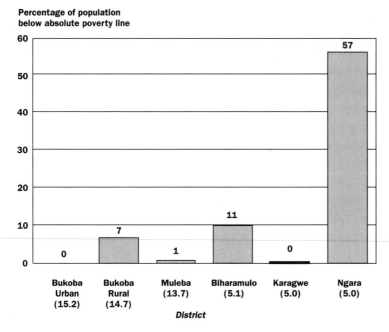

Percentage of population
below absolute poverty line

AIDS is not necessarily more wide-spread in poorer districts. Among the districts comprising Tanzania's Kagera region, some have severe epidemics but low rates of poverty; the district with the most poverty (Ngara) has a smaller epidemic.

Note: Adult mortality rate is per 1,000 for 1988.
Source: Gupta, Mujinja, and Over, forthcoming.

adults ages 15 to 50, as calculated from the 1988 census data. The fact that AIDS is not the most important cause of poverty is obvious from the fact that two of the districts where the AIDS epidemic had been extremely severe for a decade, Bukoba Urban and Muleba, had the least problem with poverty, while one of the districts with almost no AIDS, Ngara, had the most poverty.[23]

While the households affected by AIDS are not necessarily poor, the poor households that are affected are much less able to cope than the nonpoor households. Less-poor households in Kagera actually experienced an increase in consumption per capita after the death, while the poorest households experienced a sharp reduction in consumption, especially food consumption, as a result of the death. While the increased consumption in the nonpoor households will not necessarily be observed in other locales, it is reasonable to assume that nonpoor households faced with an adult death will be much better placed to smooth consumption than poor households.

Finally, we have seen that even households that lack access to formal credit and insurance markets nonetheless use a variety of measures to cope with AIDS deaths and other misfortunes. Thus, even if the shock from an AIDS death is larger than that from other misfortunes, and poor households are more vulnerable to the shock than nonpoor households, governments seeking to mitigate the impact of AIDS on poverty must ask themselves whether it is possible to design and implement, either directly or through NGOs, assistance programs that are more efficient and equitable than the informal coping strategies already in place.

These broad findings can be further distilled into three general recommendations for policymakers:

- Not all households experiencing an AIDS death need assistance.
- If survivor assistance is to be offered, it should be targeted to all very poor households that suffer a prime-age adult death, regardless of whether the death was due to AIDS.
- Assistance will do the most good immediately before and after the adult death, during the period when per capita food consumption has fallen but not yet recovered. It need not be permanent.

In addition to these three points, the findings also suggest that there is potentially important synergy between AIDS mitigation and antipoverty programs. For example, the finding that poor households are more vulnerable to the impact of an AIDS death implies that general antipoverty policies can also be AIDS mitigation policies. If general antipoverty policies are effective in reducing the number of poor households, then AIDS deaths will occur in stronger households that can cope at smaller cost to the survivors.

Similarly, the finding that an adult death depresses per capita food consumption in the poorest households by 15 percent implies that AIDS deaths that occur in poor households exacerbate poverty. Thus, when AIDS mitigation policies are targeted to households that were poor before the AIDS death, they are likely to prevent the affected household from slipping further into misery as a result of the death. In this case, AIDS mitigation policies could be effective at limiting the depth, if not the extent, of poverty.[24]

In sum, the results of these studies suggest that antipoverty programs and mitigation programs be integrated. When an antipoverty program is designed for a community with low living standards, consideration should be given to including components that specifically address the needs of the poorest households hit by AIDS deaths. For example, suppose the antipoverty program in a specific AIDS-affected community consists of a labor-intensive public works program. Components that would generate such synergy might include:

■ home-based or hostel-based care for the terminally ill to enable healthy adults who would otherwise have provided this care to take advantage of the jobs

■ day-care centers or "feeding posts" to enable single parents to take the jobs.

Examples of targeted antipoverty programs in developing countries that could be modified to use adult death in the household as an additional targeting criterion are presented in box 4.9. Conversely, when a mitigation program is established, locating it close to and combining it with a conventional antipoverty program will improve its effectiveness. In an area severely affected by an AIDS epidemic, implementing either program in isolation from the other sacrifices an opportunity for effective development policy.

How Governments Can Cope with the Impact of HIV/AIDS on Health Care and Poverty

ALTHOUGH THE SPECIFICS OF THE IMPACTS ON HEALTH and poverty differ, the analysis leads to broadly similar conclusions in both areas. Special government assistance for people infected with HIV/AIDS and their survivors must be weighed carefully against the many other pressing needs that governments face. Well-intentioned government efforts to assist individuals with HIV/AIDS and their families may divert resources from other families that have not been afflicted by HIV/AIDS but are nonetheless suffering from illness, poverty, or both. In particular, HIV-infected patients should be responsible for the same portion of the cost of their care as other patients with similar income and likelihood of infecting others.

Because of HIV's long incubation period, governments may initially underestimate the cost of programs to provide special assistance to those

Box 4.9 Using Adult Death as a Targeting Criterion for Antipoverty Programs

THE GENERALLY ACCEPTED STRATEGY FOR REDUC-
tion of long-term poverty comprises three components:
pro-growth macroeconomic policies, human capital
development, and social safety net programs. While
most countries depend primarily on the first two, some
countries, including some very poor countries, also
have substantial safety net programs that attempt to
provide assistance directly to the poorest households.
An important question for all such programs is how to
identify the households that need help most. Even a
brief summary of the extensive literature on targeting is
beyond the scope of this volume. However, it is important to note that a number of countries at different income levels and at different stages of the epidemic
already have in place targeted safety net programs that
could help poor households that suffer a prime-age
adult death. In some of these, including prime-age
adult death as a specific targeting criterion along with
other criteria may help to identify the neediest families.
Box table 4.9 describes five such programs.

Prior to the AIDS epidemic, prime-age adult death
was rare, perhaps too rare to warrant including it as a
targeting criterion. Sadly, it is now common enough
that countries with targeted poverty reduction programs should consider whether and how to include it
as a targeting criterion. Because this is a new area, impact evaluation of programs that attempt to do so
would generate important new knowledge.

Using prime-age adult death as a targeting criterion is likely to have several advantages. Compared
with providing help to families with a death from
HIV/AIDS, it is fairer, since it will include families
with prime-age adult deaths from other causes. Combining this criterion with others that identify the
household as poor may help identify the neediest families. Since the death of a prime-age adult is usually
well known to everyone in the community, using this
as a targeting criterion may help program administra-
tors identify destitute families that might otherwise be
missed. For the same reason, such a criterion may be
effective in minimizing opportunistic responses: feign-
ing a household death to obtain the benefits of the
targeted program for surviving household members
would be very difficult. Finally, including prime-age
adult death as a targeting criterion may help to increase the political acceptability of safety net programs
among those who do not benefit, since many people
will readily understand that poor households suffering
such a death—and especially the children in such
households—are likely to face severe hardship.

Sources: Besley and Kanbur 1988; Subbarao and others 1996;
van de Walle and Nead 1995.

Box Table 4.9 Social Safety Net Programs in Which Prime-Age Adult Death Could Be Used as an Additional Targeting Criterion, Five Countries

Country and state of the epidemic	Program and existing targeting criteria
Zimbabwe (generalized)	*Feeding program.* Targeted to children in drought-prone areas using nutritional sur-veillance data. Uses locally grown food and includes nutrition education.
India (concentrated)	*Food grain distribution* Through publicly operated "ration" shops, the states have dis-tributed grain to anyone requesting it, but under a new government program they are required to limit distribution to those below the poverty line.
Honduras (concentrated)	*Food stamps.* Distributed through health centers to low-income children under 5 and pregnant and lactating mothers, and through schools to poor mothers and their chil-dren grades 1–3.
Bangladesh (nascent)	*Microcredit program.* Targeted to households owning less than 0.5 acre of land; group lending and peer monitoring serve the poor without collateral and ensure repayment.
Chile (nascent)	*Cash transfer.* Targeted to rural and urban poor based on their answers to a computer-scored questionnaire.

affected by HIV/AIDS. As the number of people who get sick and die from the disease increases, these programs will absorb a growing share of resources that could have been used to address other problems. Because AIDS may divert resources from other pressing problems and commit governments to expenditures from which it will later be politically difficult to withdraw, policymakers in developing countries should be wary of programs that provide special assistance to people with HIV and their families solely on the basis of an HIV diagnosis. At a minimum, they should consider the long-term cost of such programs based on a range of likely assumptions about the course of the epidemic.

Notes

1. Impacts on other sectors may be substantial in some countries. See Ainsworth and Over (1994b).

2. The AIDS epidemic will increase costs, and thereby reduce supply in all sectors of the economy, and will reduce the domestic demand for nontradable goods. Since the value of health care output increases while that of other sectors declines, AIDS will increase the share of health care in national expenditure and product.

3. The increased number of deaths among prime-age adults is eventually offset by decreased numbers of deaths in older age groups. Since terminally ill elderly patients use little care in poor countries, this offset can be ignored in the developing countries that are the subject of this report.

4. A population that goes from 0 to 5 percent HIV infection in one year will not experience any increase in mortality the first year. Assuming the incubation period has a median of ten years, mortality from HIV would begin to rise in the second year, reach 2.5 per thousand in the tenth year and 5 per thousand in the twentieth year.

5. The calculation is: *100 x 0.25 x (5.3 /5.0).*

6. For example, one source assumes that only about 2.5 percent of those African transfusion recipients who receive blood infected with hepatitis B would contract the disease, which would then require 20 years to kill them (Beal, Bontinck, and Fransen 1992, p. 116).

7. Gertler and van der Gaag (1990) show that the poor are more responsive to price (measured as travel time) than the less poor. Lavy and Quigley (1993) and Mwabu, Ainsworth, and Nyamete (1993) provide recent evidence on the elasticity of demand with respect to quality. See Carrin, Perrot, and Sergent (1994) and Gertler and Hammer (1997) for reviews of the literature.

8. Barnum and Kutzin (1993, tables 3.3, 3.4) give occupancy rates for the developing world that range from 31 percent in Belize and 46 percent in Fiji to 116 and 129 percent in Malawi and Lesotho, respectively. But the pattern observed in Kenya (Collins and others 1996) of a positive correlation between the occupancy rate in a public facility and its degree of medical sophistication is common within many public health systems, especially if the user charge is equally low at all levels of care.

9. As part of a national health care reform process, the Kenyan government decreed a fee increase throughout the country in December 1989, which is after the 1988/89 data on Kenyatta Hospital patients were collected in the Floyd and Gilks study. Since AIDS patients were exempted from fees imposed during this period, care at the hospital simultaneously became more expensive for the HIV-negative and less expensive for the HIV-positive. This change in relative prices between the two groups probably accounts for part of the change in admission mix seen in figure 4.2. Collins and others 1996 describe health financing reform in Kenya.

10. Since Tanzania's national prevalence rate only recently attained 5 percent, the epidemic's full impact on mortality and health care expenditures are still in the future. Expenditure data on Mexico and Tanzania are from figure 4.6 and Shepard and others (*background paper,* 1996).

11. Insurance policies suffer to varying degrees from "moral hazard," when the amount of the loss incurred depends upon whether or not the individual is insured. For example, insured houses are somewhat more likely to burn than uninsured ones. The result of the problem is that insurance against a specific risk becomes more expensive per dollar of risk coverage and, in the extreme, may not be available at all (Arrow 1963). The problem arises regardless of whether the insurance is private or public and is particularly severe with health insurance, where it is controlled in practice by coinsurance provisions. These provisions typically specify higher coinsurance rates on highly price-elastic services like outpatient visits or psychiatric care than on less-elastic ones like inpatient services.

12. Areas of Africa studied include areas surrounding Lake Victoria—Rakai (Serwadda and others 1992; *background paper,* Menon and others 1996b); the Masaka districts of Uganda; Kagera (Killewo and others 1990) and Mwanza (Barongo and others 1992; Grosskurth and others 1995a,b) regions of Tanzania; and Kigali, the capital of Rwanda (Allen and others 1991).

13. One study converts consumption in developing countries to parity with the U.S. dollar (using the purchasing power parity indices). It estimates that two of every three persons in the developing world, and, in Africa, four out of five, spend fewer than two 1985 U.S. dollars a day (Chen, Datt, and Ravallion 1994), a level of consumption that suggests substantial poverty by almost any standard.

14. Surveys of survivors in the Kagera study found that, of the 264 household members age 15 to 50 who died during or in the year before the survey, 82 percent sought treatment, while 15 percent sought no medical care at all (survivors were unsure about the other 3 percent). Among men, about 90 percent of those who died of AIDS were reported to have sought medical care, compared with only 66 percent of those who died of other causes. (About 85 percent of women sought care, regardless of the cause of their death.) For both men and

women, the percentage incurring out-of-pocket medical expenses was higher for those who died of AIDS (70%) than for those who died of other causes (59%).

15. Since everyone must eventually be buried, that portion of the funeral costs that would have occurred anyway, discounted to the present, should not be attributed to the prime-age death. However, when the death occurs many years before it would have otherwise been expected, as is the case with most AIDS deaths, the discounted value of the future funeral costs is quite small in comparison with the actual cost of the funeral.

16. The authors of the Thai survey analysis did not analyze direct costs by the gender of the deceased.

17. Analysis of a single wave of data such as this does not show the direction of causality: were the expenditure patterns the result of the death or were households with certain expenditure patterns more likely to suffer a death? Figure 4.11 demonstrates from an analysis of changes in consumption over time that the differences between the two pie charts are largely due to the impact of the death.

18. In most of the 29 households, the person with AIDS died; in a few cases the AIDS patient moved to another household.

19. The difference is statistically significant at the 0.01 level on the Rakai sample, but not significant in the Kagera sample, perhaps because it is a smaller sample.

20. The initial differences between Kagera and Rakai households that later suffered an adult death and those that did not are intriguing. The former began the survey period with somewhat lower ratios of dependents to adults, slightly more household members and assets, and greater participation in ROSCAs. All of these drop after the death, causing the households suffering a death to resemble the other households more after the death than before. There are two possible explanations for this. On the one hand, it is possible that households that anticipate a death prepare by accumulating assets, recruiting additional household members, joining a ROSCA, and so on. Such adaptive coping in a risky environment is undoubtedly part of the explanation. On the other hand, there is also evidence that AIDS-affected households in the samples were on average somewhat less poor than their neighbors. The initial differences cited above between the

households that later suffered a death and those that did not could simply be indicative of the greater affluence of the average AIDS-affected households.

21. Because averages are pulled upward by extreme values and some households received as much as $5,000 in private transfers, the mean amount of assistance received in wave 4 by households that had experienced a death was $192, much larger than the median.

22. The figure of $124 is the 1996 dollar equivalent of 31,000 Tanzanian shillings at 1991 prices. See Ferreira and Goodhart (1995) and World Bank (1996b) for a discussion of poverty in Tanzania and details on the derivation of this and other poverty lines for Tanzania.

23. The recent influx of refugees to Ngara District from Rwanda and Burundi may be exacerbating the AIDS epidemic there. See CARE and ODA (1994).

24. That is, AIDS mitigation policies targeted to the poor may reduce the poverty gap if not the poverty head count.

CHAPTER 5

Working Together To Confront HIV/AIDS

I F DEVELOPING COUNTRY GOVERNMENTS, DONORS, AND multilateral organizations were already following the policies advocated in previous chapters, HIV would be a serious but manageable health challenge, perhaps not unlike cancer or other difficult-to-treat diseases, rather than a global epidemic. Unfortunately, national and international efforts to fight AIDS are far from optimal. Fifteen years into the epidemic, many developing country governments still lack adequate surveillance systems and have yet to enable a large enough share of those who are most likely to contract and spread HIV to protect themselves and others. Moreover, many countries also lack the society-wide policies to support such prevention interventions. Despite the willingness of nonprofit private groups to contribute to the fight against AIDS, some governments have difficulty providing the right mix of support and supervision. Donor governments and multilateral organizations, which provide much of the financing for national AIDS programs, have not always encouraged recipient governments to set and address appropriate priorities and they have invested too little in international public goods: knowledge and technology for fighting the epidemic in developing countries.

Why has the national and international policy response to the AIDS epidemic not been more satisfactory? Much of the explanation involves an understandable lack of knowledge. AIDS is a relatively recent challenge. Faced with a potential emergency, governments, donors, and multilateral organizations responded as best they knew how using the information then available. As knowledge about the epidemic and ways to combat it increases, public responses are likely to improve.

Like all public policies, however, AIDS policy is not made in a vacuum. Indeed, because the spread of HIV involves private behavior that many people deplore—frequent changes in sexual partners and the injection of addictive drugs—governments that adopt programs to reduce the riskiness of these activities may be accused by their constituents of facilitating socially deviant or immoral behavior. Thus HIV/AIDS policy may be subject to strong political pressures, some of which work against the policies most likely to contain the epidemic.

In considering these issues, this chapter looks beyond the national policies discussed in previous chapters to consider how the main actors in the AIDS policy arena can work together to more effectively confront the epidemic. First we examine the evolving roles of national governments, donors, and the multitude of other nonprofit and for-profit organizations which we refer to collectively as NGOs. We conclude that many low-income countries should confront the epidemic more forcefully, both directly and in collaboration with NGOs. Turning to a detailed examination of donor funding and policies, we argue that bilateral donors and multilateral organizations, despite their substantial contributions, have focused too little on fostering new knowledge and technology, such as information about costs and effectiveness of alternative prevention strategies and research on an HIV vaccine. Finally, the chapter discusses how public opinion and politics shape AIDS policy and how developing country governments can work with a variety of partners to overcome the obstacles to sound policies for fighting AIDS.

Government, Donors, and NGOs

NATIONAL GOVERNMENTS BEAR THE RESPONSIBILITY FOR protecting their citizens from the spread of the HIV epidemic and of mitigating its worst effects once it has spread. But they are not alone in the effort. Bilateral and multilateral donors have provided both leadership and major funding for national AIDS prevention programs, especially in the poorer developing countries. And both local and international nongovernmental organizations have stepped forward to help against the epidemic, sometimes prodding reluctant governments into action.

The challenge for national governments is to define their role in the struggle against the epidemic, not in isolation from or in passive response to the other actors, but in active collaboration with them. Only the government can claim to represent and act on behalf of the national population. Among the three types of actors, it has the unique ability to authorize implementation of an intervention by a donor or NGO. However, a donor cannot be commanded to finance or implement a program in which it has little interest. NGOs, too, have preferences and technical strengths or weaknesses. Thus the government cannot simply assign tasks to itself and the other actors. Instead it must learn the preferences and judge the comparative advantages of donors and NGOs. If there are important tasks that public economics considerations assign to the public sector, but that donors and NGOs either cannot or will not perform, then the government must undertake them directly or subcontract them.

What roles have the three sets of actors played in the struggle against the AIDS epidemic? How can cooperation be improved to make the most of the strengths of each? To answer these questions, this section first describes the important role that donors have played in funding AIDS interventions in most developing countries. Although the data on financing are incomplete and imprecise, they present a coherent picture of the relative roles of national governments and donors: donors have assumed the major financing burden in the poorest developing countries, and bilateral donors show a preference for countries suffering from generalized epidemics. Since the available cross-country data speak only to financing, the analysis of the NGOs' role in implementation is based on examples. Although no generalization regarding roles will apply to every country, the analysis suggests that many national governments and NGOs should assume a somewhat larger share of the funding of prevention activities, leaving donors to focus on the international public goods discussed in the next section. Furthermore, anecdotal evidence supports the claim that the donors frequently work at cross-purposes at the country level. Efforts would be more effective if donors would improve their coordination with one another and with national authorities without slowing the speed with which they deliver assistance.

Most of the National Response Is Funded by Donors

The total amount of donor funding for AIDS was estimated at approximately $300 million in 1996. The largest contributor of new funds in that year was the United States ($117 million); the European Union ($55 million) and Japan ($40 million) provided the next largest amounts of grant funding, and the World Bank provided approximately

$45 million in new loan commitments that year, most of which was at concessional rates. This seemingly large amount of money is, however, only about 6 percent of total donor health assistance to developing countries.[1] Nevertheless, since AIDS expenditures represent a substantial fraction of total public spending on health in some developing countries, observers have asked whether too large a proportion of health resources is devoted to AIDS in these countries relative to other health problems.

The WHO Global Programme on AIDS, predecessor of UNAIDS, collected data on donor, national, and NGO funding of AIDS programs in participating countries for the period 1991–93. While this database is imperfect and underrepresents funding from national and NGO sources, it provides the only detailed view of AIDS funding for a significant number of developing countries. By matching it with data on total national health expenditures from the World Bank (1993c) and data on domestic AIDS spending collected by Mann and Tarantola for *AIDS in the World II* (1996), it is possible to measure the relationship between expenditures from each source and total national health spending in a country, and to compare the importance of national government and donor funding across countries.

Average annual 1991–93 spending on AIDS by donors recorded by the GPA funding database exceeded 10 percent of 1990 public health spending in only seven countries: Uganda (59), Tanzania (36) Zambia (27), Malawi (16), Central African Republic (13), Guinea (11), and Rwanda (11). In these seven countries, all of which are in Africa and have some of the most serious AIDS epidemics in the world, international AIDS spending is large enough to overshadow all other prevention programs operated by the ministries of health. International spending was greater than 1 percent of the public health budget in 32 additional countries, including the non-African countries of Haiti (7 percent), Vietnam (3 percent), Thailand (3 percent), Lao PDR (3 percent), Bolivia (3 percent), Bangladesh (2 percent), Sri Lanka (2 percent), Pakistan (1.4 percent), Honduras (1.1 percent), and Chile (1.01 percent).

However, table 5.1 reveals that countries with large ratios of donor-funded AIDS spending to total national health spending are the exception rather than the rule. The average country in fact received a little less than 2 percent of its 1990 health budget for AIDS. Even in low-income countries with generalized epidemics, the average percentage of the national health budget received for AIDS was only 8.5 percent. Looking across the table's three country income groups (see "Average"

Table 5.1 Average 1993 Donor-Funded HIV/AIDS Expenditures, by Stage of the Epidemic and Income Level

(percentage of 1990 national health expenditure)

Stage of the epidemic	Income level (GDP per capita)			
	Low (< $725)	*Lower-middle* ($726–$2,985)	*Upper-middle* ($2,986–$8,955)	*Average*
Nascent	0.7	0.2	0.01	0.3
Concentrated	1.7	0.4	0.1	1.2
Generalized	8.5	n.a.	0.1	7.9
Unknown	0.4	0.1	0.4	0.2
Average	3.2	0.2	0.1	1.8

n.a. Not applicable.

Note: See end of chapter 2 for definitions of "nascent," concentrated," and "generalized." Donor assistance data are extracted by Pyne (*background paper,* 1997) from the GPA funding database developed by the predecessor program to UNAIDS, the WHO Global Programme on AIDS. National health expenditure data for 1990 are from World Bank (1993c).

row), we see that the average percentage of the health budget received from donors for AIDS declines from 3.2 percent among the low-income countries to one-tenth of a percent among the upper-middle-income countries. This is due to higher total public health spending in the higher-income countries, as well as to lower donor allocations to these countries. Looking across stages of the epidemic (see "Average" column), we see that donor funding as an average percentage of national health spending rose consistently from one-third of a percent in the nascent countries to nearly 8 percent in the countries with generalized epidemics. However, even the higher figure does not threaten to overwhelm ministries or to overshadow other public health programs in the *average* recipient country.

If donor-funded AIDS spending is occasionally large compared with the national health budget, national AIDS program spending never exceeds 10 percent of total national health spending and only rarely exceeds 1 percent, as judged from the survey of national program spending conducted for *AIDS in the World II* (Mann and Tarantola 1996). Figure 5.1 shows that only three developing countries reported a figure for 1993 national AIDS program spending that was above 1 percent of their 1990 public health spending: Thailand (5 percent), Mali (2 percent), and Malaysia (2 percent).[2] Twenty countries reported spending nothing

Figure 5.1 Comparison of Average Annual AIDS Spending by Donors with That of National AIDS Programs, 1991–93

(percentages of 1990 government health spending)

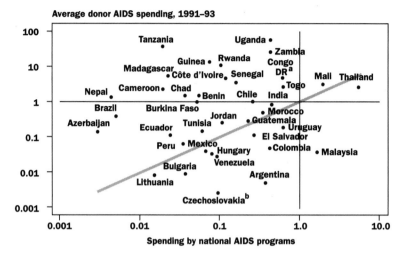

International and national AIDS expenditures are not correlated across countries and are usually small in relation to total public health spending.

a. Formerly Zaire.

b. Now Czech Republic and Slovak Republic.

Note: Since both axes are scaled in logarithms, 16 countries with zero national AIDS program spending are omitted from the scatter plot. Donor funding for AIDS is the average of 1991–93 funding from the GPA funding database. National AIDS program expenditures are from the survey conducted by Mann and Tarantola (1996) and are typically for various years between 1990 and 1993. The denominator for the ratios on both axes is 1990 public health spending as estimated in *World Development Report 1993* (World Bank 1993c). See Pyne (*background paper,* 1997) for further discussion of the data.

at all of their own funds through their national AIDS program in the relevant fiscal year, although seven of these received donor contributions for AIDS in excess of 1 percent of their national public health budgets.

Since international and national AIDS spending are expressed as a percentage of the same denominator in figure 5.1, the scatter plot would reveal any tendency for spending from these two sources to be correlated. However, the distribution of points is almost spherical: there is no relationship, either positive or negative, between donor funding and national AIDS program spending. This and further evidence presented below suggest that in the average country, the national program budget was not primarily determined by donor spending decisions.

The 45-degree diagonal line in figure 5.1 represents equal allocations of donor funding and domestic funding in response to HIV/AIDS. The thirteen countries below the diagonal received *less* in donor funds to combat AIDS than they spent of their own resources. The 26 countries above the diagonal, plus another 16 that reported spending none of their own resources for AIDS and are omitted from the figure, received *more* from donors than they spent through their national AIDS programs. Thus in roughly three-quarters of developing countries, donor spending on AIDS exceeded domestic allocations over this time period.

This analysis suggests that, although donor allocations for AIDS are not large enough to overwhelm the domestic health care system in most developing countries, these allocations are remarkably large relative to national spending on the same problem and probably in comparison with current international spending on any other disease. Perhaps only the international campaign to eradicate smallpox in the 1970s benefited from such a large preponderance of donor funds. But the amount of both types of funding varies a great deal from one country to the next. The next section seeks to explain this variation.

Donors Favor Lower-Income Countries That Have Larger Epidemics

As discussed in chapter 3, the severity of the epidemic and the availability of resources should be the two primary determinants of the extent of HIV/AIDS interventions in a developing country. Furthermore, the absolute size of a country's population will affect the scale of activities and therefore of expenditure.

How does spending vary across countries by infection rate, GDP per capita, and population size? To avoid double-counting the HIV-infected, the analysis below breaks the population into two components, the number infected and the number not infected. It then examines their separate influence, and that of GDP per capita, on aggregate levels of national and international AIDS expenditures in a country. More than 60 percent of the cross-country pattern of domestically financed AIDS expenditures can be explained by these three variables. The analysis focuses on the number of people infected and on GDP per capita to see how these two variables affect national and international allocations for confronting AIDS.

It is not surprising that national and international decisionmakers respond to the severity of the AIDS epidemic. Figure 5.2 presents the relationship between the number of HIV infected in the country and the amount of national and international AIDS expenditures in a country, after controlling for the number of people not infected and for GDP per

Figure 5.2 Relationship between the Number of HIV-infected People in a Country (in Millions) and the Amount of National and International AIDS Expenditures

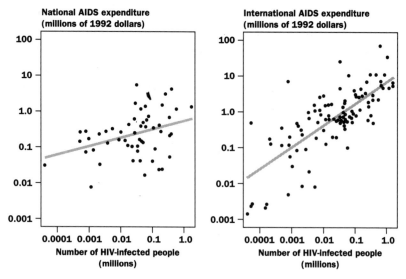

The expenditure of international donors is more responsive than that of national governments to the number of HIV-infected people in a country.

Note: The country data plotted in the graphs have been adjusted for the influences of the country's GDP per capita and the number of uninfected people in its population. See note 9 in chapter 1.

Sources: Expenditure data: see note on figure 5.1. HIV infection data: see Pyne (*background paper,* 1996).

capita. The relationships are positive (and statistically significant) in both cases, but the international donors were responding much more to the number of HIV-infected people than were the national governments. Every 10 percent increase in the number of HIV-infected people (after controlling for the other factors) is associated with a 6 percent increase in international spending in the country, while national spending rises by only 2 percent. Although it is reasonable for governments to respond to evidence of HIV infection with increased funding for both prevention and curative programs, national governments that view the epidemic with urgency might be expected to respond *more* strongly to HIV infections than would international donors, not less so.

One possible explanation for the weakness of the national governments' spending response to the epidemic might be the availability of international donor funding. If this were true, one would expect that some of the variation in national expenditure, after controlling for HIV infections and the other variables, could be explained by the receipt of donor funds. However, as we have seen in figure 5.1, there is no statistically significant correlation between national and donor funding in a country. Moreover, this is true even if we correct for the influence on national expenditure of GDP per capita, the number infected with HIV, and the number not infected.[3] Although some individual governments undoubtedly made their decisions on national funding levels based on what they were receiving from outside, this evidence suggests that this is not the case for the average country.

Turning to GDP per capita (figure 5.3), we see in the left-hand panel that national spending is extremely responsive to national income: of two countries of the same size and the same number of HIV-infected people, the one that is 10 percent poorer spends about 12 percent less on managing its AIDS epidemic. And the fit to that relationship is quite good. This striking sensitivity to income level could be explained as the rational response by decisionmakers who have full information about the danger of AIDS and the role of the public sector in confronting it but are not convinced that government intervention can slow the

Figure 5.3 Relationship between GDP Per Capita and National and International AIDS Expenditure

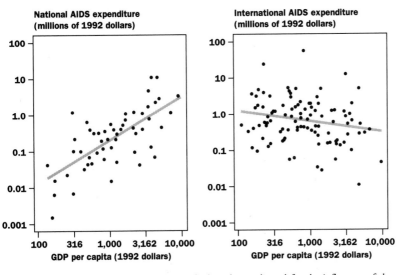

With an AIDS epidemic of a given size, countries with more national resources spend more on AIDS, while receiving less from donors.

Note: The country data plotted in the graphs have been adjusted for the influences of the country's number of HIV infections and the number of uninfected people in its population.

Source: See note on figure 5.1.

epidemic and are acutely aware of the many other demands on extremely scarce public resources. Under this interpretation, national decisionmakers view AIDS expenditures as a luxury, affordable only at higher income levels. Alternatively, it could be that decisionmakers in lower-income countries have less complete information about AIDS than those in other countries and perhaps are handicapped to a greater degree by conservative constituencies. Either of these interpretations suggests that donor assistance is acutely needed in the lowest-income countries to enable significant national activity against AIDS. But the latter interpretation, which is supported by the discussion of the political economy of AIDS in the last section of this chapter, further argues that *low-income countries should strive to increase their national effort against AIDS in order to ensure that people most likely to contract and spread HIV are able to protect themselves and others.*

The right-hand panel shows that the level of donor spending is also related to the recipient country's income, but in the opposite direction. This bias of the donors in favor of the poorer countries compensates to some degree for the much smaller national expenditures there; poorer countries receive somewhat more in donor funding than less poor ones, after correcting for population size and epidemic severity. However, donors do not fully compensate for the reduction in national spending: the country that is 10 percent poorer receives only 3 percent more donor resources. Furthermore, although the relationship is statistically significant, the fit is not very good. Thus many factors besides population size, epidemic severity, and GNP per capita influence international assistance to a country's AIDS program.[4] While donors must and should take many other considerations into account, this evidence suggests that *donors should give somewhat greater consideration to per capita income than they did from 1991 to 1993 when determining how to allocate resources to confront AIDS across countries, so that low-income countries with severe epidemics would be sure to receive the resources needed for the essential core functions of an AIDS program.*

Box 5.1 gives a detailed breakdown of AIDS funding by source for four countries and the Brazilian state of São Paulo. These detailed data from in-depth background studies performed for this report confirm the patterns discussed above (*background paper,* Shepard and others 1996). First, donor and national spending on AIDS both vary a great deal, even within this small sample of five countries: national government AIDS spending ranges from only 5 percent of AIDS spending in Tanzania to

Box 5.1 Government, Private, and Donor Expenditures on AIDS in Five Countries

SHEPARD AND OTHERS (1996) EXAMINED THE LEVEL and source of expenditures on HIV/AIDS in Tanzania, Côte d'Ivoire, Thailand, Mexico, and São Paulo State, Brazil.[1] Box table 5.1 reflects the international dollar amount and percentage of funding by source for each of the countries.

With the exception of Mexico, public funding per capita rises steadily with per capita GNP. Donor funding is by far the largest share (85 percent) of resources in Tanzania; in other countries, it is no more than 12 percent of total AIDS expenditures. The importance of donor funding outstrips its monetary value. First, it is insulated from domestic political pressures from patients and health care providers toward curative care, potentially at the expense of prevention. Second, it may play a catalytic role, showing the effectiveness of preventive expenditures and sparking contributions from other sources of funding.

The shares of public expenditures devoted to AIDS differ from total expenditure for AIDS by more than 1 percentage point only in Tanzania and Thailand. In Tanzania, the overall share is much higher due to substantial donor funding. In Thailand, the overall share is smaller because Thailand's extensive prevention program is predominantly publicly funded.

Variations in incidence explain why Tanzania, with the highest AIDS incidence (14.3 per 100,000) has moderately high expenditures per capita despite the lowest per capita GNP, while Mexico, with the lowest incidence, also has the lowest expenditures despite the second highest GNP.

Finally, political factors within the country and the donor community are also important. Tanzania's egalitarian ideals and relatively honest administration have long earned respect from the international donor community, and helped the country gain international support for its efforts to control AIDS. Thailand's openness in addressing AIDS through its National AIDS Task Force, chaired by the prime minister, has also brought support for that country's program.

---

[1] In the time available for this study, Brazilian data on AIDS expenditures could be obtained only for the state of São Paulo. With a 1991 population of 33 million, this *state* is larger than two of the five *countries* in the study (Tanzania and Côte d'Ivoire). As it contains 54% of Brazil's reported AIDS cases, the state's AIDS situation is believed to capture the reality of AIDS expenditures in Brazil. Economic and expenditure data were inferred from national statistics.

Box Table 5.1 Per Capita AIDS Expenditures Broken Down by Source of Financing in Four Countries and São Paulo State, Brazil

(1993–95 international dollars)

Source	Tanzania	Côte d'Ivoire	Thailand	Mexico	São Paulo, Brazil	Average
Public	0.20	1.34	3.45	0.76	5.78	2.31
(Percent)	(5)	(42)	(72)	(52)	(67)	(48)
Private	0.35	1.69	0.76	0.70	2.65	1.23
(Percent)	(10)	(53)	(16)	(47)	(31)	(31)
Donor	3.12	0.16	0.56	0.02	0.26	0.82
(Percent)	(85)	(5)	(12)	(1)	(3)	(21)
Total	3.68	3.18	4.76	1.48	8.69	4.36

Note: Columns arranged left to right from lowest to highest GDP.

Source: Background paper, Shepard and others 1996.

72 percent in Thailand. Second, donors clearly favor the lower-income countries over the higher-income ones, while national government spending on AIDS is positively correlated with income. And last, the low spending on AIDS in Mexico reflects the trend, discussed above, for the level of spending to be associated with the number of HIV-infected.

Bilateral and Multilateral Funding and the Stage of the Epidemic

HAVE BILATERAL DONORS AND MULTILATERAL ORGANIZA-tions responded differently to the epidemic? Figure 5.4 shows the allocation of approximately $1.2 billion in donor funding recorded in the GPA funding database over the period 1991–93 according to the type of donor and the stage of the epidemic in the recipient country.[5] While bilaterals allocated the lion's share of their AIDS assistance ($316 million, or 63 percent) to countries in the generalized stage of the epidemic, multilaterals allocated the bulk of their assistance ($379 million, or 62 percent) to countries in the concentrated stage of the epidemic. International NGOs (not shown) accounted for only a small portion of total funding reported in the database; the 16.4 million they provided was about equally split between countries with concentrated epidemics and countries with generalized epidemics.

This discrepancy between the funding patterns of bilaterals and multilaterals may have been only temporary and was perhaps partly due to two large World Bank loans made to India and Brazil during this period, both of which have concentrated epidemics. However, the pattern casts doubt on the frequent assertion that countries with advanced epidemics will encounter donor "fatigue" from bilateral donors and be forced to turn to multilaterals as the funders of last resort.

The observed pattern suggests instead that the bilaterals are particularly concerned about countries where the caseload is highest. Such behavior is consistent with two views of the motives of bilaterals. Perhaps they are altruistically responding to the suffering of countries with generalized epidemics. Or perhaps they view their self-interest as jeopardized most acutely by countries where there are large numbers of infected people. Whatever their motive, the bilateral focus on countries with generalized epidemics has left multilaterals to fund countries at the nascent

Figure 5.4 Donor Funding for HIV/AIDS Interventions in Developing Countries in 1993 by Type of Donor and Stage of the Epidemic

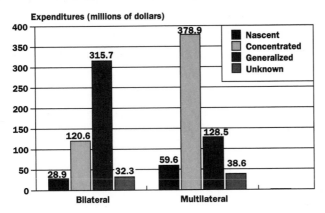

Expenditures (millions of dollars)

Source: GPA funding database as analyzed in Pyne (*background paper*, 1977, table 8).

Bilateral donors devoted the largest share of their AIDS funding to countries with generalized epidemics; multilateral institutions focused their funding on countries where the epidemic was still in the concentrated stage.

and concentrated stages. The outcome of this division of responsibility is that countries where the epidemic is nascent or concentrated pay higher costs for their external AIDS funding than those where it is generalized, but it does ensure that funds are available for all countries. It also permits donor governments to focus their resources on hard-hit countries where their constituencies are likely to most support spending.

NGOs Extend the Reach of Government and Donor Programs

Achieving the most cost-effective response to the HIV/AIDS epidemic requires cooperation between governments and NGOs, both nonprofit and for-profit. But working with NGOs can be costly for governments. Governments need to develop and apply guidelines and procedures to assure that the collaborative relationship operates with minimal friction and maximum effect.

Many of the highest-priority interventions require delivering highly differentiated services to small distinctive groups of clients, such as sex workers or poor AIDS-affected households. Effective service delivery requires the ability to learn from, and respond quickly to, the changing needs of a specific subpopulation. Because the needs of one client group are different from those of the next, unit costs are likely to rise rather than fall when the same organization attempts to deliver to multiple groups. In this situation, service delivery costs less if undertaken by many small entities rather than by one large one, such as a government agency.

When a highly differentiated service has the attributes of a public good (as chapter 1 argues is the case for many AIDS-related prevention and mitigation services), local communities often spontaneously create a nonprofit, grass roots NGO to deliver them, endowing it with finances and volunteer labor (Weisbrod 1977, James 1982). However, in developing countries many local communities lack the internal organization or resources to create their own NGOs, and few have sufficient incentives or resources to subsidize services whose benefits extend beyond their boundaries. Thus, governments cannot expect spontaneously created NGOs to tackle the epidemic alone. NGOs need the public mandate, technical information, financing, and sector coordination that governments can provide, while governments need NGOs for their diversity, flexibility, potential cost-effectiveness, and credibility with marginalized people. By working together, NGOs and governments can be a formidable force in the struggle against HIV/AIDS.

How should governments select an NGO partner to deliver an AIDS-related service? Characteristics of the service to be delivered can often indicate the type of NGO that will be most appropriate, but ultimately governments will have to judge the qualifications of competing NGO candidates for the specific service delivery contract in question.

Once government has identified an AIDS-related service that is undersupplied by the private market, it must ask the question whether it is possible to (1) *precisely specify in a written contract* the quantity and quality of the service to be provided, and (2) *monitor the contract* for compliance. It might be difficult to specify a complete contract either because the quality of the service depends on subjective aspects of its delivery (for example, the kindness and solicitude of an individual delivering home-based care), or because even the physical aspects of service delivery are difficult to measure (for example, whether the condoms reportedly sold to sex workers really went to them or were instead sold to pharmacies catering to the middle class). Monitoring might be difficult for technical reasons (for example, the presence of a government representative in the room to watch an NGO member educate prisoners or sex workers might destroy the rapport between educator and clients), or a government might simply not have sufficient trained and motivated personnel to monitor a large number of NGO contracts.

When the contract can be specified accurately and monitored for compliance, the government can choose among all the available NGO candidates, including both for-profit and nonprofit firms, depending only on their technical qualifications for the task at hand. In such circumstances the winning candidate will often be a for-profit firm with no direct ties to any client constituency, because they typically have the capability to mobilize the best expertise in the country, to produce outputs to international standards of excellence, to follow government or donor guidelines regarding records and accounts, to minimize costs, and to raise capital for expansion as needed to fulfill the contract.[6] However, since it typically has no other constituency to satisfy, a for-profit firm will reduce its costs, not only by eliminating waste, but also by reducing the quantity or quality of any unmonitored dimension of service.

When the contract for a highly differentiated service is difficult to specify or monitor, there is a strong argument for the government to favor an NGO that has its own constituency with a stake in the quality of the service. In contrast to the for-profit firm, which might divert resources in order to maximize profits, the nonprofit NGO has the incentive to divert resources toward the other services it provides or toward advocacy. Thus, governments considering delegation to an NGO as a solution to their inability to completely monitor performance must consider the conformity of the NGOs' overall objectives with the public interest.

Different types of nonprofit NGOs have different overall objectives. Broad-based public charities have large constituencies drawn from the general public and are therefore likely to have objectives in broad conformity with the general public interest. However, such broad-based public charities are likely to be less credible with the client groups than an organization composed of members of that group. Therefore, in selecting an NGO for a specific contract, governments may face a tradeoff between the degree to which the organization's objectives conform to those of the general public and the organization's effectiveness in working with the specific client group. Figure 5.5 shows the differing strengths of four stylized types of nonprofit NGOs—client affinity groups, social service clubs, nonprofit firms, and broad-based private charities—along these two dimensions. NGOs of all four types may be indigenous or may be the local affiliate of an international organization.

These tradeoffs can be best understood by considering the two types of organization at either end of the spectrum. Because the management and staff of a client affinity group is typically drawn from and selected by the client group it is serving, this type of NGO will have the most credibility with its clients. Furthermore, to the extent that the services for which the government pays the organization are perceived by its mem-

In selecting an NGO partner, a government may face tradeoffs between credibility with clients, on the one hand, and accountability to majority preference and its own objectives and procedures, on the other.

Figure 5.5 Differing Strengths of Four Types of Nonprofit NGOs

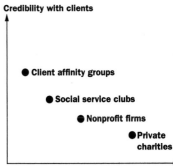

Credibility with clients

● Client affinity groups

● Social service clubs

● Nonprofit firms

●Private charities

Conformity of objectives to majority preferences

Source: Authors' construction.

bers as in their interest, the clients will themselves monitor its performance, greatly reducing government monitoring costs. For these reasons, client affinity groups can be very cost-effective in delivering highly differentiated services, such as peer counseling of sex workers.

However, the government's ability to delegate to a client affinity group is limited by the fact that the group's interests and objectives will sometimes diverge from those of the general public.[7] For example, it might not be a good idea to subcontract to a client affinity group of sex workers the collection of data on the proportion of its members who are HIV-positive, since the group might perceive that publication of such a number would be against its best interest. Moreover, some of the members of the affinity group may have socially unacceptable objectives that could be cross-subsidized from government resources. Thus, the divergence of objectives between the client affinity group and the government implies that a service contract with this type of organization will entail a risk of resource diversion toward the group's own objectives. Addressing this problem will increase monitoring costs.

Examples of client affinity groups that attain international recognition and receive international funding are multiplying. Perhaps the two earliest and best known such groups are WAMATA of Tanzania and TASO of Uganda. Founded by female relatives of people with AIDS, these organizations began as grass roots self-help groups providing basic home care services to home-bound or bedridden AIDS patients. Later, with outside support, they began to offer counseling for other HIV-infected individuals, as well as other services.

At the other end of the spectrum are the broad-based private charities. These organizations may be religious or secular but typically represent a large body of dues-paying members. Therefore, their constituency mirrors an important portion of the entire public, those who are willing to contribute regularly for charitable causes. Except for divergences due to religious belief, the interests of this mainstream constituency are likely to conform quite closely to those of the general public. However, private charities may not have credibility with all of the subpopulations that the government wishes to reach with its message and thus might be less effective delivering services to them. An example of a broad-based charity is the Thai Red Cross Society, which organized the first HIV/AIDS support group for affected individuals and their families in 1991 and only later developed the expertise to reach out to sex workers. That first support group was a powerful example that led the Red Cross and many different types of NGOs to create 80 such groups by mid-1996 (Phoolcharoen and Phongphit 1996).

Two other types of nonprofit NGOs fall between the affinity group and the large private charity. Social service clubs are *local* charitable organizations typically composed of middle class and elite community members who volunteer their time in order to improve their community. Their direct constituency, to whom they are primarily accountable, is their peer group within their own community. Social service clubs may be able to establish trust in client groups because service providers live in the same community with the clients and are volunteering their time. The members of such social service clubs have skills and education that can enhance the NGO's utility as a delivery organization. Although the interests of the typical social service club will conform to those of the local elite society, they may not exactly match those of the government or general public. For example, a social service club NGO providing AIDS information to truck drivers in Lahore, Pakistan, mentioned only one possible source of infection: blood transfusion ("Signs of Change . . ." 1996).

Nonprofit firms constitute the majority of NGOs in most countries. The distinction between a nonprofit and a for-profit firm varies from country to country and depends both on the tax laws of the country and on the vigor with which they are enforced. The most sophisticated nonprofit firms are like for-profits in that they can draw on the best national expertise and be held accountable to international standards. But nonprofit firms can, with greater ease and legitimacy than for-profit firms, develop a constituency, independent funding sources, and their own agenda of objectives. However, the nonprofit firm typically is constrained by a much smaller constituency, perhaps consisting only of the members of its board of directors and their immediate acquaintances. The rapid proliferation of nonprofit firms observed in some countries in response to the availability of service contracts suggests at least some degree of profit motive. For example, in the four years after the Brazilian government initiated a grant program for AIDS-related services, the number of NGOs registered with the Ministry of Health jumped from 120 to 480. A 1996 evaluation report that compared the earlier with the later group of NGOs found a change toward a more consolidated and formal organization structure, greater dependence on government funding, and a tighter focus on service delivery at the expense of public advocacy. This change suggests that the profile of the average Brazilian NGO working against AIDS is now closer to that of a nonprofit firm than to either a client affinity group (which would typically be less formally organized) or a broad-based charity (which would be less dependent on government funding).

Of course an NGO can embody the characteristics of more than one stylized type and some have objectives that conform closely to the public interest while also having high credibility with its clients. Box 5.2 describes one such program in Sonagachi, one of the largest red-light districts in Calcutta, India. The program combines the characteristics of the nonprofit firm and the social service club.

How well have governments done in delegating preventive or mitigating service delivery to NGOs? A program in Burkina Faso, one of the four West African nations with a generalized epidemic, offers an example of how government and NGOs acting in concert can extend the reach of their AIDS prevention and mitigation efforts, achieving better quality and access than if either had acted alone (Van der Gaag 1995). The project, which is supported by the World Bank, seeks to increase the use of condoms and other contraceptives and change behaviors that facilitate the spread of STDs. The government and NGOs share responsibility and

---

Box 5.2 Helping Calcutta Sex Workers Avoid AIDS

IN 1992 THE INDIAN GOVERNMENT, INTERNA-tional donors, three local NGOs, and sex workers in Sonagachi, one of the largest red-light districts in Calcutta, joined together to launch a remarkably successful STD/HIV Intervention Program. The program, known as SHIP, has trained sex workers as peer educators, providing them with knowledge about STDs, the use of condoms, and negotiation skills, which are essential if sex workers are to convince their clients to use condoms without the support of pimps and brothel owners.

The success of this approach can be seen in several indicators. The number of condoms distributed through the program per month rose from 1,500 at the start of the program to 65,000 at the end of 1995. The number of abortions and the STD rate among sex workers in Sonagachi have declined significantly. And strikingly, HIV prevalence among the sex workers has remained at less than 1.5 percent.

Much of the program's success is credited to the sex workers who have become peer educators, since other sex workers regard them as trustworthy advocates of behavior change. Moreover, their employment in the program has brought community recognition, self-respect, and dignity, which have encouraged other sex workers to become peer educators, thus helping to ensure that the program will continue.

SHIP has been expanded into four other red-light districts in Calcutta; by 1997 it was reported to cover areas that include more than 80 percent of the sex workers in the city.

Source: Singh 1995.

costs. Government roles include providing supplies at a subsidized rate; launching a national media campaign to promote the purchase of condoms; and teaching traditional healers to fill prescriptions, diagnose STDs, and refer cases to health clinics. Treatment of STDs will be handled primarily by NGOs, which are both nonprofit and for-profit firms; NGOs will also provide training to traditional healers. Both NGOs and public clinics will provide free condoms to people with high-risk behavior. The government has also provided encouragement and access to grant money for NGOs to enable them to offer additional services. This type of collaboration lays the foundation for increased coordination in the future between the two actors and fosters an environment of trust.

The largest and most elaborate effort to subcontract AIDS services to NGOs is probably the annual competition for service grants in Brazil. Supported by a World Bank loan, the program has funded all four types of nonprofit NGOs, including client affinity groups such as an association of transvestites in Rio de Janeiro, and nonprofit firms such as a university-affiliated research center in São Paulo. Clients have included children, hemophiliacs, pregnant women, feminists, transvestites, prostitutes, drug users, prisoners, truck drivers, and men who have sex with men. While grant competitions are managed centrally by an NGO liaison office attached to the Ministry of Health in Brasilia, state and municipal as well as federal government agencies provide complementary funding to, and collaborate actively in the execution of, funded programs. In the recent evaluation of this program, only 7 percent of the 111 current grantee NGOs were deemed to be falling short of their project objectives, and only 2 percent were having serious difficulties reaching their target populations. The financial control mechanisms employed by the liaison office, which include an annual visit to each grantee and audit of its accounts, has identified serious mismanagement in connection with less than 1 percent of projects. While the NGO liaison office constitutes a substantial new and expensive function for the Ministry of Health, in the four years of its existence it has facilitated the funding of 308 projects and disbursed a total of $14 million. Although the total impact of this activity on HIV infection rates in Brazil has not been assessed, it is clear that no government agency could have carried out directly so many diverse and precisely focused activities with these resources.

Unfortunately, to our knowledge there is no systematic study that compares the merits of alternative government procedures for evaluating

NGO proposals for an AIDS-related service delivery contract; nor are we aware of any study that compares ways for governments to monitor NGO performance under such a contract (National Research Council 1996, appendix to chapter 6). A starting place for such a study would be a comparison of the lessons learned in the recent service contracting experiences of Brazil, Burkina Faso, and Thailand. The availability of a set of standard, transparent, internationally recognized procedures for governments to follow in delegating service provision to NGOs could greatly facilitate government-NGO cooperation and minimize the disappointment of all parties concerned. AIDS donors, NGOs, government policymakers, and indeed the entire international health community would benefit from studies of the costs and effectiveness of alternative procedures for identifying effective NGOs to be service providers and for monitoring their performance.

Such studies are but one example of an urgently needed international public good, the topic of our next section.

Who Will Invest in New Knowledge and Technology?

DONOR SUPPORT FOR NATIONAL AIDS PROGRAMS IS IMPORTANT and, in a nascent epidemic, often critical; yet there are other crucial activities in which donors have a greater comparative advantage and a clearer public economics mandate. Because the benefits of prevention programs accrue primarily to a country's own population, all but the poorest national governments can and should finance a significant share of these costs. In contrast, donors are in a unique position to mobilize international support for the creation and dissemination of knowledge and technology that is transferable across countries. This section first discusses the organizational response and financial contributions of bilateral donors and multilateral organizations since the start of the epidemic. It then explains why knowledge and technology should be regarded as international public goods that the donor community alone is likely to provide. Finally, it discusses the need for specific types of knowledge and technology, including a vaccine, and organizational innovations for tapping the creative energy and resources of private firms.

The Evolution of Donor Policy

Although AIDS was first diagnosed in 1981, a systematic international and national response to the epidemic was not evident until the late 1980s. In many parts of the world, NGOs led the way in providing care and prevention services for individuals and communities affected by the epidemic (Mann and Tarantola 1996; *background paper*, Pyne 1997; Sittitrai 1994). The incremental and relatively limited response of WHO in the early years has been attributed to resistance by many member states to addressing the problem of HIV/AIDS (Panos Institute 1989). The establishment of the WHO Global Programme on AIDS (GPA) in 1987 helped to generate momentum for global prevention and mitigation efforts; that same year the U.N. General Assembly adopted a resolution encouraging U.N. agencies and other members of the U.N. family to initiate their own HIV/AIDS activities (Mann and Tarantola 1996).

During its early years GPA focused on helping national governments develop strategies to curb the spread of the epidemic. The year GPA was established, 170 countries requested assistance; by 1989 GPA had helped 151 countries to establish national AIDS programs, 102 countries to develop short term (6 to 12 month) plans, and 30 countries to develop medium-term (3 to 5 year) plans (Panos Institute, 1989). Largely as the result of the prodigious efforts of GPA, almost all countries today have national AIDS programs; most of them were formed between 1985 and 1990.

Meanwhile, in response to the U.N. General Assembly Resolution, UNDP, UNICEF, UNFPA, and UNESCO developed a joint HIV/AIDS strategy document that specified the resources and staffing that each would allocate to combating the epidemic. UNDP played the most prominent role, devoting 2.1 percent of overall agency resources and 0.43 percent of overall agency staff (Garbus 1996, as cited in *background paper*, Pyne 1997). Other multilaterals also initiated AIDS programs. In 1987 the European Union established the AIDS Task Force in order to fund AIDs-related programs in developing countries. The World Bank, which made its first loan exclusively to combat AIDS in 1986, had financed 61 projects in 41 countries, for a total commitment of $632 million by the end of 1996 making it the largest source of funds for confronting HIV/AIDS (*background paper*, Dayton 1996; World Bank 1996a).

In the late 1980s, the wealthier donor countries, in addition to making contributions to the GPA and providing support through the other multilaterals, also launched their own bilateral HIV/AIDS programs. By 1993 the largest of these was the U.S. program; launched in 1988, it includes the centrally funded AIDS Control and Prevention Project (AIDSCAP) as well as other activities initiated and funded by country USAID missions.[8] Other countries with large bilateral AIDS programs include Canada and Norway (launched in 1987); Denmark, Germany, the Netherlands, Sweden, and the United Kingdom (1988); Japan (1989); Belgium and France (1990); Australia (1991); and Switzerland (1993). Table 5.2 gives the total amount spent in 1993 by twelve major donor countries.

Under the leadership of the GPA, many national plans were written, many AIDS interventions were launched, and many national leaders became aware of the severity of the AIDS epidemic. For the first time, senior policymakers discussed high-risk sexual behavior and how gov-

Table 5.2 International AIDS Expenditures through Bilateral and Multilateral Channels, by Major Donor Countries in 1993 and Net Immigration in 1992

(millions of dollars except as indicated)

Country	Bilateral	Multi-lateral	Both	Total	Net immigration (thousands)
United States	82.0	34.0	1.0	117.0	793
France	18.5	1.4	0.1	20.0	86
United Kingdom	7.8	8.4	n.a.	16.2	147
Germany	7.8	0.9	4.1	12.8	788
Canada	8.2	3.1	0.3	11.6	195
Sweden	3.7	5.1	1.0	9.8	20
Norway	4.6	2.5	2.3	9.4	10
Denmark	2.1	2.7	4.1	8.9	12
Australia	7.1	0.5	0.3	7.9	48
Netherlands	2.7	2.4	0.9	6.1	43
Japan	1.0	4.5	n.a.	5.5	48
Luxembourg	1.0	0.3	n.a.	1.2	6
Total for 12 donors	146.4	65.9	14.1	226.3	2196

n.a. Not applicable.

Note: Funding totals exclude the AIDS share of national contributions to the multilateral lending agencies.

Sources: Laws 1996, table 35-1; and OECD 1995, table I.1, p. 24.

ernments should respond. However, the epidemic continued to spread. In the early 1990s a group of member states, especially the donor governments then funding the GPA, became concerned that, as a part of WHO, it had insufficient mandate to coordinate the expanding efforts against the epidemic across the U.N. system. The donor community perceived that the GPA was unable to restrain donors from competing vigorously with one another instead of cooperating around a mutually agreed plan of action and came to believe it necessary to create a specialized international institution with an explicit mandate to coordinate the work of the other U.N. agencies at the country level. As a result they worked with UNDP, the World Bank, and other multilaterals to create a new special-purpose U.N. program dedicated uniquely to combating AIDS. The Joint U.N. Programme on AIDS, widely known as UNAIDS, officially began operations on January 1, 1996. It is based in Geneva and works most closely with its six cosponsoring agencies: WHO, UNDP, UNICEF, UNFPA, UNESCO, and the World Bank. It is governed by a Programme Coordinating Board (PCB) of 22 member states and 6 cosponsors, plus, for the first time in the U.N. system, 5 rotating nonvoting representatives of NGOs.

The PCB has assigned UNAIDS four roles: first, *policy development and research*, which are to account for a larger share of UNAIDS' activities than it did of the GPA's; second, like the GPA before it, UNAIDS is to take the lead among U.N. agencies in providing *technical support* to national AIDS programs around the world; third, the program is committed more formally to *advocacy* on behalf of HIV/AIDS prevention and mitigation than was the GPA; and finally, UNAIDS is charged with the difficult task of *coordination* of its cosponsors and of other U.N. agencies. In this last role, it can potentially address the needs described in the next section by serving as a forum within which multilateral and bilateral donors can agree to donate more to AIDS research, prevention, and control than they otherwise would. Since cooperation with other donors at the country level entails substantially increased costs to each donor and deprives each of being able to claim sole credit for supporting the government on a specific activity, the incentives for such cooperation are weak. Since UNAIDS lacks the power to enforce cooperation from its co-sponsoring multilaterals, much less from the bilaterals, the hope for this form of donor cooperation lies in the good will of the staff of the various donors working at the country level—perhaps reinforced by the insistence of the national government.[9]

Donors Should Focus More on International Public Goods

One explanation for international assistance to help developing countries combat their AIDS epidemic is altruism. Just as famine and flood overseas can elicit an outpouring of generous assistance from more favored countries, the disease problems of low-income countries have often been the cause of generous government and private contributions.

However, in the case of an infectious disease that even the most sophisticated medical technology cannot always cure, like drug-resistant tuberculosis, the Ebola virus, or HIV, it is also in the self-interest of higher-income countries to help poorer ones combat the disease. Chapter 1 argues that there is a compelling role for government in the prevention and control of infectious disease. Figure 5.6 illustrates that in industrial countries HIV is estimated to have caused 65 percent of adult deaths from infectious disease in 1990 and, unless new antiretroviral therapies are effective and become widely available and affordable, is projected to account for over 96 percent of such deaths in 2020.[10] This is much higher than the HIV share of deaths from infectious disease in developing countries (see chapter 1).

The current and future magnitude of HIV's contribution to the infectious disease burden within industrial countries' borders provides them with two reasons to spend money on HIV control in the low-income countries. First, any lessons learned about how to slow the spread of the epidemic, whether through behavioral modification or technological advances, are potentially applicable at home. Second, because HIV is infectious and the higher-income countries exchange thousands of tourists and attract thousands of legal and illegal immigrants to their shores every year, a reduction in HIV prevalence in low-income countries has a secondary effect of protecting the citizens of higher-income countries. Evidence suggests that countries are already aware of these arguments: the five countries that provide the most support to the global effort against AIDS also receive the most immigrants.

Assuming that self-interest is at least part of the explanation for the high-income countries' contributions to AIDS prevention in the developing world, will this be sufficient to generate the globally optimal expenditure on AIDS control in developing countries? Recall the discussion in chapter 1 of the difficulty in coordinating the contributions to mosquito control of all the individuals who inhabit mosquito-infested

Already the major cause of adult death from infectious disease in industrial countries, HIV could account for twice as many adult deaths by 2020, unless new treatments prove to be effective and widely affordable.

Figure 5.6 Deaths of Adults from HIV and Other Infectious Diseases in the Established Market Economies, 1990 and Projected to 2020

Annual deaths
(thousands)

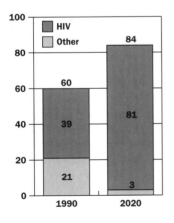

Source: Murray and Lopez 1996.

land. Once the mosquitoes are gone, even people who have contributed nothing to the effort will benefit. Since each individual can hope to "free-ride" on the others' contributions, each holds back from giving as much as he would be willing to pay to end the mosquito infestation. A similar free-rider problem threatens to prevent the donor countries from voluntarily donating as much to the AIDS effort in developing countries as the abolition of the epidemic would in fact be worth to them. Because it suffers from this international free-rider problem, the effort to combat AIDS can be viewed as an international public good.

Another good on which it is easy to free-ride is new technical information, such as that generated by frontier medical research on treatment of AIDS and opportunistic illnesses, AIDS vaccines, or, to the degree that the results are transferable across countries, by operations research on the best way to market condoms to the people most likely to contract and spread HIV.

The solution to problems of local or national public goods is typically government intervention. At the local level it is in the interests of all the individuals concerned to support a government that taxes them all and uses the taxes to control mosquitoes and fight other infectious diseases. A similar argument can be made for an international government with the power to tax countries and spend the proceeds on international public goods such as the control of HIV/AIDS. However, since countries are unlikely to surrender their sovereignty to a supranational body for this or any other reason, another solution to the international free-rider problem must be found.

As an alternative to government, the individuals who live on the mosquito-infested land could negotiate with and persuade one another ("I agree to give more if you will") until sufficient money was raised among them all to solve their joint problem. While requiring more time and effort from individuals than the simple solution of a tax, the negotiated solution is potentially workable. At the international level, the United Nations is a forum for such negotiation and persuasion. Through it, countries can potentially be persuaded to donate their "fair share" to international public goods, such as AIDS control.

Thus from the public economics point of view, it is not surprising that donor countries have been willing to donate to AIDS control and to research on AIDS. However, given the free-rider problem, it is unlikely that the donor countries have committed as much as it would be in their joint best interest to provide.

Investments in International Public Goods

Information that can be generalized beyond the country in which it is produced can originate in either the social or the physical sciences. This section discusses both types of knowledge and a third type of international public good: international institutions.

The medical and social sciences of epidemiology, sociology, economics, and operations research are necessary to track the epidemic and to learn what sort of interventions prevent the most secondary cases of HIV infection per government dollar spent. Applied social science research offers the greatest hope for immediately slowing the spread of AIDS and of improving the well-being of the hardest-hit survivors.

The biological sciences, including microbiology, immunology, and virology, are making slow progress toward a vaccine and a cure. However, market imperfections mean that only a small share of biomedical research is designed to produce products or knowledge that will benefit low-income countries. WHO's Ad Hoc Committee on Health Research estimates that 95 percent of spending on health research and development is directed toward solving health problems that mostly affect the richest 10 percent of the world's population; only 5 percent of such spending is directed toward the diseases that account for most of the disease burden of the remaining 90 percent of the world's population (Ad Hoc Committee 1996, p. 102). An important role of governments, especially of donors, is to tilt incentives for medical research somewhat more in favor of the low-income countries.

A third important type of public good is the international institution that enables a group of countries to coordinate their efforts in their mutual best interests. Two types of international institutions are relevant to the AIDS epidemic: those among low-income countries in a region, and those that bring poor and high-income countries together in a common struggle against HIV/AIDS.

Information from the social sciences on behavioral interventions. Any successful preventive intervention among individuals who are very likely to spread the virus will produce positive spillover effects for the host country, in the form of reduced secondary transmission, which to some degree will also benefit other countries. But the most valuable output of such an intervention for the outside world is knowledge that can be applied in other countries. Donors who fund behavioral interventions have a responsibility to ensure that the opportunities for the generation of new knowledge that arise from such programs are not wasted.

Although the imperative to learn from interventions seems self-evident, surprisingly little is being done in this regard. Recent literature reviews found that publicly available written evaluations exist for only about 10 percent of donor-funded interventions. Worse, of the few hundred published studies, very few were conducted with sufficient thoroughness to determine whether or not the intervention actually changed the risk behavior or HIV incidence (Choi and Coates 1994; Oakley, Fullerton, and Holland 1995; National Research Council 1991).[11]

The reviewers noted many deficiencies in the available studies. In some cases the lack of baseline data made it impossible to know whether a measured difference between a control and experimental group was due to differences in the two groups present before the intervention. In others, baseline data were collected but there was no control group against which the intervention group could be compared. Few studies attempted to determine whether changes in behavior were due to the intervention or to a placebo effect arising from the existence of the study. To be sure, ethical considerations and the complexities of research with human subjects often make it impossible to use a true experimental approach. An alternative is to have copious baseline data and implement quasi-experimental research designs (Moffitt, 1991). However, very few studies attempted such an approach.

Differences between the standards of knowledge for pharmaceutical products and those for behavioral interventions against HIV are striking. Since pharmaceutical products can be patented, private firms have a strong incentive to win the race to the market with a new drug. Governments have responded by requiring that companies prove the safety and the efficacy of new drugs, typically at a cost of millions of dollars. These sums are spent even on such relatively minor drugs as a new headache pill in order to ensure very high standards. The government does not hesitate to require such expenditures, knowing that firms will spend this money on any drug they think will pass the market test.

In contrast, preventive interventions that have the potential of producing far more public benefits, in the form of secondary HIV infections averted, are held to much weaker standards. Since these kinds of interventions can not be patented and they produce positive externalities, the public sector typically must finance them. If governments held themselves to standards as rigorous as those they set for pharmaceutical man-ufacturers, HIV prevention interventions would be forced to meet stan-dards of rigorous design and data collection methods that would enable the public to learn whether an intervention will be safe and efficacious in subsequent application.

Although it might seem that safety would not be an issue, the exam-ples of needle exchange programs and counseling and testing for HIV infection suggest the contrary. It is precisely the fear that the provision of clean needles might encourage injecting drug behavior and that an HIV testing program, even with the accompanying counseling, might reduce the propensity to practice safe sex among those who are told they are positive that often undermines public support for these programs. The public has an interest, and indeed a right, to know the magnitude of any such "side effects," as well as the efficacy of the intervention, before it finances its continuation or expansion.

Information from the biological sciences on medical interven-tions. With potential profits protected by the patent system and a large potential market in industrial countries for an AIDS cure, research by both private firms and nonprofit institutes has been intense in the in-dustrial countries. The most recent product of that research is the triple-drug therapy discussed in chapter 4. As shown, the high costs of provid-ing this therapy mean that it will not be of immediate benefit to the 90 percent of HIV-infected people who live in low-income countries.

Some observers, aware of these prohibitive costs and pessimistic about the prospects for successful behavioral interventions, believe that the only hope for reducing the impact of HIV on low-income countries is a vaccine. But vaccine research of all types faces serious impediments.[12] These include the increasing complexity and expense of vaccine research, the need to sell perhaps 40 million doses before production processes attain economies of scale; the inability of people in developing countries to afford expensive vaccines; and, perhaps most serious, companies' vul-nerability to damage claims in the millions of dollars, if even one dose of a vaccine causes the disease it was designed to prevent (Ad Hoc Com-mittee on Health Research 1996, Robbins and Freeman 1988). Partly as a result of these impediments, worldwide public and private sector in-vestment in vaccine development totaled a mere $160 million in 1993, compared with an estimated $1.3 billion spent on other approaches to prevent HIV infection and about $5 billion spent on HIV-related health care (FitzSimmons 1996).

In order to achieve the substantial international public benefits of vaccines for the diseases of the developing world, governments must play a role. The May 1997 announcement of a U.S. goal to produce an effective AIDS vaccine within ten years as a U.S. national goal is welcome news not only for people in the U.S. but for people everywhere, including developing countries. His choice of a ten-year target date, which some experts believe to be too optimistic, is a sobering reminder that no vaccine will solve the AIDS problem in the developing world in the near future (see box 5.3.)

The need for government involvement is apparent not only for an AIDS vaccine, but also for other medical advances, which would substantially benefit people in the developing world who lack the purchasing power to motivate the pharmaceutical companies of the industrial countries. Examples include vaginal microbicides and simple inexpensive diagnostic kits for classic STDs such as chlamydia and chancroid that are currently difficult and expensive to properly diagnose (Ad Hoc Committee on Health Research 1996; Elias and Heise, 1994).

As the example of the hepatitis B vaccine discussed in box 5.4 makes clear, once a vaccine or other drug has been invented, tested, and produced on a large scale, its price is likely to fall to the point at which commercial firms can profitably manufacture and distribute it in large quantities at prices that are affordable in developing countries. Thus, the need for government involvement is likely to be temporary, but critical.

International institutions can produce international public goods. We noted above that the United Nations and other multilateral organizations can provide forums in which countries can persuade one another to contribute more than they otherwise would to the production of an international public good. Two additional types of international institutions that could solve specific kinds of international free-rider problems are private-public alliances for health research and regional cooperation bodies.

Public-private alliance for health research. WHO's Ad Hoc Committee on Health Research has recently proposed a "Health Product Development Alliance" between the public and private sectors whose mandate would be tightly focused on the development of a limited number of products for major causes of disease burden that are currently neglected by existing efforts (1996, p. 101). Such an alliance would use a variety of approaches to improve the incentives for private firms to develop pharmaceutical and other health care products urgently needed in developing

Box 5.3 Challenges To Be Overcome in Developing an HIV Vaccine

THIS BOOK ARGUES THAT DONOR COUNTRIES AND multilateral institutions have a comparative advantage in creating incentives for HIV vaccine research and that doing so would be in their own self-interest, as well as the interest of developing countries. Policymakers asked to provide such support, either directly or by supporting mechanisms to generate appropriate incentives, have a right to ask: Is an HIV/AIDS vaccine really possible? What challenges must be overcome?

The short answer is that many scientists believe that a vaccine is indeed possible, but that the challenges are very substantial. The most basic challenge involves the question of whether human immune responses can prevent HIV infection or prevent illnesses in a person infected with HIV after vaccination. Although most people infected with HIV develop a broad range of anti-HIV immune responses (antibodies are one example), these responses are generally not capable of eliminating the infection or preventing progression to disease. Nobody knows whether these same immune responses would be more effective if they were induced by a vaccine, before exposure to HIV.

Intriguingly, some individuals do seem to have protective responses that enable them to ward off either HIV infection or the effects of the virus. Examples include the apparent absence of HIV in half to three-quarters of babies born to HIV-infected mothers and apparent resistance to HIV infection in a few individuals who remain uninfected, despite repeated exposure to the virus. Similarly, a few individuals, called long-term nonprogressors, have carried the virus for ten or more years but have not become sick with AIDS. In addition, trial HIV vaccines appear to have been effective in protecting chimpanzees from HIV, while other vaccines appear to protect monkeys from the simian immunodeficiency virus or SIV. All of these responses could be due at least in part to a strong immune response.

A second set of challenges involves the high degree of genetic variability in HIV: there is no guarantee that a vaccine developed to protect against one strain would necessarily protect against the others. HIV strains from different parts of the world have been grouped into ten genetic subtypes: A, B, C, D, E, F, G, H, I and O. Most of these subtypes

---

countries. These mechanisms, some of which require changes in the tax codes and legislation of participating countries, include:

- direct support for the costs of the early stages of product development
- analysis of the potential market for a specific new product that would primarily benefit people in the lowest-income countries[13]
- tax relief and or streamlined regulatory controls for the development of products for low-income countries
- worldwide tax breaks for pharmaceutical companies and extended periods during which they have the exclusive right to sell the drug (provisions similar to those of the U.S. Orphan Drug Act of 1983)

are present in Africa, subtype B is most common in developed countries. Encouragingly, recent research indicates that the genetic differences among subtypes may not necessarily affect the way that they respond to a vaccine. Nonetheless, the issue remains high on the vaccine research agenda and is of particular importance for those developing countries where several subtypes are present.

The third set of challenges involves the need for human trials; and a related need to ensure that these are conducted according to accepted standards of medical ethics. Despite the progress in testing vaccines on chimpanzees and monkeys, human trials are essential to determine the safety and effectiveness of an HIV vaccine. More than 20 candidate HIV vaccines have been tested in Phase I and Phase II trials with more than 2,000 HIV-negative volunteers, mostly in the United States. These trials have indicated that candidate vaccines are safe (Phase I) and that at least some of them induce HIV-specific immune responses (Phase II) which could confer protection against HIV infection or disease. However, because deliberately exposing trial volunteers to HIV is unthinkable, information on protective efficacy can only be obtained from large-scale Phase III field trials.

Multiple Phase III trials will be necessary to evaluate the protective efficacy of different vaccine concepts, against different HIV subtypes, against different routes of transmission, and under different health, nutritional, and/or genetic conditions which may be present in different countries where the vaccine is to be used. In order to gather the necessary information, these trials must be conducted in industrial and developing countries. The United States has announced its intention to proceed with a U.S. Phase III trial within the next two years, and discussions are underway to conduct Phase III trials in selected developing countries. The results of these trial may become available early next century. Of course, there is no guarantee that these trials will lead to an effective vaccine. However, without proceeding to Phase III trials, an HIV vaccine will never be available.

Sources: Esparza, Heyward, and Osmanov 1996; FitzSimmons, 1996; Gold 1996; International AIDS Vaccine Initiative 1996; Johnston 1996; and Osmanov 1996.

■ advance guarantee of a market for a health product that meets certain objectively verifiable criteria.

The last idea is a particularly innovative approach to solving the incentive problem. One way to implement the idea would be for one or more low-income countries with a specific disease problem not being adequately tackled by biomedical research to offer to buy a large quantity of the first drug or other medical product that meets precise specifications, regardless of the identity of the developer. To be credible, this offer would be guaranteed by a consortium of international donors and lenders. The total financial package could include, for example, grants from bilateral donors and a mixture of soft or market-rate loans from multilateral institutions or even commercial lenders. In the purest form

Box 5.4 Can Companies Make a Reasonable Return from the AIDS Vaccine?

"I CAN ONLY TELL YOU ABOUT THE EXPERIENCE with hepatitis B vaccine, which was developed 20 years ago. In the first couple of years the price of the vaccine was as high as $25 to $40 a dose, with three doses needed [US$50 to $80 at 1997 prices]. So companies aimed for the upper end of the market and the market was stalled. The price could not go too high despite the global need for the vaccine. In china alone, with 1.2 billion people, the hepatitis carrier rate is 10 percent. Yet many countries were practically out of the purchasing market.

"When the recombinant hepatitis B vaccine was developed, the price decreased a little. Now the price has come down to probably US1.00 per dose, [less than two percent of the initial price]. And four years ago, Thailand put the hepatitis B vaccine on the general program of immunization. So all babies in our country now receive the vaccine.

"Companies must realize that the potential market for an HIV vaccine in the developing world is tremendous, but it can only be captured by using two or three price tiers. The high price would be for industrialized countries, while developing countries would have another price. Companies must profit from their investment. And developing countries must be able to afford the vaccine. Figuring out how to work this out is a very important challenge for government, business, scientists and international organizations."

Dr. Natth Bhamarapravati, chairman, Subcommittee on HIV Vaccine Trials, developer of a vaccine against dengue hemorrhagic fever, and former president of Mahidol University, Bangkok, Thailand. From an interview published in IAVR (1997b).

of the guarantee, none of this financing would be released until the desired product was approved by independent testing laboratories. Only then, as the culmination of a period of research and development that might last as long as five or ten years, would the financial instruments be executed, the donor contributions delivered, the international loans disbursed, and the delivery and distribution of the product initiated.

The most urgently needed anti-AIDS technology—a vaccine to protect against HIV infection—is already the subject of a public-private partnership. Established in 1996, the International AIDS Vaccine Initiative (IAVI) is the first attempt to organize a health product development alliance along the lines recommended by the Ad Hoc Committee on Health Research. First proposed by the Rockefeller Foundation, the IAVI has attracted support from the Merieux Foundation, UNAIDS, and the World Bank, and Until There's A Cure, an NGO with links to the AIDS community. Its mandate is to accelerate the development of HIV vaccines appropriate for worldwide use by reducing obstacles to vaccine development and filling gaps in the current effort. In 1997, IAVI's first full year of operation, participants expected to devote a total of $2 million to $4 million in direct support of research on an AIDS vaccine (IAVR 1997a). In light of the U.S. initiative to produce an AIDS vaccine, the challenge to IAVI will be to assure that vaccine development programs do not neglect the needs of low-income countries, where 90 percent of HIV infections occur.

Other important potential applications for a public-private alliance in the AIDS arena lie in the development of vaginal microbicides and virucides, which would allow a woman to protect herself from HIV infection without asking her partner to wear a condom. Poor women in developing countries are often at a particular disadvantage in negotiating condom use with their partners; yet these same women are also most likely to lack the purchasing power to buy a vaginal virucide. By guaranteeing the market, a public-private alliance would provide an incentive for pharmaceutical companies to develop such products. Public-private partnerships could also offer incentives for development of reliable, inexpensive diagnostic tests for STDs such as chlamydia, which are prevalent in developing countries and accelerate the spread of HIV, and to stimulate research that would lower the cost of antiretroviral therapies.

Regional cooperation. When an AIDS epidemic first comes to public attention, many people and some governments blame neighboring countries or "foreigners" generally for the introduction of the virus. But every infectious event, whether through sex, or needle sharing, or transfusion, involves two people. One of them must be a national resident if the epidemic is to enter the national population. For the epidemic to then spread within the country, there must be additional infectious events involving residents. Thus, in any country with a serious AIDS problem, the national population must have taken an active role in its spread.

Blaming foreigners for the spread of the disease within the national population is not only illogical, it also undermines efforts to confront the epidemic. First, blaming foreigners gives people who are not in direct contact with foreigners a false sense of security, thus hindering efforts to encourage safer behavior. Second, blaming foreigners may worsen relations with neighboring countries, making it more difficult for governments to coordinate prevention and mitigation efforts. Soured relations may also threaten other benefits of economic cooperation, such as those from migrant workers or from commerce.

Rather than casting blame, a more productive approach would be for governments in neighboring countries to discuss ways in which they can cooperate to overcome the *shared* problem of AIDS. For example, neighboring countries can agree not to attempt to screen out HIV-positive migrants; to share information about prevention measures and the course of the epidemic; to coordinate policies on AIDS-related social issues such as prostitution and drug addiction; and to offer similar levels of subsidies for AIDS treatment and assistance to affected households, to avoid creating incentives for HIV-positive people to migrate in search of higher subsidies. Donor governments and multilateral institutions can play a useful role in supporting such regional dialogues.

Overcoming Political Impediments to Effective AIDS Policy

THE POLICY MESSAGES OF THIS REPORT ARE NOT STARTLING new findings. The call for preventing infections among people most likely to contract and spread the virus is a reiteration of arguments for the control of the sexually transmitted diseases that were already recognized 20 years ago (Brandt 1987). The warning that subsidizing AIDS treatment more generously than, say, cancer treatment endangers the quality and accessibility of health care for everyone is familiar from discussions of health sector reform (World Bank 1993c). The finding that the poorest households are most vulnerable to the shock of an AIDS death is consistent with previous work demonstrating that poor households have difficulty weathering other kinds of shocks. The conclusion that "survivor assistance" provided by the government or NGOs should be targeted to the poorest AIDS-affected households follows logically. The advantages of decentralizing and privatizing government service programs are well established. At the level of international public goods, the need for better knowledge and technology for developing countries has been glaringly apparent for years.

If these messages are familiar, why are they not being followed in countries around the world? The answers clearly lie outside the technical discussions that have occupied this book so far and fall instead into the domain of political science, a less-developed discipline than either epidemiology or economics, with fewer guiding principles. However, the examples of countries that have achieved modest success in confronting AIDS suggest some lessons.

Interest Groups and AIDS Policy

Many groups with divergent interests affect the design and implementation of HIV/AIDS policy, and the mix of groups and their relative strength changes over the course of the epidemic. At the outset, few groups are concerned. However, as the epidemic progresses, the number of interest groups increases and the politics of AIDS becomes ever more complex.

Early in the epidemic, physicians and medical suppliers have an interest in learning how to treat AIDS and how to protect the safety of health care workers from needle sticks and other accidental infection on the job. A group that emerges at about the same time is people infected with HIV. Although the number of these individuals is initially small, and they may at the outset lack political influence, they are often highly motivated to lobby government, since their very lives may depend on persuading the government to subsidize AIDS treatment and care. As the epidemic spreads, the size of this group and its potential to influence government policy increase. Often overlapping with this group are individuals who practice high-risk behavior but who are not infected—or hope they are not infected. Although these people have a strong interest in government-subsidized prevention for themselves, in the early stages of the epidemic they are rarely well organized enough to lobby on their own behalf. Yet increasingly the AIDS epidemic has induced people with the highest risk behavior to organize in order to promote their interests. Furthermore, NGOs working on HIV prevention and AIDS care become advocates for the populations they serve. Finally, as the number of AIDS cases increases, insurance providers and employers will become concerned about rising health care costs and increased sickness and death among employees.

At all stages of the epidemic, the largest interest group is the one *least* motivated to learn about the issues or lobby on its own behalf: the general public of HIV-negative individuals who rarely practice risky behavior. Like most of those with riskier behavior, these people at low risk have an interest in marriage, in conceiving and raising healthy children, and in seeing them married, all without the risk of HIV infection. Although not themselves suffering from AIDS, in a generalized epidemic these people find that the price of health care has greatly increased because of higher demand and increased costs. Some of these people are the poor who have never suffered an HIV infection or death but nevertheless need help in order to escape poverty. Some suffer dangerous chronic diseases other than HIV, such as cancer, kidney disease or diabetes, and cannot afford the treatment to keep themselves alive.

To be truly democratic, a society must find ways—for example, opinion polls or elections—for the many with a small interest in an issue to express their views inexpensively and influence the course of events. Politicians facing a ballot box have an incentive to seek the opinions of ordinary people and consider these together with the views of smaller, more vocal interest groups. A government that is responsive to the nation's political leadership will follow suit.

However, in the case of HIV/AIDS, the policies that will best protect the average citizen are not necessarily popular. Politicians and government officials, who may themselves be unsure of the best policies for confronting the epidemic, have the difficult task of explaining to the public why taxes should be spent subsidizing condoms and STD treatment for prostitutes and clean needles for injecting drug users. Conservative social and religious groups, perhaps not fully appreciating the great harm that can arise from failing to prevent the spread of HIV, may oppose efforts to reduce the risks involved in commercial sex or injecting drug use, or to encourage condom use generally, out of concern that these efforts will encourage behavior they regard as immoral. Business interests, having immediate profits in mind, may apply the kind of pressure to government that was dramatized in Henrik Ibsen's 1883 play *An Enemy of the People:* a physician who discovers that his Norwegian town's polluted public baths are a threat to tourists' health is pressured to keep silent by the democratically elected mayor and his supporters, and ultimately declared to be an "enemy of the people" himself.

Mexico and Thailand offer two dramatic examples of AIDS policy-making in the midst of all these conflicting pressures. The former coordinator of Mexico's National Committee for the Prevention and Control of AIDS (CONASIDA), Dr. Jaime Sepulveda, has summarized the responses of government, NGOs and mass media during three periods from 1985 to 1992 (Sepulveda 1992). As shown in table 5.3, the government response evolved from "erratic and medicalized" in 1985–86 to "reactive and participatory" in 1989–92. Strikingly, organizations of homosexual and bisexual men and liberal NGOs were initially silent and then actively opposed to the AIDS control program. Through continued efforts to engage these interest groups, government policymakers eventually won them over; by the third period they were active participants

Table 5.3 Responses to the AIDS Epidemic in Mexico: Government, NGOs, and Mass Media

| Government response | Nongovernmental response | | Mass media response |
	Gay and liberal NGOs	Pro-Vida and other right-wing groups	
Erratic, medicalized, 1985–86	Silence	Slight opposition	Alarmist
Planned technocratic, 1987–88	Anger, protest	Strong opposition	Reactive only to "sensational" news
Reactive, participatory, 1989–92	Protest, participation	Lawsuits, marches	Fatigue

Source: Sepulveda 1992.

in carrying out prevention programs. Meanwhile, Pro-Vida, a conservative religious group, and other right-wing organizations became increasingly outspoken, if ultimately ineffective, in their opposition.

Sepulveda includes the mass media among the actors in the shaping of Mexican AIDS policy, but he describes their role as only occasionally helpful. As late as 1992 he characterizes media coverage as continuing to focus on the number of AIDS cases, while neglecting other crucial information about the disease: "In spite of the constant presence of information about AIDS in the mass media, specific aspects of the disease are not addressed so that collective accurate knowledge about AIDS is not generated nor is participatory discussion promoted." He points out that television and radio do a somewhat better job than print media, sometimes using live programs with interviews, phone-in questions, and audience participation to generate discussion (Sepulveda 1992, p. 143). However, he concludes that, by the third period covered in the table, the media have passed from "alarm" to "fatigue" without ever providing the information that the public needs to understand the epidemic.

An authoritative case study of Thailand highlights other political problems that can arise in designing and implementing an effective response. In the second half of the 1980s, as evidence that HIV was spreading rapidly among Thai sex workers and injecting drug users accumulated, a government official insisted that the situation was under control: "The general public need not be alarmed. Thai-to-Thai transmission is not in evidence." In keeping with this sanguine view, the government spent only $180,000 on HIV prevention in 1988 (the GPA committed $500,000 to Thailand that same year). The study suggests that during this period of democratic rule, in a pattern reminiscent of that described by Ibsen in Norway 100 years before, "high-level cabinet pressure was brought to bear on the ministry of public health not to publicize the emergence of increasing HIV in the population" (Porapakkham and others 1996, p. 8).

Although Thai national funding increased to $2.6 million by 1990 (and donor funding reached $3.4 million), the government did not initiate a high-profile, aggressive campaign to control HIV until 1991–92, when the country was led by Premier Anand Panyarachun, who had been appointed by the leaders of a military coup. The new prime minister took several important steps that have since been credited with helping to slow and perhaps reverse the epidemic in Thailand. First, he shifted control of the AIDS control program out of the Ministry of Public Health to the Office of the Prime Minister, giving it added political clout. Second, he increased the budget almost 20-fold, to $44 million in 1993. Perhaps most important, he initiated the "100 percent condom program" focused on brothels, as described in chapter 3. Since then Thai funding to AIDS control has continued to increase, reaching more than $80 million in 1996, a sum equivalent to more than one-quarter of the *entire* international donor commitment to AIDS control in developing countries that year.

The high-profile campaign was initially unpopular with the influential tourism industry, and tourism indeed temporarily declined. However, once AIDS had a prominent place on the national agenda, opposition to the measures gradually faded—and support increased. "There were too many vested interests in maintaining the high status of the national AIDS program to make a policy reversal," the case study noted. "In particular, the enormous budget allocated to the HIV/AIDS prevention and control campaign was vigorously coveted by a wide-range of participants" (Porapakkham and others 1996, p. 17). Thus, the policy situation in Thailand had come full circle, from one in which special interest groups used their influence to oppose a vigorous prevention policy, to one in which the participants in the prevention program assumed the role of vested interests in sustaining it. Since all programs that involve significant public expenditure develop their own constituencies, policymakers must be careful at the outset to initiate programs that are in the interest of the general public, as appears to have been the case in Mexico and Thailand.

Donor Assistance and Public Consensus

Although the politics of AIDS will differ greatly across countries, bilateral donors and multilateral organizations can help to encourage public consensus on effective, low-cost responses to HIV through direct funding and through a judicious use of encouragement and conditionality. For countries that are still in the nascent stage, where citizens are not sufficiently aware of the epidemic to support funding activities from public revenue, donor funding can be critical in gathering surveillance data or establishing a demonstration project. Sometimes donors can require certain actions as a condition of the receipt of an aid package. However, the leverage afforded by conditionality is often quite limited and may depend on all donors agreeing to the desirability of a given condition. Conditionality is more likely to work if the government (or important elements of it) intends to carry out the action in any case but has not yet made it high enough priority to get it accomplished.

One example of the effective application of conditionality occurred during negotiations of the $70 million World Bank loan to India. In 1991 the government's initial posture was that there was no need for specific interventions with sex workers and their clients in Indian cities. One influential government figure asserted that "in India AIDS is not sexually transmitted." As a result of a position taken jointly by GPA and the World Bank, the government of India agreed to double the size of its proposed AIDS program to include interventions with those most likely to contract and spread HIV, to be implemented by NGOs. Since then the extent of the sexually transmitted AIDS epidemic in India has become obvious to the highest levels of government, as evidenced in a 1997 speech by Prime Minister Deva Gowda. Attention has turned from whether interventions with those who practice the riskiest behavior are necessary to how best to implement them.

These instances suggest that donors can significantly improve the timing and quality of country-level responses to HIV/AIDS. However, the evidence cited in chapter 3 and earlier in this chapter suggests that donors have often waited until AIDS has moved beyond the nascent stage before providing support. Although the data suggest that multilateral institutions are more likely than bilateral donors to direct resources to countries at the concentrated stage of the epidemic, neither supports countries sufficiently at the nascent stage, when the largest benefits can be achieved with the smallest expenditure. We return to this issue in the policy recommendations in chapter 6.

Individuals Who Make a Difference

Although this chapter, and indeed most of the book, has focused almost exclusively on national governments, donors, or groups, sometimes a courageous individual changes the way an entire nation or society thinks about HIV/AIDS, opening the way for a more effective and compassionate response. These individuals may be national political leaders or other well-known figures, such as athletes or movie stars, who are not themselves infected. Or they may be individuals, famous or not, who are infected with HIV and summon the strength and courage to serve as advocates for a sound national response.

Examples of such individuals in the industrial countries are known worldwide. Actress Elizabeth Taylor has made fundraising for AIDS a nearly full-time occupation. Others, such as the late Princess Diana of Britain, have reduced prejudice and fear simply by being photographed embracing a child with AIDS. Among U.S. athletes, diver Greg Louganis, the late tennis star Arthur Ashe, and basketball's Magic Johnson have each helped to raise awareness of the disease by coming forward with the news of their infection.

But while these figures are widely known and often admired around the world, the fact that they are from industrial countries means that their high-profile activities have only a limited ability to overcome denial in developing countries. People in a poor country learning that a movie star or athlete in a rich country has become infected may continue to think, "It can't happen here"—even though 90 percent of HIV infections occur in developing countries. Because of this, every country and all societies need local individuals with the courage to advocate an effective response to HIV/AIDS. Where such individuals have stepped forward, their efforts have often had a significant positive impact on public awareness and attitudes.

Fortunately, as the understanding of the epidemic increases, a growing number of individuals in developing countries are demonstrating such leadership. To mention just three examples: the speech by Indian Prime Minister Deve Gowda naming HIV/AIDS as a national health problem helped to overcome the idea that India was somehow not threatened by the virus. Zambian President Kenneth Kaunda, in acknowledging publicly that his son had died of AIDS, helped to energize his country's response to the ravages of a widespread epidemic. Finally, Marina Mahathir, daughter of Malaysian Prime Minister Mahathir Mohamad and the president of the Malaysian AIDS Council, a nongovernmental organization, has spoken out in her own country and internationally for greater political commitment to mobilizing the resources necessary for effective prevention.

Some of the most compelling advocates of an effective response to the epidemic are people who are themselves infected with HIV. Philly Lutaaya, an enormously popular Ugandan singer and songwriter, became the first prominent African to acknowledge that he was infected with HIV. He spent his remaining healthy time writing songs about his battle with AIDS and touring churches and schools throughout Uganda to spread a message of prevention and hope. After Lutaaya's death at age 38, the Philly Lutaaya Initiative continued his work. With assistance from UNICEF, the Initiative sponsors lectures in schools and communities across Uganda highlighting personal testimonials of hundreds of people infected with HIV. A 90-minute television documentary on Lutaaya's struggle with AIDS released in 1990 reached millions of television viewers around the world (Graham 1990, Kogan 1990, McBrier 1995).

But a person need not be a celebrity prior to infection for personal testimony to have a powerful impact. Perhaps the most courageous individuals are otherwise ordinary people who, after becoming infected, step forward to acknowledge their disease and, in the face of discrimination and persecution and with very limited personal financial resources to draw upon, speak out for a more effective public response. All these individuals serve as a powerful example to those who meet them, a few become nationally known. Box 5.5 describes how one such individual, a factory watchman, raised awareness about HIV/AIDS in Thailand.

■ ■ ■

This chapter has analyzed the roles of governments, donors, and NGOs in financing and implementing effective policy responses to HIV. It has argued that each of these types of organizations has particular strengths and that for an effective global response to HIV/AIDS, all of these groups, plus countless exemplary individuals, must work toward a common goal of overcoming the epidemic.

As the chapter relates, much has already been done; yet the analysis also identified some key shortcomings. Governments have the unique

Box 5.5 Someone with AIDS Who Made a Difference

WHEN CHA-ON SUESUM CONTRACTED HIV FROM A blood transfusion he was fired from his job as a factory watchman; his wife, who worked in the same factory, was also fired. In 1987, Cha-on decided to make his case public and accepted a job as an AIDS educator with the Population and Community Development Association of Thailand, a nongovernmental organization.

Cha-on soon appeared on national television talk shows and on the front pages of Thailand's biggest newspapers. The result was an outpouring of public sympathy for his own hardship, and a turning point for society as a whole in dealing with the epidemic. While Cha-on was still well, his activities focused attention on discrimination against those infected with HIV. However, as the entire nation witnessed his rapid progression to clinical AIDS and finally his death, another critical issue was brought to the fore. Thais began to understand that AIDS was real and that they themselves could become infected and die. Cha-on's lasting legacy was stronger support for and receptivity to HIV prevention efforts throughout Thailand (Porapakkham and others 1996).

responsibility for coordinating their country's overall response to the epidemic. As part of that responsibility, many governments, especially in developing countries, should take on greater responsibility for basic epidemiological surveillance and prevention activities. NGOs have often played an important role in prodding governments into action; governments that select appropriate NGO partners can often greatly increase their reach, especially in working with marginalized groups to help people who practice the riskiest behavior to protect themselves and others. Donors and the multilateral institutions they support have provided significant financing and other assistance for all of these efforts. But donors need to do a better job of focusing attention and resources on countries where the epidemic has yet to attract policymakers' attention, especially countries with nascent epidemics, where prevention is most cost-effective. Moreover, international donors have the unique ability to mobilize financing and other support for international public goods, such as evaluation of alternative approaches to preventing HIV and mitigating the impact of AIDS, as well as research on a vaccine that would work in developing countries. Such efforts are in the donors' own best interest, as well as the interest of developing countries, and deserve much greater attention and support. Finally, donors have the responsibility to coordinate their activities at the country level, both among themselves and with the national government.

Although there are no easy solutions to the technical and political problems posed by the HIV/AIDS epidemic, examples from countries around the world offer hope that people of good will, working together, can overcome this global epidemic.

The next and final chapter in the book summarizes its main policy recommendations and looks toward the future.

Notes

1. This calculation uses the estimate of $4.8 billion for total health assistance in 1990 (World Bank 1993c, p. 166).

2. Although national AIDS program spending under-represents total national spending on the AIDS program, it probably captured most of it in 1993.

3. The effect of donor spending on national spending was estimated under the maintained hypothesis that national funding does not affect donor funding by including donor funding as a fourth regressor in the equation to predict national funding. After controlling for (the logarithms of) GDP per capita, population and the number of HIV-infected people, the coefficient of (the logarithm of) donor spending is 0.01 with a t-statistic of 0.08. The instruments necessary to identify a model of simultaneous causation between national and international funding are not available; therefore, such a pattern cannot be ruled out.

4. These two relationships do not appear to be the result of a national decision to reduce AIDS funding in response to the perception that donors are providing such funding already. See note 3.

5. According to the available data, multilateral funding totaled $605.7 million during this period, 22 percent more than bilateral funding. The World Bank made two large loans, one to India for $70 million and one to Brazil (the total size of the Brazil project was $250 million, of which $160 million was borrowed from the World Bank). The loan to India was at a concessional interest rate accorded to the lowest-income countries and thus approximately equivalent to a $50 million grant (Arias and Servén 1997). The loan to Brazil was at higher rates accorded to less-poor countries, so that the equivalent grant would be significantly smaller. For the purposes of this analysis, the difference between loans and grants has not been taken into account.

6. Tax law typically forbids nonprofit firms from selling shares of the firm to raise capital, as for-profit firms are free to do.

7. See, for example, "NGOs Flout AIDS Control Policy" (1994). We set aside the fact that the government may imperfectly represent the interests of the public.

8. In late 1997, USAID was preparing programs to succeed AIDSCAP.

9. Given a constant total donor AIDS budget in a recipient country, the country would benefit if its AIDS programs were developed as a coherent whole and all donors agreed to pay a share of the total. However, experience shows that the amount of any given donor's expenditure in a country is not usually fixed. In those cases where the donor's budget for the country is fixed in the short run, it is still likely to be fungible across sectors. Thus, the amount of AIDS financing from a given donor depends upon how much its representatives want to fund the AIDS projects that the government allows it to fund. It is often alleged that donors prefer to "put their flag" on a project, so that they can claim credit for it in the international community and to their domestic constituency. These incentives lead to a situation in which no donor wants to fund the overhead costs of an AIDS programs or a portion of any part of the program. Any attempt to coordinate donors, whether bilateral or multilateral, must struggle against these perverse incentives.

10. This projection is based on the assumption that the number of incident cases will stabilize in every region of the world once incidence falls to half of its peak value. "The choice of an equilibrium value for incidence that is 50 percent of peak incidence is entirely arbitrary and does not take into account advances that may be achieved in behavior modification or technological breakthroughs such as a vaccine or more effective chemotherapy. Consequently considerable caution is required in interpreting these HIV projections, particularly for the years beyond 2005" (Murray and Lopez 1996, p. 347).

11. See the examples in chapter 3 and the summary of rigorous evaluations of preventive interventions in developing countries in appendix A of this report.

12. A "decrease in the willingness of pharmaceutical companies to become involved in vaccine research, development and manufacturing" was observed as early as 1985 (Institute of Medicine 1985, p. viii).

13. The behavioral sciences can also contribute information that will increase the profitability of a medical intervention. For example, the European Commission is sponsoring a "market perspectives study" on vaginal microbicides in Brazil, Côte d'Ivoire, Egypt, India, Kenya, the Philippines, Poland, and South Africa. A finding that women are willing to pay for this drug would improve the incentives for private pharmaceutical firms to invest in their development (AIDS Analysis Africa 1996).

CHAPTER 6

Lessons from the Past, Opportunities for the Future

O VER THE PAST TWO DECADES THE HUMAN immunodeficiency virus has spread silently throughout the world, profoundly affecting the lives of men and women, their families, and societies. It has not respected international boundaries or spared the elite. By the time that researchers understood how HIV spreads, how it can be prevented, and the behaviors that put people at risk, HIV had already infected millions of adults in the industrial and developing world. In the hardest-hit countries in Sub-Saharan Africa, poverty, illiteracy, poor health, low status of women, and political instability fueled its spread. By the time East African health authorities identified the mysterious "slim" disease as AIDS in the early 1980s, HIV had already widely infected those with the riskiest behavior and had a firm foothold in the general population.

On the medical frontier there have been many advances, but there is still no vaccine for HIV and no cure for AIDS. Medical researchers have succeeded in substantially prolonging the lives of some people living with HIV and AIDS in industrial countries. However, these treatments are still very expensive, they are not always successful, and no one knows for how long they prolong life. The costs of these new therapies are so high and the requirements for their implementation are so demanding that they are simply not feasible in low-income countries and would bankrupt the health systems of middle-income countries.

Lessons from Two Decades of Experience

STILL, WE HAVE LEARNED MUCH IN THE PAST TWO DECADES that is cause for optimism as we confront the epidemic. We now know that HIV is not spread easily and that it can be prevented through behavioral change. Other STDs signal risky behavior, and preventing or treating these STDs can slow HIV transmission rates. Low-cost, cost-effective interventions to prevent HIV/AIDS in poor countries are now known to exist. Behavior change has reduced incidence of HIV among specific groups in countries as diverse as Australia, Thailand, and Uganda. And there are many opportunities to alleviate suffering and prolong the lives of HIV-infected people in developing countries, for example, through low-cost treatments of common opportunistic infections, particularly tuberculosis.

We can also learn from the policy mistakes of the past. No country, rich or poor, is insulated from the risk of HIV. Governments should intervene as soon as possible; if policymakers wait until AIDS is killing many people, HIV already will have spread widely, interventions will be less cost-effective, reducing infection will become more difficult and, absent a cure, the epidemic and its terrible impact are likely to persist for decades. Behavioral change must focus first on people with high-risk behavior who are most likely to become infected and unknowingly infect others. But discrimination against such individuals makes behavioral change more difficult and inhibits efforts to cope with the impact of AIDS.

The Role of Government

EXPERIENCE HAS ALSO SHOWN THAT ACTIVE GOVERNMENT involvement is crucial if AIDS is to be overcome. Only governments have the means and mandate to finance the public goods necessary for the monitoring and control of the disease—epidemiological surveillance, basic research on sexual behavior, information collection for identifying high-risk groups, and evaluation of the costs and effectiveness of interventions. Private individuals left to their own devices would not invest adequately in these activities. Governments also have a unique responsibility to intervene to reduce the negative externalities of high-risk behavior, while preventing discrimination that would inhibit behavioral change. Without these government efforts, those at high risk of contracting and spreading HIV are unlikely to reduce risky behaviors enough from the perspective of the rest of society. The government role extends to ensuring equity in access to HIV prevention and treatment for the most destitute.

Other key functions that most governments are already attempting to perform can also make an important contribution to slowing the spread of HIV: promoting labor-intensive economic growth to reduce poverty; assuring basic social services, law and order, human and property rights; and protecting the poor. Investing in female schooling and ensuring equal rights for women in employment, inheritance, divorce, and child custody proceedings are part of this broader mandate. These policies yield large development benefits in their own right but are also important for preventing an HIV epidemic and coping with its impact. Reform of health systems, as outlined in the *World Development Report 1993* (World Bank 1993c), will improve the efficiency of health care delivery, including HIV and STD prevention, and will reduce the impact of AIDS on the health system. In areas where there is a severe epidemic and targeted poverty reduction programs already exist, these can often be combined with efforts to ease the impact on surviving household members, especially children, in the most destitute families that suffer a prime-age adult death.

We know that certain policies can work, yet developing countries face many financial, political, and managerial obstacles to implementing them. Financial resources are scarce. In low-income countries, annual health spending from public and private sources averages only $16 per person.[1] This is one-tenth the resources available in middle-income countries and only 0.7 percent of the $2,300 per capita annual health spending in high-income countries. Many developing country governments also have limited capability to implement complex or multifaceted programs. The *World Development Report 1997* (World Bank 1997a) makes a compelling case that the government role must be matched with its capability. In fighting the spread of HIV and mitigating the impact of AIDS, developing country governments will be most effective if they focus their financial and other resources on a limited set of feasible activities that have the potential to be highly cost-effective. Pressures from the public and from international donors can lead governments to try to do too much with too few resources, reducing the effectiveness of programs. Governments can sometimes expand their effectiveness by involving the private sector, reputable NGOs, those most severely affected, and decentralized community organizations in the design and implementation of high-priority HIV/AIDS prevention activities. However, coordination and management of these activities can also stretch the capabilities of government.

Opportunities To Change the Course of the Epidemic

ONCE LAUNCHED, AN HIV/AIDS EPIDEMIC CAN TAKE decades to unfold. Epidemiological models predict that between 1996 and 2001, between 10 and 30 million new infections will occur in developing countries. But the future of the epidemic is not carved in stone. One reason that such projections are very uncertain is that nobody knows the extent to which individuals, especially those most likely to contract and spread HIV, will change their behavior in response to the virus. Concerted, focused action in developing countries, where more than 90 percent of HIV infections occur, can save millions of lives (box 6.1).

Preventing the Expansion of Nascent Epidemics

Public action can make the greatest difference for the 2.4 billion people who live in areas where the epidemic is nascent. Developing areas with nascent epidemics include half of the world population, two-thirds of the population of developing countries, and nearly 40 percent of the population of low-income countries (table 6.1). Half of India, all of China except Yunnan Province, Indonesia, the Philippines, most of Eastern Europe and the former Soviet Union, North Africa, and a third of the countries in Latin America and the Caribbean are at this stage. In these areas, HIV has not yet spread widely even among those whose behavior puts them at risk. But countries with nascent epidemics cannot assume that they will never be affected; every country that now has a generalized AIDS epidemic went through a phase of denial that gave the virus time to gain a foothold.

These nascent-stage areas present an enormous opportunity for governments and donors to prevent an HIV epidemic by intervening actively and early. Epidemiological surveillance of those who practice the

Box 6.1 Estimating the Power of Prevention in Three Countries

THE PREVIOUS CHAPTERS HAVE HIGHLIGHTED simulations of the epidemic in hypothetical populations. What might happen in a real country?

Modeling the potential benefits of interventions for a specific country requires detailed information about behavioral and biological characteristics of the population—the types and distribution of risk behavior, the number of people involved, sexual mixing patterns, and the prevalence of other STDs in specific population groups. Such information is rarely available and is urgently needed. Efforts are underway to calibrate the STDSIM model for Nairobi, Kenya, for example, and the iwgAIDS and SimulAIDS models have been calibrated to predict the impact of interventions in Kampala, Uganda (Bernstein and others 1997).

Nevertheless, we can get some sense of the likely impact of interventions that change high-risk behavior by applying limited country-specific parameters to existing models. Simulation results for three countries at different stages of the epidemic— Indonesia (nascent), Brazil (concentrated), and Côte d'Ivoire (generalized)—have been derived from a model developed at the World Bank.[1] Like the STDSIM model used earlier in this report, the World Bank model simulates the spread of HIV through heterosexual contacts and from mothers to their children, and it takes into account factors, like the presence of STDs and condom use, that affect the probability of HIV transmission. In addition, it models transmission through blood transfusion, needle sharing, and homosexual contact.

Country-specific parameters for these simulations were based on information from surveys and other studies in each country, as well as on informed estimates. We show below the simulated impact of raising condom use among two segments of the population with high rates of partner change: women who have 500 new partners per year (1 percent or less of all women), and women (and men, in countries where homosexual transmission is modeled) who have one new partner per month (5 to 10 percent of

the population). The simulations show the impact of raising condom use in these two groups with the most partners from 20 to 80 percent and from 5 to 15 percent, respectively, between 1997 and 2000. In Brazil, where needle sharing has played an important role in spreading HIV, the share of injections with clean needles among injecting drug users is assumed to rise from 20 to 80 percent. Finally, for comparison, we show the effect of raising condom use among women in stable relationships, from 1 to 3 percent in Côte d'Ivoire and from 5 to 10 percent in Indonesia between 1997 and 2000. The simulations show the results of these interventions through 2010.[2]

- In Indonesia, HIV prevalence is still very low—less than 0.01 percent of the population is infected. Among sex workers, homosexuals, and transvestites, however, HIV prevalence is as high as 3 percent. A rapid expansion of condom use among the two groups with the highest rates of partner change can prevent the level of HIV infection in the general population from reaching above 0.2 percent. Increased condom use among women in stable relationships has very little impact.

- In Brazil, with a concentrated epidemic, an increase in condom use among the two groups with the highest rates of partner change is sufficient to bring HIV prevalence down to about 2 percent in 2010. An increase in the use of clean needles accelerates this trend, but by itself is insufficient to reduce prevalence substantially.

- In Côte d'Ivoire, where HIV prevalence in the general population has already reached 13 percent, the epidemic would, in the absence of any behavior change, continue to increase, reaching 16 percent of the population by 2010. Interventions to dramatically increase condom use among those with the highest rates of partner change would lower preva-

(Box continues on the following page.)

Box 6.1 *(continued)*

lence to 9 percent by 2010. In contrast, increased condom use among women in stable relationships would have a negligible impact.

These results are suggestive of the impact that programs can have if they succeed in changing the behavior of the population with the highest rates of partner change. However, they understate the impact to the extent that other segments of the population may also change their behavior, either spontaneously or as the result of interventions. More detailed information on sexual behavior in these countries is necessary to generate more accurate models.

---

1. The model called "Projecting AIDS," or PRAY, is described in Bulatao (1991).

2. Condom use continues to climb at the same rate until 2020, the end of the simulation period. It reaches 98 percent and 70 percent among the groups with the highest and next-highest rates of partner change. Among the women with the fewest partners, it reaches only 10–25 percent.

Box Figure 6.1 Projected Impact of Behavioral Interventions in Three Countries

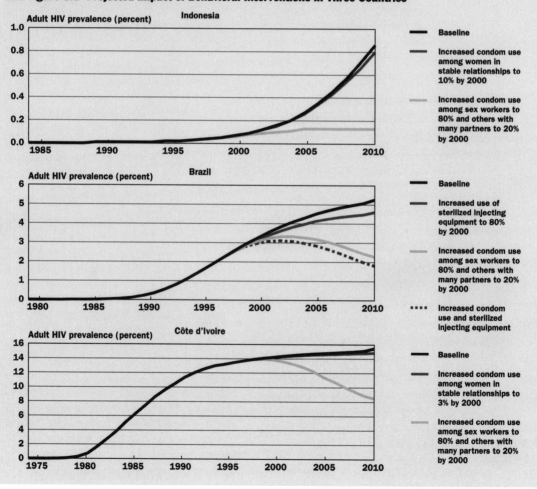

Table 6.1 Distribution of Developing Country Population by Stage of the Epidemic and Income

Stage of the epidemic	Low income[a] Population (millions)	%	Lower-middle income Population (millions)	%	Upper-middle income Population (millions)	%	Total Population (millions)	%
Nascent	1,735	37	503	11	28	1	2,265	49
Concentrated	1,008	22	320	7	311	7	1,640	35
Generalized	181	4	3	0	42	1	226	5
Unknown	151	3	307	7	42	1	500	11
Total population[b]	3,075	66	1,133	24	422	9	4,630	100
Number of countries	60		46		17		123	

a. The populations of China and India, both low-income countries, have been distributed between nascent and concentrated stages of the epidemic, based on the stage in specific provinces and states, respectively.

b. Any discrepancies in totals are from rounding numbers.

Sources: Income groups are from the *World Development Report 1997* (World Bank 1997a). Stage of the epidemic and 1995 population are from table 2 of the statistical appendix to this report.

riskiest behavior and studies of risky behaviors in the general population and specific subgroups have high payoffs at this stage. An HIV/AIDS epidemic can be pre-empted, for little cost, by promotion of safe injecting behavior among injecting drug users and of safe sex and STD prevention through condom use among those with high levels of sexual activity. We know that this can be done. In chapter 3 we highlighted the example of five cities in which early intervention kept infection levels among injecting drug users below 5 percent, even as HIV prevalence soared among injecting drug users in nearby cities. Experience has shown that early interventions focused on groups at high risk of sexual transmission can be equally effective.

Prevention efforts focused on those who practice the riskiest behavior may be politically controversial, especially if such efforts are perceived by some constituencies as facilitating antisocial or immoral behavior. Policymakers who encounter such opposition have an obligation to make clear that preventing infections among those with risky behavior is the best way to protect everyone.

Containing Concentrated Epidemics

Developing countries with concentrated epidemics—where HIV prevalence exceeds 5 percent in one or more groups with high-risk behavior but not in the general population—are a diverse group of low- and middle-income countries, with a variety of risk factors. In Latin America, Ukraine, Yunnan province of China, much of Indochina, and the northeast of India, the epidemic has reached concentrated levels among injecting drug users; in many countries in Latin America, the epidemic has also reached concentrated levels among homosexual and bisexual men. In addition, HIV has infected more than 5 percent of high-risk heterosexuals, among them sex workers, in southern India, Indochina, and much of Africa.

Once HIV has reached high levels among those who are most likely to contract and spread the virus, containing the epidemic is difficult and requires drastic action—but is nonetheless possible. Thailand undertook such a massive effort when injecting drug users and prostitutes were discovered to have high infection rates. A policy of heavily subsidized condom promotion and STD treatment programs for prostitutes and others with high-risk behavior, supplemented by widespread dissemination of information to the general population, brought down the prevalence of HIV among military conscripts within a few years. Not all countries have the same institutional setup or implementing capacity as Thailand. Each country will have to find its own way. But whatever tactics are adopted, the underlying strategy of massive interventions to change the behavior of those most likely to contract and spread HIV is crucial.

Adapting this strategy successfully will require better information about the cost-effectiveness of alternative interventions to prevent the spread of HIV. Research documenting the effectiveness of such interventions in preventing secondary infections can be very valuable in generating and sustaining support for these measures. Governments also have a role in ensuring that basic information about HIV is presented to the general public in ways that will minimize irrational fear and persecution of individuals who are infected with HIV or thought to engage in high-risk behavior, since such responses make it harder to reach those with risky behavior and encourage safer behavior.

As people infected early in the epidemic begin to get sick and die from AIDS, governments will face growing pressure to spend public resources on care and treatment. Responding to these needs compassionately, while keeping them in perspective with the many other pressing human needs and demands upon public resources, is one of the most difficult challenges posed by the epidemic. Pressure for spending for AIDS care and treatment will be stronger in a generalized epidemic, when the disease has spread into the general population and people infected with HIV are a large and highly motivated constituency. By then, subsidies begun during the concentrated stage may be unsustainable and yet very difficult politically to withdraw. The concentrated stage of the epidemic is therefore the time when policymakers and their constituents need to consider how government can best respond to the medical needs of people with HIV.

The fair response in terms of health care, advocated in chapter 4, is to offer the same level of subsidy for the care and treatment of people with AIDS as for the care and treatment of people with other diseases that are expensive and difficult to treat. Denying care to individuals simply because they have HIV/AIDS is unjust to those who are infected and to their families. By the same token, providing a higher level of subsidy for AIDS care than for other illnesses is also unfair to the majority of people who are not infected with HIV. Choices about the appropriate overall level of public subsidies for health care will vary across societies. Governments and their constituents should be aware, however, that high subsidy levels will be extremely difficult to sustain in the face of a large epidemic. Since it is unfair and impractical to deny care and treatment subsidies to people with HIV while providing them to people with other illnesses, any changes in subsidy levels should apply equally to the HIV infected and the uninfected.

Policymakers need also be aware that the care and treatment of AIDS, in sharp contrast to preventive interventions focused on those most likely to spread HIV, is primarily a private rather than a public good: most of the benefits of the care and treatment of AIDS accrue to the person who receives the care. There are important exceptions to this general rule. Treating tuberculosis, STDs, and other infectious diseases in people with HIV can prevent these infections from spreading to others, including people who are HIV-negative; these "externalities" are a sound rationale for public funding of such treatments, regardless of whether the recipient of the treatment has HIV. Similarly, outreach programs that include care for those infected with HIV who practice high-risk behavior may be a justifiable use of public funds if the program results in behavior changes that reduce the spread of the virus. Often, however, demands for publicly funded care and treatment threaten to drain scarce resources that could have been used for preventing new infections.

Maintaining Focused Prevention in Generalized Epidemics

Countries with a generalized epidemic will face two related sets of challenges: establishing or maintaining prevention programs focused on those most likely to contract and spread HIV, while expanding prevention efforts to those with somewhat lower risk of transmitting the virus; and mitigating the impact of AIDS sickness and death, especially among the poor.

Except for Botswana and South Africa, all countries that currently have generalized epidemics are low-income, with 1995 per capita income of $765 or less. Scarce financial and managerial resources mean that these governments must be especially vigilant in implementing the most cost-effective prevention programs. Although prevention measures for the general population become increasingly cost-effective as prevalence rises, interventions for those practicing the riskiest behavior continue to have the greatest impact on incidence per dollar spent and must be maintained even as prevention programs are expanded to others. Condom social marketing programs and other forms of prevention subsidies aimed at poor people who would otherwise be unable to afford to protect themselves are an appropriate government response at this stage, where resources are available. But these programs are no substitute for reaching the highest-risk groups. Indeed, one of the greatest threats to effective prevention in generalized epidemics is pressure to divert resources from highly targeted cost-effective interventions to politically popular interventions with lower cost-effectiveness.

Even where prevention measures are very effective, declines in prevalence will occur only gradually, as people already infected die and are succeeded by younger cohorts. But declines in incidence—the number of new infections—can be achieved relatively quickly, even in the face of a generalized epidemic. Recent declines in HIV incidence among young people in Uganda are an encouraging sign that even the worst-hit countries can make progress against the epidemic.

The second challenge to governments in a generalized epidemic is mitigating its impact, especially on the poor. A widespread epidemic will greatly increase the number of households that suffer a prime-age adult death. In poor households, such deaths can have a severe and lasting impact on surviving children, who may suffer further declines in already inadequate nutrition and schooling. But not all households that suffer a prime-age adult death are poor. Indeed, in many of the countries hardest hit by AIDS, while most of those infected may be poor, it is still the case that nonpoor people are more likely to be infected than the poor.

Confronted with demands to finance programs to help households affected by AIDS, policymakers need to balance the needs of poor households hit by AIDS with the needs of other poor households that are more numerous and often poorer. In approaching this task, they should ask two questions: Which households need help most? How best can they be helped? If many households are very poor and children are malnourished and not in school, government's priorities must include such basic development policies as fostering labor-intensive economic growth, improving nutrition levels, and increasing school enrollments, especially of girls. Where targeted poverty reduction programs are already in place, modifying these programs to make assistance available to very poor families that suffer a prime-age adult death can help to improve the targeting of assistance to the households that need help most.

Challenges for the International Community

INTERNATIONAL DONORS HAVE BEEN GENEROUS IN THEIR support for AIDS prevention in developing countries, but their support has not always gone to those interventions that are most cost-effective from the perspective of government. To have the largest impact now on the pandemic, donors need to consider two main strategies.

First, in terms of bilateral and multilateral assistance, donors should support major interventions in countries at the nascent stage of the epidemic, including epidemiological surveillance, surveys of risky behavior, and programs to change behavior among those who practice the riskiest behavior. Among countries at the concentrated and generalized stages, particularly the low-income countries, ensuring prevention of infection among those who practice the riskiest behavior would be the most cost-effective strategy. Moreover, donor funds could help promote such programs when they might be politically unworkable if openly sponsored by government. With respect to mitigating the tragic impact of AIDS on society, donors must not lose sight of the myriad development problems faced by the low-income countries with generalized epidemics. The AIDS epidemic will increase poverty and will undermine household investments in human capital. Countries with generalized epidemics are therefore likely to need renewed support for core public programs to raise levels of human capital and reduce poverty. In addition, there may be some scope in specific hard-hit areas for assistance in integrating targeted poverty reduction efforts and AIDS mitigation. However, governments and donors need to be careful that such assistance does not displace household and community efforts to cope or, worse, drain time, energy and money from prevention measures focused on people who practice high-risk behavior.

The second important strategy for international donors is to finance key international public goods that poor countries cannot afford to support collectively. Two important public goods stand out: knowledge about the costs and impact of interventions on the incidence of HIV in differing environments; and development of vaccines and low-cost preventive medical technologies that will be effective under conditions prevailing in developing countries.

■ ■ ■

The poet and philosopher George Santayana said, "Those who cannot remember the past are condemned to repeat it." This maxim is nowhere truer than with the AIDS epidemic. Country after country responded to evidence of the first infections by saying "We are different. AIDS cannot strike us." Each has been proven wrong. When countries discovered that they indeed did have a fatal, sexually transmitted disease spreading rapidly in their midst, one after another responded by cleaning up the blood supply or conducting general awareness campaigns, while avoiding or devoting insufficient resources to efforts to encourage safer behavior among people most likely to contract and spread the virus.

But recent history also offers valuable examples of success. Experience demonstrates that enabling people who practice the riskiest behavior to protect themselves and others can be extraordinarily effective. National policymakers now face the challenge of applying this strategy in the cultural and political context of their own countries.

Note

1. When China and India are excluded, average health spending in low-income countries is even lower—$11 per person per year (1994 World Bank data).

Appendix A

Selected Evaluations of Interventions
To Prevent Transmission of HIV
In Developing Countries

Appendix A. Selected Evaluations of Interventions To Prevent Transmission of HIV in Developing Countries

Author (year)	Country	Intervention	Study design[a]	Sample size
Intervention: Condom promotion/safe sex				
Bhave and others (1995)	India	Condom distribution and HIV testing and counseling	QE	541 sex workers and 37 madams
Ford and others (1996)	Indonesia	HIV education, pimp training, and condom sales and distribution	QE	300 sex workers and 300 clients (3 sites: 2 case, 1 control)
Fox and others (1993)	Honduras	Condom distribution and HIV education	PC	134 female sex workers
Ngugi and others (1988)	Kenya	Targeted condom promotion	QE	366 female sex workers
Pauw and others (1996)	Nicaragua	Community-wide AIDS education	QE	Residents ages 15–45: 2,160 at baseline and 2,271 at follow-up
Intervention: STD treatment, treatment only				
Cohen and others (1997)	Malawi	Antibiotic treatment for men with urethritis	E	135 HIV-positive men; 86 with urethritis and 49 controls without urethritis
Grosskurth and others (1995a)	Tanzania	STD treatment	E	1,000 adults in each of two randomly selected communities
Wawer and others (1996b)	Uganda	Mass STD treatment	E	Over 5,700 adults ages 15–59 per arm; 58 villages in 10 clusters randomly assigned to intervention or control arm
Intervention: Combined STD treatment and condom promotion				
Ettiegne-Traore and others (1996)	Côte d'Ivoire	Two STD treatment methods: standard and intensive; all received health education, free condoms, and free treatment.	E	Group 1: 21 sex workers; Group 2: 23 sex workers (selected from a CS of sex workers and randomized to group)
Jackson and others (1997)	Kenya	STD screening and treatment and condom promotion	PC	556 HIV-negative male employees of trucking company
Laga and others (1994)	Congo DR (formerly Zaire)	STD screening and treatment; condom promotion	PC	531 initially HIV-negative female sex workers
Levine and others (1996)	Bolivia	STD treatment and condom promotion	PC	150 initially HIV-negative female sex workers

Location	Length of observation	Results[b]
Urban	24 months	Among the intervention group, an increase in "always" using condoms from 3 to 28% ($p<0.001$), as compared with no change in the control group; increase in "sometimes" using condoms from 31 to 70% ($p<0.001$) and from 36 to 53% ($p<0.01$), respectively, in the two groups.
Urban	6 months	Condom use with clients increased from 18 to 75% and 29 to 62% ($p<0.01$) in the two intervention sites and from 47 to 60% ($p<0.05$) in the control site.
Urban	6 months	Increase in mean condom use from 64 to 70% ($p<0.05$); condom use reported in diaries during program was even higher (90%).
Urban	12 months	Those who received individual and group counseling (group 1) increased occasional condom use from 10 to 80%; those with group counseling (group 2) increased from 9 to 70%; and control group (group 3) from 7 to 58%; mean condom use was 39%, 35%, and 27%, respectively (1 vs. 2, $p<0.002$; 2 vs. 3, $p<0.005$); condom use resulted in three-fold reduction of risk (OR = 0.34, $p<0.05$).
Urban	12 months	Condom use increased from 9 to 16% ($p = 0.003$) among intervention women, but only from 9 to 11% ($p = 0.5$) in control women; condom use among men increased from 31 to 41% ($p<0.001$) and from 30 to 37% ($p = 0.06$) among intervention and control groups, respectively.
Urban	2 weeks	Men with HIV and urethritis have HIV-1 RNA concentrations in seminal plasma eight-fold higher than men without urethritis (and the same CD4+ T-cell count). Two weeks after antibiotic treatment for urethritis, the level of RNA in semen decreased significantly. Gonorrhoea caused the greatest increase in viral shedding.
Rural	24 months	HIV seroconversion rates were 1.2% in intervention communities and 1.9% in control communities; risk ratio for seroconversion was 0.58, 95% CI 0.42 to 0.79, ($p = 0.0007$).
Rural	6–9 months (ongoing)	Six to 9 months after the mass treatment there was a statistically significant decline in STD symptoms and prevalence in the intervention arm and not in control arm. (Ongoing research)
Urban	6 months (ongoing)	Individuals were randomized to either traditional STD treatment (treated only when symptomatic) or intensive treatment (examined every month and treated according to an intensive therapy). To date, no significant differences in mean number of visits or in STD prevalence have been detected.
Urban	12 months	Decline in extramarital sex from 49 to 36% ($p<0.001$); decline in sex with sex workers from 12 to 6% ($p = 0.001$); significant decrease in incidence of gonorrhoea, nongonococcal urethritis, and genital ulcer disease, but no change in reported condom use.
Urban	36 months	Decline in HIV seroconversion from 11.7 per 100 woman-years to 4.4 per 100 woman-years ($p = 0.003$); increase in regular condom use with clients from 10 to 68%.
Urban	42 months	Self-reported condom use increased from 36 to 74% ($p<0.001$); STD prevalence decreased: gonorrhea from 21 to 10% ($p<0.001$); syphilis from 15 to 7.4% ($p = 0.003$); genital ulcer disease 4 to 2% ($p = 0.03$).

(Table continues on the following page.)

Appendix A. *(continued)*

Author (year)	Country	Intervention	Study design[a]	Sample size
Intervention: Voluntary counseling and testing				
Allen and others (1992b)	Rwanda	HIV counseling for discordant couples	PC	53 cohabiting discordant couples
Allen and others (1992a)	Rwanda	AIDS education, HIV testing and counseling, free condoms and spermicides	PC	460 HIV-positive and 998 HIV-negative women
Allen and others (1993)	Rwanda	HIV counseling and testing	PC	1,458 childbearing women
Deschampes and others (1996)	Haiti	Counseling, testing, and free condoms	PC	475 discordant heterosexual couples
Kamenga and others (1991)	Congo DR (formerly Zaire)	HIV counseling for discordant couples	PC	149 discordant married couples
Moore and others (1996)	Uganda	HIV counseling and testing	PC	3,000 clients at the AIDS Information Centre in Kampala
Pickering and others (1993)	The Gambia	HIV counseling and testing	PC	31 CSW (12 HIV-positive and 19 HIV-negative)
Ryder and others (1991)	Congo DR	HIV counseling and testing	PC	238 HIV-positive and 315 HIV-negative women
Temmerman and others (1994)	Kenya	HIV counseling and testing	PC	24 HIV-positive and 33 HIV-negative childbearing women
Intervention: Harm reduction				
Peak and others (1995)	Nepal	Needle and syringe exchange	CS/PC	424 injecting drug users

CI   Confidence interval
OR   Odds ratio

Note: Studies were included if they had at least a 3-month follow-up, and if they reported the statistical significance of the results, with the exception of the clinical experiment by Cohen and others (1997).

a. E=experimental; QE=quasi-experimental; PC=prospective cohort; RC=retrospective cohort; CC=case-control; CS=cross-section; RCS=repeat cross-section.

b. Statistical significance is indicated where the authors have reported it.

Source: Studies compiled by Julia Dayton and Michael Merson, Yale University School of Medicine, Department of Epidemiology and Public Health.

Location	Length of observation	Results[b]
Urban	26 months	Condom use increased from 4 to 57% after one year; condom use less common among sero-converted (100% vs. 5%, $p = 0.01$ in men; 67% vs. 25%, $p = 0.14$ in women).
Urban	24 months	Reports of "ever using condoms" increased from 7 to 22% after 1 year; HIV-positive women were more likely to use condoms (36% vs. 16%; $p<0.05$); HIV seroconversion rates decreased (13 to 6%; $p<0.04$) in women whose partners were tested and counseled.
Urban	24 months	Two-year incidence of pregnancy was 43% in HIV-positive women and 58% in HIV-negative women ($p<0.05$). HIV-positive women with fewer than 4 children were more likely to become pregnant than those with 4 or more children.
Urban	6 months	Safe sex practices or abstinence were followed by 45% of couples; HIV incidence was 1.0 per 100 person-years (95% CI: 0.80 to 1.19) among those practicing safe sex; 55% of couples continued to have unprotected sex, with HIV incidence of 6.8 per 100 person-years (95% CI: 6.53 to 7.14).
Urban	18 months	Before notification of serostatus, less than 5% of couples reported using condoms; one month after notification, 71% of couples reported condom use in all sexual intercourse; 18 months after notification, condom use rose to 77%. Intensive counseling following notification of HIV status led to low seroconversion in partner (3.1% during 100 person-years observed).
Urban	6 months	Significant increase between baseline and 3- and 6-month follow-up in risk reduction strate-gies: e.g., refraining from sexual activity, faithfulness, and condom use.
Urban	2–5 months	Overall percentage increase in condom use in first month was 1.9% (95% CI:−2.8 to 6.6) and between 1 and 2–5 months fell by 6.4% (95% CI: −14 to 1.2).
Urban	36 months	Condom use rates after 36 months differed according to serostatus: 17% for HIV-positive and 3% for HIV-negative ($p<0.01$). Adjusted fertility rates also differed: 245 live births per 1,000 women for HIV-positive and 316/1,000 for HIV-negative ($p<0.05$).
Urban	12 months	Use of condoms was infrequent (8% for HIV-positive and 6% for HIV-negative) and not sig-nificantly different according to serostatus; pregnancy rates were 16% and 18%, respectively, with no statistical difference by serostatus.
Urban	4 years	HIV seroprevalence remained low: 1.6% in 1991 and 0% in 1994; unsafe injection reduced; no change in unsafe sex.

Appendix B

Selected Studies of the Cost-Effectiveness of Preventive Interventions in Developing Countries

THE TABLE BELOW SUMMARIZES STUDIES OF THE cost-effectiveness of interventions to prevent HIV in developing countries and several other studies that have measured only the costs of program outputs, without measuring their effects.[1]

Readers are advised to use this table cautiously. The cost-effectiveness of a particular intervention is not a constant; the costs, effects, and ranking of different interventions are very likely to differ across countries because of the degree to which the intervention is targeted to those with high rates of partner change, the prevalence of HIV in high- and low-risk groups, the length of time that an intervention has been in the field, the labor-intensity of the intervention, and the local cost of labor and other inputs (*backgroud paper*, Mills and Watts 1996). Thus, the results of the different studies in the table are not directly comparable with each other. Ideally, we would like to have measures of cost-effectiveness across multiple interventions for a single country (see box 3.9 of the text).

The cost per HIV infection averted is available for only four interventions—one targeted to people with very high rates of partner change (sex workers) in Nairobi and three others addressed to those with lower rates of partner change. As discussed in chapter 3, government has a strong interest in supporting interventions that prevent the most secondary infections per dollar spent. However, except for one study, secondary infections were not included in the benefits.

- The annual operating costs of the Nairobi, Kenya, sex worker program came to roughly $70,000, or $140 per sex worker per year (Moses and others 1991). At the beginning of the program, 80 percent of the participants were infected with HIV and they had a mean of four clients per day. The annual cost per case of HIV averted came to $8, under the assumption of 80 percent condom use, or $12, under the assumption of 50 percent condom use. The number of cases of HIV prevented among the clients of sex workers and among sex workers themselves were included in the calculations, but infections prevented among the partners of clients were not included. Had they been, the intervention might have been even more cost-effective. Reportedly, the largest share of the program's costs was for STD treatment, although most of the benefits arose because of increases in condom use (Mills and others 1993). However, the availability of STD treatment may have been a major factor in obtaining the cooperation of participants.

- The Mwanza, Tanzania, STD intervention is the only one in the table for which the cost per DALY saved has been calculated—$10–11 (Richard Hayes, personal communication). The cost-effectiveness of this intervention is understated because the authors did not include prevention of any secondary infections in their analysis. The intervention might also have been more cost-effective had it been implemented in an urban area, where the number of secondary infections prevented might have been greater for each primary case. Of the total cost of $10.08 per treated case of STD, $2.11 was for drugs (Richard Hayes, personal communication). The incremental annual cost of this intervention, which served a catchment population of about 150,000 people, was $59,000, or $0.39 per capita. By comparison, the recurrent health budget of Tanzania in 1993 amounted to $2.27 per capita.

- The cost-effectiveness of safe blood programs is strongly dependent on the level of HIV prevalence in the population and on the extent of risky behavior among transfusion recipients. The Ugandan study included only averted primary infections, that is, infections due directly to transfusions (European Commission 1995a,b). It assumed a prevalence rate of 16 percent among blood donors and of 40 percent and 9 percent, respectively, among adult and child

Table B.1 Annual Costs per Infection Averted, per Condom, and per Contact for Interventions To Prevent HIV

Intervention	Location, implementing agency and year of launch	Cost per HIV infection averted	Cost per condom distributed	Cost per contact
Interventions targeted to people with high-risk behavior				
Information, condoms, STD treatment for female sex workers	Nairobi Kenya, research project (1985–91)	$8.00–$12.00 [a]		
Peer education and condoms, female sex workers	Prostitute peer education project, Yaoundé, Cameroon; Ministry of Health (1989)		$0.34 [b]	
Education and condoms, male sex workers aged 11–23	Pegação programme, Rio de Janeiro, Brazil; Social Health Guidance Unit (NOSS, 1989)		$0.70 [b]	$3.73
Peer education and condoms, female sex workers and clients, others[d]	Bulawayo peer education project, Bulawayo, Zimbabwe; Bulawayo City Health Department, University of Zimbabwe, AIDSTECH (1989)		$0.10 [b,c]	$0.47
Needle exchange, bleach, education, condoms, health care to IDUs	Kathmandu, Nepal; Lifesaving and Lifegiving Society (1992)			$3.21
Population-level interventions				
Treatment of symptomatic STDs	Research project, six rural communities of Mwanza Region, Tanzania, early 1990s	$234		$10.08
	Maputo city and province, Mozambique			$9.46
	Johannesburg, South Africa			$10.16
Condom social marketing	Ten programs (Bolivia, Congo DR,[e] Côte d'Ivoire, Dominican Republic, Ecuador, Ghana, Indonesia, Mexico, Morocco, Zimbabwe)		$0.02–$0.30 [b]	
Safe blood supply	Uganda	$172		$30 (per unit of blood)[f]
Short-course AZT therapy to prevent mother-to-child transmission	Hypothetical intervention in Sub-Saharan Africa, assuming perinatal transmission is reduced from from 25% to 16.5%.	$3,748		

a. Lower figure assumes 80% condom use, higher figure assumes 50% condom use. b. Includes the value of donated condoms. c. Cost excluding technical assistance from AIDSTECH is $0.07 per condom distributed. d. Also included people in bars, workers, STD patients. e. Formerly Zaire. f. $27–29 per unit collected, $33–35 per unit used in 1993. Mills and others (1993) find a cost of $51 per unit produced for the same program but possibly for an earlier year.

Sources: Nairobi study, Moses and others (1991); Cameroon, Brazil, Zimbabwe sex worker peer counseling and condom programs, condom social marketing programs, and STD treatment in Mozambique and South Africa, Mills and others (1993); Mwanza STD treatment, Richard Hayes, personal communication, and Gilson and others (1996); Ugandan safe blood, European Commission (1995a,b); AZT therapy in Sub-Saharan Africa, Mansergh and others (1996).

transfusion recipients. The calculations concerning the number of primary infections averted are in box 4.2 of this report. The cost per HIV infection averted was obtained by dividing the total additional costs of HIV screening in 1993 ($319,894) by the total number of infections averted (1,863).

■ The effectiveness of short-course zidovudine (AZT) therapy to prevent mother-to-child transmission is not known as of this writing; clinical efficacy trials are under way in a number of countries. The cost-effectiveness numbers in the table are, therefore, hypothetical. The calculations assume that the therapy would reduce transmission from 25 percent to 16.5 percent, or half the effect of longer-course therapy. Program costs were estimated from the literature and are based on those in Sub-Saharan Africa, where most mother-to-child transmission occurs (Marsergh and others 1996). The authors calculated that a national program in a country with a 12.5 percent HIV seroprevalence rate would lower incidence of HIV by 12 percent. Since infants and young children are very unlikely to transmit HIV to others, there are virtually no secondary cases generated by this intervention. Preventing infection of children is one of the important external benefits of preventing infection in their mothers (see box 4.6 of the report).

The cost per case of HIV averted or per DALY has not been calculated for the other studies in the table; only the costs are available. A needle exchange and bleach program serving injecting drug users in Katmandu, Nepal, cost $3.21 per contact after only one year of observation and was organized with community-based outreach. A second program in Lubljana, Slovenia (not shown), was based in a fixed facility and had been operating only 5 months when costed at $12.59 per contact (Mills and others 1993). The cost per condom distributed varied from $0.10 to $0.70 for three highly targeted programs that had peer education and condoms for sex workers. Costs were much lower for ten condom social marketing programs—from $0.02 to $0.30 per condom distributed, including the value of donated condoms.

Note

1. For examples of studies of cost-effectiveness and cost-benefit analysis of HIV/AIDS interventions in industrial countries, see Holtgrave, Qualls, and Graham (1996) and National Research Council (1991). Key principles of the economic analysis of health projects are reviewed in Hammer (1997).

Statistical Appendix

This statistical appendix assembles information about the levels and determinants of the HIV/AIDS epidemic and selected policy variables for low- and middle-income countries as classified in the *1997 World Development Report* (World Bank 1997a).

Table 1: HIV Infection Rates by Subpopulation

Table 1 assembles the most recent information on HIV prevalence by subpopulation for all low- and middle-income countries as defined in the *1997 World Development Indicators* (World Bank 1997b). Country-level adult HIV-1 seroprevalence estimates are for adults age 15–49, as estimated by the World Health Organization for December 1994. The remaining data in table 1 on HIV seroprevalence by subpopulation, for all regions except Eastern Europe and Central Asia, are from the HIV/AIDS Surveillance Database of the U.S. Bureau of the Census (U.S. Bureau of the Census [database], 1997).[1] Except where otherwise noted, data for Eastern Europe and Central Asia are from the WHO/EC Collaborating Centre on AIDS (1996).

---

[1] The complete HIV/AIDS Surveillance Database can be obtained from: International Programs Center, Population Division, U.S. Bureau of the Census, Washington, D.C. 20233-8860 USA. The e-mail address is ipc-hiv@census.gov.

The second through fourth columns of table 1 show the U.S. Bureau of the Census's "best estimates" of HIV-1 prevalence in "high-risk" and "low-risk" subpopulations in the capital or major cities and outside major cities. Their summary estimates are based on the detailed data they have compiled from published and unpublished sources in developing countries in the January 1997 version of their HIV/AIDS Surveillance Database, and usually represent the results of a specific study that is judged to be of good quality. As a rule, "high risk" is defined as sex workers and clients, STD patients, or other people with known risk factors. "Low risk" is defined as pregnant women, blood donors, or other people with no known risk factors. The figures for urban high- and low-risk groups in the capital/major city were used in the cross-country analysis of the societal determinants of HIV infection in chapter 1 of this report.

The data in columns 6 through 17 on HIV prevalence among specific high- and low-risk groups of men and women are derived from analysis of the approximately 24,000 individual data records, drawn from 3,100 publications and presentations, contained within the Census Bureau's HIV/AIDS Surveillance Database. The figures in these columns represent either the result of a specific study, if there was only one suitable study for a given year, or an average of prevalence rates from several suitable sources or sites within the same country for the most recent year available. This table uses only seroprevalence surveys measuring HIV-1, except where otherwise noted as including HIV-1 and/or HIV-2. The data in these columns were used to define the stage of the epidemic for chapter 2 of this report, as summarized in table 2 of the statistical appendix, below.

Whenever possible, the entries for columns 6 to 17 represent the results of one or more studies with a sample size of at least 100 people. If more than one study satisfied these criteria for the same year and covered comparable geographic areas, an unweighted average was taken. Surveys with exceptionally small sample sizes (<100) were not taken into account, except as a last resort where they were the only estimate available. Estimates based on small samples have been noted and should be used with caution because of their potential unreliability.

Because of the difficulty of establishing samples of individuals with certain characteristics, studies based on samples of "high-risk" subpopulations may be unrepresentative. The same caveat applies to samples of pregnant women, when such data are collected from a non-random sample of clinics. Further, self-selection of some of the individuals in these groups—such as women attending antenatal clinics and STD patients—may be serious and the HIV prevalence rates should not be taken to be representative of individuals who do not use these services. Similarly, while military populations are characterized by tighter-than-usual health surveillance in general, these populations are selected by age and other characteristics. For these reasons, results from the various high- and low-risk groups should not be considered to be indicative of prevalence in the general, low-risk population.

Table 2: Indicators of Socioeconomic Development and Government AIDS Policies by Stage of the Epidemic

Table 2 presents indicators of socioeconomic development and government AIDS policies for 123 low- and middle-income countries with at least 1 million population. The countries in this table have been ordered alphabetically according to the "stage" of the HIV/AIDS epidemic—the extent to which it has spread among those practicing high-risk behavior and outward to low-risk populations—as used in chapter 2. In countries with a *nascent* epidemic, HIV prevalence in high-risk subpopulations is less than 5 percent. A *concentrated* epidemic is defined as one in which HIV prevalence in high-risk subpopulations is 5 percent or higher, but is still less than 5 percent among women attending antenatal clinics. A *generalized* epidemic is one in which HIV prevalence among women attending antenatal clinics is 5 percent or higher. Classification of individual countries is based on the data in statistical appendix table 1, columns 6–17, generally using data from 1990 onward. If the only available data on high-risk subpopulations were from before 1990 and indicated low prevalence, then the country was classified as "unknown." Countries were also classified as "unknown" if there were no data on high-risk subpopulations and data on antenatal women were either missing or below 5 percent. Note that there are many countries with missing data, particularly on high-risk subpopulations, and many with outdated data. Therefore, these are conservative estimates of the spread of HIV; many countries may in fact be further along than indicated by this typology. China and India have been classified as having concentrated epidemics, based on a concentrated epidemic in at least one province or state, respectively.[2]

The first ten columns of table 2 are socioeconomic factors likely to affect the spread of HIV.[3] The mid-1995 population, the 1995 gross national product (GNP) per capita, and the growth rate in GNP per capita are from the *World Development Report 1997* (World Bank 1997a). The Gini index is a measure of income inequality; an index of zero indicates perfect equality, while an index of 100 indicates perfect inequality. Data are for various years in the 1980s, from Deininger and Squire (1996), and are based on either income or consumption data from household surveys. The urban population as a percentage of the total and the growth rate in urban population are from the *World Development Indicators 1997* (World Bank 1997b), table 3.6. The 1990 urban male/female ratio is the ratio of adult men ages 20 to 39 to women of the same age in urban areas, calculated from United Nations (1993). Foreign-born as a percentage of the total population in 1990 is from United Nations (1995). The 1995 adult illiteracy rates are for adults ages 15 and older and come from the *World Development Indicators 1997* (World Bank 1997b), table 1.1.

The last six columns are indicators of HIV/AIDS policies on information and prevention. Year of the first reported AIDS case is as of December 31, 1995, from WHO/GPA data (UNAIDS/Country Support 1996). Note that in virtually all countries AIDS cases occurred before they were first reported and that the first case of HIV would have occurred years before the first AIDS case. The four categories of sentinel surveillance implementation as of 1995 are: planned sites, limited sites, many sites, and extensive sites. This information comes from Sato (1996), in Mann and Tarantola (1996); unfortunately, the definition of these categories was not provided in the original source. Government condom distribution in 1992 is from the *AIDS in the World II* survey, appendix table D-7.3, column PN5, of Mann and Tarantola (1996). The presence of a condom social marketing program in 1996 is from

---

[2] Countries with population of less that 1 million that do not appear in the table but could be classified by stage of the epidemic include: Bahrain, Cape Verde, and St. Lucia (nascent stage); and Djibouti and Swaziland (concentrated stage).

[3] Many of these variables were used directly or in some transformation in the national-level regressions of urban HIV infection in chapter 1. Note, however, that for the regressions in chapter 1, data for the same variables for earlier years were often used. Other variables used in those regressions are not reported in this table, such as the purchasing-power-parity–adjusted GNP per capita. The complete data set used for the chapter 1 regressions is available on request from the authors.

personal communications with Philip Harvey (DKT) and Guy Stallworthy (PSI). The number of socially marketed condoms sold per adult in 1995 is derived from the data on total condom sales for 1995 from table 3 of the statistical appendix (see below) and the number of adults 15 to 49 from the *World Development Indicators 1997* (World Bank 1997b) on compact disk.

Table 3: Socially Marketed Condom Sales in Developing Countries, 1991–96

The 1991–95 data are from DKT International (1992–96) and 1996 data are from personal communications with Philip Harvey (DKT) and Guy Stallworthy (PSI). Several countries have now or had in the past more than one social marketing program:

- India I is the government program offering *Nirodh* brand condoms; India II is implemented by PSI/India (*Masti* and *Pearl* brands); India III is operated by Parivar Seva Sanstha (*Sawan, Bliss,* and *Ecroz* brands); India IV is implemented by DKT (*Zaroor* and *Choice*). The total for India I includes *Nirodh* condoms sold by PSI/India.
- Indonesia I is the project that sells *Blue Circle* and *Gold Circle* condoms, and is implemented by the National Family Planning Coordinating Board (BKKBN); it has devolved to the private sector; Indonesia II is implemented by SOMARC.
- Nigeria I is currently supervised by PSI; Nigeria II was started by Family Planning International Assistance and Sterling Products. They were consolidated into a single program in 1993.
- Philippines I is operated by DKT and Philippines II by SOMARC.

Statistical Appendix Table 1 HIV Infection Rates by Subpopulation

| | | Summary: HIV-1 Prevalence | | | | Women | | | |
| | | Capital/major city | | Outside major city | | | | | |
Country	Adult HIV prevalence, 12/94	Low risk	High risk	Low risk	High risk	Antenatal clinic, urban	Year	Sex workers	Year
Africa									
Angola	1.0	1.0		0.5		1.0[a]	95		
Benin	1.2	1.4	50.8[b]	4.9[b]		0.4[c]	93	53.3[a]	93–94
Botswana	18.0	32.4	41.6	16.0		34.2[c]	95		
Burkina Faso	6.7	12.0	60.4			12.0	95	60.4[a]	94
Burundi	2.7	20.0		1.8		17.2	93		
Cameroon	3.0	5.7[c]	45.3	2.9	9.0	1.9	96	21.2[a]	94
Cape Verde								2.0[a,d]	87
Central African Republic	5.8	16.0	31.0	6.5		10.0[c]	93	17.0	89
Chad	2.7	4.1				4.5[c]	92		
Comoros	0.1								
Congo, DR	3.7	5.0[c]	30.3	2.9	25.4	4.6	93	30.3	95
Congo, Rep.	7.2	7.1	17.6[c]	2.6		7.1	94	49.2[c,d]	87
Côte d'Ivoire	6.8	12.5[b]	77.0[b]			11.6[a,c]	95–96	67.6[a]	94–95
Djibouti	3.0		43.0					26.9[c]	91
Equatorial Guinea	1.1								
Eritrea	3.2	1.6						5.8	89
Ethiopia	2.5	6.7	54.2	8.6	65.6	4.9[c]	91	67.5[c]	91
Gabon	2.3	1.7				1.7	94		
Gambia, The	2.1	0.6[b]	13.6[b]			1.7[a]	93–95	34.7[a]	93
Ghana	2.3	3.2[b]	5.2[b]	1.0[b]		2.2[a,c]	95	30.8[c]	86–87
Guinea	0.6	0.7[b]	36.6	0.3		0.7[a]	90–91	36.6	94
Guinea-Bissau	3.1	2.6[b]		0.5[b]		6.9[a]	95	36.7[a,d]	87
Kenya	8.3	18.1	85.5	10.3		13.7[c]	95	85.5	92
Lesotho	3.1	6.1	11.1	4.2	21.3	6.1	93		
Liberia	1.3							0.0[d]	87
Madagascar	0.1	0.1	0.3	0.0	0.4	0.1	95	0.2	95
Malawi	13.6	32.8	70.4	11.8		32.8	96	78.0[c]	94
Mali	1.3	4.4	55.5	3.4[c]	52.8[c]	3.5[c]	94	55.5[a]	95
Mauritania	0.7	0.5[b]	0.9			0.5[a]	93–94		
Mauritius	0.1		0.8			0.0	86		
Mozambique	5.8	2.7	7.6[d]	1.5[b]		10.5[c]	94		
Namibia	6.5	17.6		10.3		17.6	96		
Niger	1.0	1.3[b]	12.6[b]	1.4		1.3[a]	93	15.4[a]	93
Nigeria	2.2	6.7[c]	29.1[c]			3.8[c]	93–94	22.5[c]	93–94
Rwanda	7.2	25.4	73.2			25.3[c]	95	87.9[d]	84
São Tomé & Principe									
Senegal	1.4	1.7[b]	10.1[b]	0.6[b]	9.2[b]	1.1[a,c]	94	22.1[c]	94
Seychelles									
Sierra Leone	3.0		26.7			0.8[a,e]	90	26.7	95
Somalia	0.3					0.0[d]	85–86	2.4	90

Men				Both			
Military	Year	Homo-sexual/bisexual	Year	Injecting drug users	Year	STD patients	Year
0.8[a]	95					24.7[a]	88
						3.7[a]	93
						42.8[c]	95
						41.8	92
						18.5[d]	86
6.2[a]	93					5.4[a]	94
2.0[a,d]	87						
15.0[a]	88					34.0	94–95
0.0	88					0.1	87–88
						9.1[d]	86
						12.0[a]	92
						20.1[a]	92
		10.0[d]	90			10.4	91
2.6	91					43.6	90
						4.2a	88
						6.8[a,c]	91
						5.6a	94
0.0[a]	86					4.5	95
11.8[a]	86–87					6.0	87
						16.0	94
						15.2[c]	93
						0.0[d]	87
0.0	89	0.0[d]	85–86			0.3	95
						54.8[c]	95
						0.9	93–94
				0.0	88–91	0.8	88–91
3.7	90					24.0[c]	94
						7.2[e]	92
						8.2[c]	93–94
						61.1	88–91
0.8[a]	93					1.8	89–91
0.4[a]	91					3.3	92
0.0[d]	90					0.0	90

(Table continues on the following page.)

Statistical Appendix Table 1   *(continued)*

Country	Adult HIV prevalence, 12/94	Summary: HIV-1 Prevalence Capital/major city Low risk	High risk	Outside major city Low risk	High risk	Women Antenatal clinic, urban	Year	Sex workers	Year
South Africa	3.2	18.2[d]	20.1[c]	6.4		10.4[f]	95	3.2	89
Sudan	1.0		3.0[d]			3.0[e]	95	7.6	89
Swaziland	3.8	21.9[d]	11.1			3.9	92		
Tanzania	6.4	13.7	49.5	15.0[d]	34.3[d]	13.9[c]	95–96	49.5	93
Togo	8.5		3.1	3.0	7.3[d]				
Uganda	14.5	18.5[c,d]	38.5[d]	6.5[d]		21.2	94–95	86.0	87
Zambia	17.1	27.9	58.0	12.7	36.0[d]	27.9	94		
Zimbabwe	17.4	32.0	86.0	16.0[d]	46.0	35.2[c]	95	86.0[a]	94–95
Latin America/Caribbean									
Argentina	0.4	2.8[d]	9.1[d]			2.8[e]	95	4.2[c]	91
Barbados	2.8	1.2[c]				0.0[h]	93		
Belize	2.0								
Bolivia	0.1					0.0[h]		0.0	88
Brazil	0.7	2.7	13.3	0.3	3.5	1.7[c,h]	95	11.2[i]	92–93
Chile	0.1	0.1	1.3			0.1	94		
Colombia	0.2	1.1	3.4			0.5[c]	94	1.1[a]	94
Costa Rica	0.5	0.0	1.6			0.0	92	0.7[c]	90
Cuba	0.02	0.0	0.0			0.0[h]	96		
Dominican Republic	1.0	2.0	6.2[c]			2.8[c]	95	7.0[c]	95
Ecuador	0.3	0.3	3.6			0.3	92	0.0	93
El Salvador	0.6	0.5	6.0			0.0[h]	94–95	2.0	95
Grenada		0.0	2.4[d]			0.0[h]	91		
Guatemala	0.4	0.0	8.5			0.0	90–91	0.2[c]	89
Guyana	1.3	6.9	25.0			6.9	92	25.0	93
Haiti	4.4	15.7	70.0[d]	4.0		8.4	93	41.9	89
Honduras	1.6[d]	4.1[d]	20.5[d]			1.0[e]	96	20.5[e]	95
Jamaica	0.9	0.7	24.6			0.7	96	24.6	94–95
Mexico	0.4	0.6	5.0[d]			0.0	96	0.1	95–96
Nicaragua	0.1	0.0	1.6[d]					1.6[d]	90–91
Panama	0.6	0.8				0.3[h]	94	0.0	86
Paraguay	0.1	0.0				0.0	92	0.1	87–90
Peru	0.3							0.7	89–90
St. Kitts & Nevis		2.0[c]				1.8[c,d]	92		
St. Lucia		0.0[c]	1.2[c]			2.0[c,h]	92		
St. Vincent & the Grenadines		0.2	1.4[c]			0.2	92		
Suriname	1.2					0.8[h]	91	2.7[d]	92
Trinidad & Tobago	0.9	0.3	14.7			0.3[h]	90	13.0	88
Uruguay	0.3	0.1[c]	10.7[c]			0.0	91	0.3[a,c]	96
Venezuela	0.3	0.1	6.1[d]					1.3[c]	94

Men				Both			
Military	Year	Homo-sexual/bisexual	Year	Injecting drug users	Year	STD patients	Year
		7.7c	86	0.6	86	20.1	94
0.8	89	7.7d	89			19.1c	89
						11.1	92
12.9g	94	0.0d	86			24.0	93–94
3.1	93					7.3a,e	93
27.1	95–96					36.1c	95
						58.0	92–93
						61.0c	95
0.6	92	14.9e	92	41.4	96	9.1e	95
						4.7	88
0.0	88	5.1c,d	88	0.0d	88		
		8.9	96	40.4c	94–95	6.1c	95
						0.7c	94
		26.2c	94			8.8c	85–87
		4.9	94	0.0d	90	1.0c	92
						0.0	93
		7.7	94	0.0d	84–85	6.7	95
0.0	88	28.8c,d	88			3.6	93
0.4	88–93					6.0	95–96
						1.2c,d	91
0.4	94					5.3c	90–93
						26.5c	92
						21.4	92
		30.0e	92			11.2	91
		15.0	86			6.2	93
0.5	86	32.7	94	1.3d	94		
0.0	90						
		3.1	84–86				
		8.8	87–90				
0.5	89–90	41.0	89–90	28.1d	89–90	18.7	89–90
						1.2c	92
						1.6c	92
						1.1	90
		40.0	83–84	4.7	88–89	12.0c	92
		3.3a,c	95	13.0a	96	0.6c	91
		25.0e	94				

(Table continues on the following page.)

Statistical Appendix Table 1   *(continued)*

Country	Adult HIV prevalence, 12/94	Summary: HIV-1 Prevalence				Women			
		Capital/major city		Outside major city		Antenatal clinic, urban	Year	Sex workers	Year
		Low risk	High risk	Low risk	High risk				
Asia									
Bangladesh	0.03	0.0	1.2[d]					0.6[d]	96
Bhutan	0.01	0.0							
Cambodia	1.9	3.2	41.6	0.9	27.3	3.2	96	43.0[c]	96
China	0.0	0.0	73.2[d]			0.0[j]	93	0.3[j]	93
Fiji	0.04	0.0	0.1						
India	0.4	2.5[d]	28.6[b]			0.3[k]	95	51.0[e]	94
Indonesia	0.05		0.3			0.0	86–87	0.3	94
Kiribati		0.1							
Korea, Dem. Rep.	0.0	0.0							
Korea, Rep.	0.01	0.8	0.0					0.1	88
Lao PDR	0.03	0.8	1.2					1.2	90–93
Malaysia	0.3	0.0	29.5					1.4	91
Maldives	0.6	0.0							
Mongolia	0.01	0.0	0.0			0.0	87–93	0.0	87–93
Myanmar	1.5	2.2[d]	18.0			1.3	95	18.2[e]	95
Nepal	0.05	0.0[d]	1.3[d]			0.0[e,h]	92	0.9	93
Pakistan	0.06	0.6	3.7			0.2[c,e]	95		
Papua New Guinea	0.2	0.0	0.3	0.0	0.0[c]	0.0	92		
Philippines	0.05		0.5					0.6[c]	93
Solomon Islands		0.0	0.0						
Sri Lanka	0.05	0.0	0.5[d]					0.5[e]	93
Thailand	2.1	2.8	21.6			2.4[c,h]	95	18.8[c,n]	95
Vietnam	0.07	0.2	43.6			0.0[c]	95	0.24[c]	95
Middle East/North Africa									
Afghanistan	0.001								
Algeria			0.0[d]					0.0	91
Bahrain	0.2								
Egypt, Arab Rep.	0.03	0.0	5.3	0.0		0.0	92	0.0	90–91
Iran, Islamic Rep.	0.003								
Iraq	0.003								
Jordan	0.02								
Lebanon	0.09								
Libya	0.06								
Morocco	0.04	0.2	1.4			0.2	93	7.1[d]	90
Oman	0.1								
Saudi Arabia	0.01								
Syrian Arab Rep.	0.01								
Tunisia	0.04	0.0				0.0	91	0.0	87
Yemen, Rep.	0.01								

Men				Both			
Military	Year	Homo-sexual/bisexual	Year	Injecting drug users	Year	STD patients	Year
6.5[d]	96			12.0	95	8.5[h]	94
		0.0[d]	90	66.5[j]	94	0.0[j]	93
						0.1	91–93
0.7[g,l]	86–91	1.1[m]	91–92	67.2[l]	92	28.3[c]	95
0.3	92–93	0.0[d]	88–89	0.0	91–92	0.0	94
				29.5	92	1.9	94
		0.0	88–90			0.0	87–93
0.7	95	0.0	85–90	56.5	95	7.6[e]	95
				0.0	94	0.7[c,e]	92
				11.5	95	2.0[c]	95
						0.1[c]	92
0.0[c]	88–89	0.0	92	0.0	92	0.0	92
						0.0	91–93
						0.1	86–91
3.0[o]	95	14.0[c]	92	34.4[c]	95	9.7[c]	95
0.0	95			7.5[c]	95	0.3[c]	95
						0.0	81–89
				2.9	90	0.0	90
				7.6	94	0.8	89–90
0.1	93					0.7	92–95
						0.0[d]	85–87
						0.0[d]	92

(Table continues on the following page.)

Statistical Appendix Table 1 *(continued)*

Country	Adult HIV prevalence, 12/94	Summary: HIV-1 Prevalence				Women			
		Capital/major city		Outside major city		Antenatal clinic, urban	Year	Sex workers	Year
		Low risk	High risk	Low risk	High risk				
Eastern Europe/Central Asia[p]									
Albania	0.01								
Armenia	0.0								
Azerbaijan	0.0								
Belarus	0.0								
Bulgaria	0.01					0.0	93		
Croatia	0.01								
Czech Republic	0.04					0.0	93		
Estonia	0.01								
Georgia	0.02								
Hungary	0.06								
Kazakhstan	0.0								
Kyrgyz Republic	0.0								
Latvia	0.01								
Lithuania	0.01					0.0	92–93		
Macedonia, FYR	0.04								
Moldova						0.0	93		
Poland	0.05								
Romania[r]	0.0								
Russian Federation	0.0					0.0	95		
Slovak Republic	0.01					0.0	92		
Slovenia	0.02								
Tajikistan	0.0								
Turkey	0.0	0.0				0.0	87–88	0.0	87
Turkmenistan	0.0								
Ukraine	0.01					0.0	93		
Uzbekistan	0.0								
Yugoslavia, FR (Serb./Mont.)	0.09								

Blank cells = Data not available.

a. HIV-1 and/or HIV-2.
b. Rate represents infection with HIV-1 only and dual infection (HIV-1 and HIV-2).
c. Data averaged.
d. Data are best available but are not necessarily reliable due to small sample size (<100).
e. Sample size unknown.
f. National data.
g. Police.
h. Not specifically urban.
i. Urban male prostitutes in São Paulo had a 48.6% prevalence rate in 1993.
j. For Yunnan Province.
k. Tamil Nadu State.

	Men				Both		
Military	Year	Homo-sexual/bisexual	Year	Injecting drug users	Year	STD patients	Year
						0.0	93
						0.0	93
				0.0	93	0.0	93
						0.0	93
						0.0	92
				4.7q	95		
0.0	95	0.05	95	0.0	95	0.0	95
				0.0	92	0.0	93
						0.0	93
0.0	85–90						
				13.0t	95	0.0	93

l. For Manipur State.
m. For the city of Madras.
n. Includes direct (brothel-based), indirect (non-brothel-based), and male prostitutes.
o. Division of Epidemiology/Public Health, Thailand, and Army Institute of Pathology, Royal Thai Army.
p. WHO-EC Collaborating Centre on AIDS 1996.
q. UNAIDS data.
r. In Romania, two 1990 studies of orphaned children found HIV prevalence rates of 7.8 and 21.6% (Hersh and others 1991).
s. Pokrovsky and others 1996.
t. UNAIDS 1996a.

Statistical Appendix Table 2   Classification of Countries by Stage of the Epidemic, with Selected Economic and Policy Variables Affecting the Spread of HIV

Country	Population, mid-1995 (millions)	GNP per capita, 1995 (dollars)	Economic growth Avg. annual GNP per capita growth (%), 1985–95	Gini index (1980s)	Urbanization Urban pop. 1995 (%)	Avg. annual urban growth rate (%), 1990–95	Urban male/female ratio, 1990	Foreign-born, 1990 (% of total pop.)
Nascent Epidemic								
Algeria	28.0	1,600	−2.4	0.39	56	4.0	1.03	1.5
Azerbaijan	7.5	480	−16.3[b]	—	56	1.6	—	—
Bangladesh	8.0	240	2.1	0.35	18	5.0	1.55	0.7
Bulgaria	8.4	1,330	−2.6	0.23	71	−0.1	1.00	0.2
Chile	14.2	4,160	6.1	0.52	86	1.9	0.96	0.8
Costa Rica	3.4	2,610	2.8	0.46	50	3.3	0.91	5.9
Cuba	11.0	c	—	—	76	1.5	0.97	0.6
Czech Republic	10.3	3,870	−1.8	0.27	65	0.1	1.01	—
Ecuador	11.5	1,390	0.8	0.43	58	3.6	0.97	0.8
Estonia	1.5	2,860	−4.3[b]	—	73	−0.9	0.97	—
Indonesia	193.3	980	6.0	0.33	34	3.9	1.04	0.1
Lao PDR	4.9	350	2.7	0.30	22	6.5	0.96	0.4
Lithuania	3.7	1,900	−11.7[b]	—	72	1.0	0.96	—
Madagascar	13.7	230	−2.2	0.43	27	5.7	0.97	0.3
Mauritania	2.3	460	0.5	0.43	54	5.5	1.34	3.3
Mauritius	1.1	3,380	5.4	0.41	41	1.3	1.01	0.8
Mongolia	2.5	310	−3.8	—	60	2.9	—	0.5
Morocco	26.0	1,110	0.9	0.39	49	3.0	1.06	0.2
Nepal	21.5	200	2.4	0.30	14	7.5	1.20	2.1
Nicaragua	4.4	380	−5.4	0.50	62	4.1	0.71	2.1
Papua New Guinea	4.3	1,160	2.3	—	16	3.7	—	0.7
Philippines	68.6	1,050	1.5	0.48	53	4.4	0.98	0.1
Poland	38.6	2,790	1.2	0.26	65	1.0	0.96	3.6
Russian Federation	148.2	2,240	−5.1[b]	—	73	−0.2	—	—
Slovak Republic	5.6	2,950	−2.8	—	59	1.1	—	—
Slovenia	2.0	8,200	—	—	64	1.0	—	—
Somalia	9.5	c	—	—	—	—	—	7.2
Sri Lanka	18.0	700	2.6	0.42	22	2.1	1.11	0.1
Suriname	0.4	880	—	—	—	—	—	2.1
Yemen	15.3	260	—	—	34	9.4	1.29	0.6
Concentrated Epidemic								
Angola	10.8	410	−6.1	—	32	5.9	—	0.3
Argentina	34.7	8,030	1.8	—	88	1.7	0.99	5.1
Brazil	159.2	3,640	−0.8	—	78	2.5	0.97	0.7

Women's status		Prevention policy				
Male adult illit., 1995 (%)	Female adult illit., 1995 (%)	Year of 1st reported AIDS case	Sent. Surv. Implem. status, 1995[a]	Gov't condom distri- bution, 1992	CSM for HIV preven- tion, 1996	SM condoms sold per adult, 1995
26	51	86	1	Y		..
—	—	94	2	N		..
51	74	90	1	—	Y	2.91
—	—	87	1	—		..
5	5	84	3	Y		..
5	5	83	2	Y	Y	3.28
4	5	86	2	—		..
—	—	87	2	N		..
8	12	86	2	N	Y	0.12
—	—	92	2	Y		..
10	22	87	2	—	Y	..
31	56	87	1	—		..
—	—	88	2	Y		..
—	—	89	2	Y		..
50	74	88	2	N		..
13	21	87	—	Y		..
—	—	—	1	—		..
43	69	86	2	Y	Y	0.18
59	86	88	2	Y	Y	0.81
35	33	88	2	—		..
19	37	87	1	Y		..
5	6	84	4	—	Y	0.30
—	—	86	2	—		..
—	—	86	2	—	Y	..
—	—	87	2	—		..
—	—	87	2	N		..
—	—	87	—	—		..
7	13	87	2	Y	Y	0.80
5	9	84	—	Y		..
—	—	90	—	—		..
—	—	85	1	Y		..
4	4	82	2	—		..
17	17	80	4	Y	Y	0.21

(*Table continues on the following page.*)

319

Statistical Appendix Table 2 *(continued)*

| Country | Population, mid-1995 (millions) | Economic growth | | | | Urbanization | | | |
		GNP per capita, 1995 (dollars)	Avg. annual GNP per capita growth (%), 1985–95	Gini index (1980s)	Urban pop. 1995 (%)	Avg. annual urban growth rate (%), 1990–95	Urban male/ female ratio, 1990	Foreign-born, 1990 (% of total pop.)
Cambodia	10.0	270	—	—	21	6.5	0.95	0.3
Cameroon	13.3	650	−6.6	0.49	45	5.3	1.05	2.4
Chad	6.0	180	0.6	—	21	3.5	1.21	0.3
China	1,200.2	620	8.3	0.33	30	3.8	1.11	0.0
Colombia	36.8	1,910	2.6	0.52	73	2.6	0.91	0.3
Congo, DR	43.8	120	—	—	29	4.0	1.05	2.8
Dominican Republic	7.8	1,460	2.1	0.47	65	3.4	0.97	2.5
Egypt	57.8	790	1.1	0.38	45	2.5	1.07	0.3
El Salvador	5.6	1,610	2.8	0.48	45	2.7	0.81	1.0
Eritrea	3.6	c	—	—	17	—	—	—
Ethiopia	56.4	100	−0.3	—	13	3.4	0.92	1.6
Gabon	1.1	3,490	−8.2	0.61	50	5.5	1.33	8.9
Gambia	1.1	320	—	—	26	6.8	1.13	11.2
Ghana	17.1	390	1.4	0.35	36	4.3	1.01	0.9
Guatemala	11.0	1,340	0.3	0.56	42	4.0	0.94	0.5
Guinea	7.0	550	1.4	0.47	30	5.8	1.25	1.7
Honduras	5.9	600	0.1	0.54	48	4.9	0.90	0.7
India	929.4	340	3.2	0.33	27	2.9	1.24	1.0
Jamaica	2.5	1,510	3.6	0.43	55	2.1	0.89	0.8
Malaysia	20.1	3,890	5.7	0.56	54	4.0	1.01	4.2
Mali	9.8	250	0.8	—	27	5.7	1.10	1.2
Mexico	91.8	3,320	0.1	0.54	75	2.7	0.97	0.4
Myanmar	45.1	c	—	—	27	3.3	0.99	0.2
Niger	9.0	220	—	0.36	23	6.9	—	1.5
Nigeria	111.3	260	1.2	0.39	39	5.3	1.19	0.3
Pakistan	129.9	460	1.2	0.32	35	4.7	1.23	6.1
Paraguay	4.8	1,690	1.2	—	54	4.4	0.96	4.3
Peru	23.8	2,310	−1.6	0.48	72	2.7	1.02	0.3
Senegal	8.5	600	—	0.54	42	4.0	1.05	2.5
Sierra Leone	4.2	180	−3.6	0.61	39	4.9	—	5.0
Sudan	26.7	c	—	0.39	26	4.6	1.35	3.3
Thailand	58.2	2,740	8.4	0.45	36	2.3	0.92	0.6
Togo	4.1	310	−2.7	0.34	31	4.8	1.23	4.1
Trinidad & Tobago	1.3	3,770	−1.7	0.46	68	1.8	0.95	5.0
Ukraine	51.6	1,630	−9.2[b]	—	70	1.0	—	—

Women's status		Prevention policy				
					CSM	
Male adult illit., 1995 (%)	Female adult illit., 1995 (%)	Year of 1st reported AIDS case	Sent. Surv. Implem. status, 1995[a]	Gov't condom distribution, 1992	CSM for HIV prevention, 1996	SM condoms sold per adult, 1995
20	47	93	3	—	Y	1.06
25	48	86	3	Y	Y	1.27
38	65	86	2	Y	Y	..
10	27	85	3	N	Y	..
9	9	86	2	Y	Y	0.28
13	32	86	2	Y	Y	0.43
18	18	83	2	Y	Y	0.18
36	61	86	2	—	Y	0.36
27	30	85	2	Y	Y	0.57
—	—	88	3	—		..
55	75	86	3	Y	Y	0.83
26	47	89	2	Y		..
47	75	89	2	—		..
24	47	86	3	N	Y	0.44
38	51	86	2	Y	Y	0.60
50	78	89	3	Y	Y	0.94
27	27	85	3	—	Y	0.32
35	62	86	3	N	Y	0.38
19	11	82	—	Y	Y	1.62
11	22	87	3	N	Y	0.71
61	77	85	3	Y	Y	0.71
8	13	87	3	Y		..
11	22	91	3	—	Y	..
79	93	87	2	—	Y	0.29
33	53	86	3	—	Y	1.09
50	76	87	1	—	Y	0.96
7	9	86	2	—		..
6	17	86	2	N	Y	0.35
57	77	86	3	Y	Y	..
55	82	87	2	—		..
42	65	86	1	—		..
4	8	84	4	Y		..
33	63	87	2	Y	Y	2.40
1	3	83	—	Y		..
—	—	88	2	N		..

(Table continues on the following page.)

Statistical Appendix Table 2 *(continued)*

| Country | Population, mid-1995 (millions) | Economic growth | | | Urbanization | | | |
		GNP per capita, 1995 (dollars)	Avg. annual GNP per capita growth (%), 1985–95	Gini index (1980s)	Urban pop. 1995 (%)	Avg. annual urban growth rate (%), 1990–95	Urban male/ female ratio, 1990	Foreign-born, 1990 (% of total pop.)
Uruguay	3.2	5,170	0.5	0.39	90	0.9	31.30	3.0
Venezuela	21.7	3,020	—	0.44	93	2.9	1.01	5.3
Vietnam	73.5	240	—	0.36	21	3.1	0.90	0.0
Generalized Epidemic								
Benin	5.5	370	−0.3	0.57	42	4.9	1.09	1.0
Botswana	1.5	3,020	6.1	0.55	31	7.4	1.24	1.8
Burkina Faso	10.4	230	−0.2	—	27	12.6	1.44	4.7
Burundi	6.3	160	−1.3	—	8	6.5	1.50	6.1
Central African Rep.	3.3	340	−2.4	0.55	39	3.3	0.93	2.0
Congo, Rep.	2.6	680	−3.2	—	59	5.1	1.15	5.9
Côte d'Ivoire	14.0	660	—	0.39	44	4.9	1.31	29.3
Guinea-Bissau	1.1	250	2.0	0.56	22	4.2	—	1.8
Guyana	0.8	590	—	0.48	—	—	0.92	0.4
Haiti	7.2	250	−5.2	—	32	3.9	0.66	0.3
Kenya	26.7	280	0.1	0.54	28	6.2	1.51	0.7
Lesotho	2.0	65	1.2	0.56	23	6.2	0.71	1.4
Malawi	9.8	170	−0.7	0.60	13	5.9	1.49	12.1
Mozambique	16.2	80	3.6	—	38	7.3	1.33	0.1
Namibia	1.5	2000	2.9	—	38	6.2	1.65	0.6
Rwanda	6.4	180	−5.4	0.29	8	4.5	1.67	1.0
South Africa	41.5	3,160	−1.1	0.62	51	2.9	1.08	3.1
Tanzania	29.6	120	1.0	0.40	24	6.5	1.31	2.3
Uganda	19.2	240	2.7	0.37	12	5.6	1.13	1.9
Zambia	9.0	400	−0.8	0.47	45	3.7	1.07	4.1
Zimbabwe	11.0	540	−0.6	0.57	32	5.2	1.39	8.0
Unknown Stage of Epidemic								
Afghanistan	23.5	c	—	—	—	—	1.12	0.2
Albania	3.3	670	—	—	37	−0.9	1.35	0.4
Armenia	3.8	730	−15.1	—	69	2.1	—	—
Belarus	10.3	2,070	−5.2[b]	—	71	1.6	—	—
Bolivia	7.4	800	1.8	0.42	58	3.2	0.94	1.0
Croatia	4.8	3,250	—	—	64	1.5	—	—
Georgia	5.4	440	−17.0[b]	—	58	0.7	—	—
Hungary	10.2	4,120	−1.0	—	65	0.6	0.99	0.3
Iran	64.1	d	—	0.43	59	4.0	1.04	6.2

Women's status		Prevention policy				
					CSM for HIV prevention, 1996	
Male adult illit., 1995 (%)	Female adult illit., 1995 (%)	Year of 1st reported AIDS case	Sent. Surv. Implem. status, 1995[a]	Gov't condom distribution, 1992	CSM for HIV prevention, 1996	SM condoms sold per adult, 1995
3	2	83	2	Y		..
8	10	83	2	Y	Y	0.05
4	9	93	2	Y	Y	0.55
51	74	85	3	Y	Y	1.10
20	40	88	4	Y	Y	3.10
71	91	86	3	Y	Y	1.45
51	78	86	4	—	Y	0.44
32	48	87	3	—	Y	1.59
17	33	86	3	Y	Y	..
50	70	85	3	Y	Y	1.73
32	58	89	3	Y	Y	..
1	3	88	2	—		..
52	58	83	4	—	Y	1.47
14	30	86	3	—	Y	0.49
19	38	86	3	Y	Y	..
28	58	85	4	—	Y	1.08
42	77	86	3	—	Y	0.26
—	—	86	4	—		..
30	48	83	4	Y	Y	0.29
18	18	82	3	Y	Y	0.07
21	43	—	4	Y	Y	0.80
26	50	85	4	Y	Y	0.73
14	21	85	3	Y	Y	1.48
10	20	87	4	Y	Y	0.11
53	85	..	—	—		..
—	—	94	—	—	Y	..
—	—	92	2	—		..
—	—	91	2	N		..
10	24	85	1	Y	Y	0.38
—	—	87	2	N		..
—	—	92	2	—		..
—	—	86	—	Y		..
22	34	87	2	—		..

(*Table continues on the following page.*)

Statistical Appendix Table 2: *(continued)*

Country	Population, mid-1995 (millions)	Economic growth		Gini index (1980s)	Urbanization			
		GNP per capita, 1995 (dollars)	Avg. annual GNP per capita growth (%), 1985–95		Urban pop. 1995 (%)	Avg. annual urban growth rate (%), 1990–95	Urban male/ female ratio, 1990	Foreign-born, 1990 (% of total pop.)
Iraq	20.1	d	—	—	78	3.5	1.11	2.8
Jordan	4.2	1,510	−4.5	0.39	72	7.9	1.11	26.4
Kazakhstan	16.6	1,330	−8.6b	—	60	1.2	—	—
Korea, Dem. Rep.	23.9	d	7.7	—	61	2.4	—	0.2
Kyrgyz Republic	4.5	700	−6.9b	—	39	1.1	—	—
Latvia	2.5	2,270	−6.6b	—	73	−0.6	0.96	—
Lebanon	4.0	2660	—	—	87	2.8	0.88	12.2
Liberia	2.7	c	—	—	—	—	1.31	5.0
Libya	5.4	e	—	—	86	4.4	1.12	12.3
Macedonia, FYR	2.1	860	—	—	60	1.6	—	—
Moldova	4.3	920	—	—	52	1.5	—	—
Oman	2.2	4,820	0.3	—	13	8.6	—	33.6
Panama	2.6	2,750	−0.4	0.52	56	2.7	0.95	2.6
Romania	22.7	1,480	−3.8	0.26	55	0.0	1.03	0.6
Saudi Arabia	19.0	7,040	−1.9	—	79	4.0	—	25.8
Syrian Arab Rep.	14.1	1120	0.9	—	53	4.3	1.10	6.6
Tajikistan	5.8	340	—	—	32	2.1	—	—
Tunisia	9.0	1,820	1.9	0.43	57	2.8	1.06	0.5
Turkey	61.1	2,780	2.2	0.50	70	4.7	1.14	2.0
Turkmenistan	4.5	920	—	—	45	5.7	—	—
Uzbekistan	22.8	970	−3.9b	—	42	2.6	—	—
Yugoslavia, FR (Serb./Mont.)	10.5	d	—	0.33	57	1.5	0.98	1.7

— Not available.

. . Not applicable.

SM Socially marketed.

Y = Yes, N = No.

a. 1 Planned. 2 Limited. 3 Many sites. 4 Extensive.

b. Estimates for economies of the former Soviet Union are preliminary.

c. Estimated to be low income ($765 or less).

d. Estimated to be lower-middle income ($766 to $3,035).

e. Estimated to be upper-middle income ($3,036 to $9,385).

Women's status		Prevention policy				
Male adult illit., 1995 (%)	Female adult illit., 1995 (%)	Year of 1st reported AIDS case	Sent. Surv. Implem. status, 1995[a]	Gov't condom distri- bution, 1992	CSM for HIV preven- tion, 1996	SM condoms sold per adult, 1995
29	55	91	1	—		..
7	21	86	2	N		..
—	—	93	2	—		..
1	3	—	—	—		..
—	—	..	2	Y		..
—	—	90	2	—		..
5	10	86	2	—		..
46	78	86	2	—		..
12	37	89	—	—		..
—	—	—	—	—		..
—	—	—	2	Y		..
—	—	86	2	N		..
9	10	86	2	Y		..
—	—	85	3	—		..
29	50	86	—	—		..
14	44	87	1	—		..
—	—	..	2	—		..
21	45	86	1	Y		..
8	28	85	1	—	Y	0.30
—	—	92	2	Y		..
—	—	92	2	—	Y	..
—	—	85	2	N		..

Statistical Appendix Table 3   Socially Marketed Condom Sales in Developing Countries, 1991–96

Country	Year program launched	Socially marketed condom sales (thousands)					
		1991	1992	1993	1994	1995	1996
Albania	1996	..	..	..	..	..	57
Bangladesh	1975	82,676	117,360	138,248	150,994	161,538	150,999
Benin	1989	634	881	1,348	1,685	2,663	2,506
Bolivia	1988	377	543	684	632	1,338	2,534
Botswana	1993	..	..	1,016	1,283	2,233	1,625
Brazil	1991	406	3,084	6,758	11,568	18,272	26,886
Burkina Faso	1991	2,795	2,252	3,176	5,171	6,583	7,570
Burundi	1990	165	1,255	1,142	559	1,273	755
Cambodia	1994	..	..	..	99	5,032	9,516
Cameroon	1989	3,194	5,111	5,756	7,205	7,563	9,254
Central African Republic	1991	310	673	1,381	1,892	2,392	2,000
Chad	1996	..	..	..	..	..	779
China	1996	..	..	..	..	..	11,778
Colombia	1974	6,548	7,015	5,976	6,227	5,310	6,390
Congo, DR	1988	18,302	7,140	2,385	3,150	8,572	1,766
Congo, Rep. of	1996	..	..	..	..	..	215
Costa Rica	1993	2,937	3,800	4,289	5,749	5,808	—
Côte d'Ivoire	1990	1,828	6,375	5,857	8,929	10,807	12,371
Dominican Republic	1986	869	1,584	1,810	1,242	758	1,700
Ecuador	1986	214	828	244	—	694	1,494
Egypt	1979	14,668	12,379	1,092	14,095	10,353	—
El Salvador	1976	1,769	2,243	2,172	1,512	1,585	1,585
Ethiopia	1990	3,782	7,076	11,788	17,293	19,833	20,662
Ghana	1986	3,748	3,996	4,289	4,386	3,411	4,665
Guatemala	1983	1,600	2,250	2,464	3,072	2,902	1,993
Guinea	1991	132	2,519	1,988	1,398	2,777	3,179
Guinea-Bissau	1996	..	..	..	..	..	495
Haiti	1990	..	1,784	3,979	3,727	5,047	4,426
Honduras	1981	600	724	921	734	890	698
India I[a]	1968	293,240	239,430	248,170	111,508	134,054	196,410
India II	1988	28,449	25,963	28,824	21,179	34,844	41,380
India III	1988	5,510	8,768	8,878	10,830	3,362	15,821
India IV	1992	..	—	747	2,727	6,150	7,984
Indonesia I[a]	var. dates	5,929	2,728	2,979	676[b]	5,996[b]	3,536[b]
Indonesia II	1996	..	..	..	..	..	1,453
Jamaica	1974	1,918	2,167	2,056	1,595	2,168	1,325
Kenya	1989	498	558	1,769	4,792	6,009	4,762
Lesotho	1996	..	..	..	82	210	28
Madagascar	1996	..	..	..	992	..	1,051
Malawi	1994	..	..	..	..	4,643	5,817
Malaysia	1991	..	1,258	3,640	5,653	7,152	8,583
Mali	1992	..	..	959	1,873	3,051	3,053
Morocco	1989	..	2,107	2,145	2,326	2,497	1,958

Statistical Appendix Table 3 *(continued)*

Country	Year program launched	Socially marketed condom sales (thousands)					
		1991	1992	1993	1994	1995	1996
Mozambique	1995	..	..	..	..	2,103	4,086
Myanmar	1996	..	..	..	..	..	368
Nepal	1976	4,585	4,676	5,688	7,203	8,146	6,710
Niger	1994	..	..	..	—	1,126	236
Nigeria I[a]	1988-90	1,876	7,051	23,614	45,243	55,654	34,185
Nigeria II	mid-80s	—	3,012	[c]	..	..	..
Pakistan	1986	73,385	34,049	39,023	48,340	58,759	88,543
Peru	1984	—	—	—	1,368	4,294	4,336
Philippines I[a]	1990	1,266	2,962	5,576	7,836	9,016	11,832
Philippines II	1992	..	—	—	1,611	1,584	—
Russian Federation	1996	..	..	..	..	..	2,212
Rwanda	1993	..	..	1,063	563	1,039	2,859
Senegal	1996	..	..	..	..	..	1,511
South Africa	1992	..	104	480	1,359	1,541	1,021
Sri Lanka	1973	6,769	6,490	7,366	7,590	7,852	—
Tanzania	1993	..	..	—	3,754	10,791	11,927
Togo	1992	..	—	—	2,272	4,403	2,979
Turkey	1991	2,398	5,877	6,326	7,743	9,694	10,500
Uganda	1991	—	1,318	1,812	4,082	5,980	9,951
Uzbekistan	1996	..	..	..	..	..	9
Venezuela	1992	..	—	—	243	425	—
Vietnam	1993	..	..	3,559	7,202	20,148	30,745
Zambia	1992	..	435	4,709	6,601	6,273	7,440
Zimbabwe	1996	1,182	1,272	1,062	863	601	769

— Not available.

.. Not applicable, program not in operation.

a. Country has or had more than one condom social marketing program.

b. Program devolved to the private sector.

c. Nigeria I and II were consolidated into a single program in 1993 and are recorded under Nigeria I.

Selected Bibliography

Background Papers

Ahlburg, Dennis, and Eric Jensen. 1996. "Economics of the Commercial Sex Industry in Developing Countries."

Ainsworth, Martha, and Innocent Semali. 1997. "Socioeconomic Correlates of Adult Deaths in Tanzania."

Béchu, Natalie. 1996. "Les Effets du Sida sur L'Economie Familiale en Côte d'Ivoire : Analyse empirique de l'évolution de la consommation des ménages touchés en Côte d'Ivoire."

Biggs, Tyler, and Manju Kedia Shah. 1996. "The Impact of the AIDS Epidemic on African Firms."

Dayton, Julia. 1996. "World Bank AIDS Interventions: Ex-ante and Ex-post Evaluation."

Deheneffe, Jean-Claude, Michel Caraël, and Amadou Noumbissi. 1996. "Socioeconomic Determinants of Sexual Behaviour and Condom Use: Analysis of WHO/GPA Surveys."

Filmer, Deon. 1997. "Socioeconomic Correlates of Risky Behavior: Results from the Demographic and Health Surveys."

Janjaroen, Wattana. 1996. "The Economic Impact of AIDS on Households in Thailand."

Jones, Christine, and Allechi M'bet. 1996. "Does Structural Adjustment Cause AIDS: One More Look at the Link between Adjustment, Growth, and Poverty."

Kremer, Michael. 1996a. "AIDS: The Economic Rationale for Public Intervention."

————. 1996b. "Optimal Subsidies for AIDS Prevention."

Menon, Rekha, Maria J. Wawer, Joseph K. Konde-Lule, Nelson K. Sewankambo, and Chuanjun Li. 1996a. "The Impact of Adult Mortality on Households in Rakai District, Uganda."

————. 1996b. "The Socioeconomic Correlates of HIV Infection among Household Heads in Rakai District, Uganda."

Mills, Anne, and Charlotte Watts. 1996. "Cost-Effectiveness of HIV Prevention and the Role of Government."

Morris, Martina. 1996. "Sexual Networks: What Is Their Relevance to Modeling the Spread of HIV?"

Over, Mead. 1997. "Societal Determinants of Urban HIV Infection: An Exploratory Cross-country Regression Analysis."

Perriëns, Joseph H. 1996. "Treatment of HIV Infection."

Pyne, Hnin H. 1997. "National and International Responses to the HIV/AIDS Epidemic in Developing Countries."

Riehman, Kara. 1996. "Drug Use and AIDS in Developing Countries: Issues for Policy Consideration."

Saba, J., and J. Perriëns. 1996. "Mother-to-Child Transmission of HIV."

Shepard, Donald S., Justine Agness-Soumahoro, Richard N. Bail, Charles S. M. Cameron, Antonio, C. C. Campino, Roberto F. Iunes, José Antonio Izazola, Tiécoura Koné, Sukhontha Kongsin, Phare Mujinja, Jeffrey Prottas, Jorge Saavedra, Adèle Silué, Laksami Suebsang, Paula Tibandebage, and Samuel Wangwe. 1996. "Expenditures on HIV/AIDS: Levels and Determinants, Lessons from Five Countries."

Stover, John. 1997. "The Future Demographic Impact of AIDS: What Do We Know?"

Van Vliet, Carina, King Holmes, Burton Singer, and Dik Habbema. 1997. "Effectiveness of HIV Prevention Strategies under Alternative Epidemiological Scenarios: Evaluation with the STDSIM Model."

Bibliography

Adler, Michael, Susan Foster, John Richens, and Hazel Slavin. 1996. "Sexually Transmitted Infections: Guidelines for Prevention and Treatment." Overseas Development Administration Health and Population Occasional Paper. Overseas Development Administration, London.

Ad Hoc Committee on Health Research Relating to Future Intervention Options. 1996. *Investing in Health Research and Development.* Geneva: World Health Organization.

Agha, Sohail. 1997. "Sexual Activity and Condom Use in Lusaka, Zambia." PSI Research Division Working Paper 6. Population Services International, Washington, D.C.

Ahituv, Avner, V. Joseph Hotz, and Thomas Philipson. 1995. "The Responsiveness of the Demand for Condoms to the Local Prevalence of AIDS." *Journal of Human Resources* 31(4): 869–97.

AIDS Analysis Africa. 1996. "Vaginal Microbicides: How Do You Kick-Start a Global Market?" Vol. 6 (5): 1.

AIDSCAP/Family Health International, Harvard School of Public Health, and UNAIDS. 1996. "The Status and Trends of the Global HIV/AIDS Pandemic." Final Report of a Satellite Symposium of the Eleventh International Conference on AIDS, Vancouver, B.C., Canada. July 7–12. Family Health International.

Ainsworth, Martha, and Godlike Koda. 1993. "The Impact of Adult Deaths from AIDS and Other Causes on School Enrollment in Tanzania." Paper presented at the annual meetings of the Population Association of America. Cincinnati, Ohio. April 1–3.

Ainsworth, Martha, and Mead Over. 1994a. "AIDS and African Development." *World Bank Research Observer* 9 (2): 203–40.

————. 1994b. "The Economic Impact of AIDS on Africa." In Max Essex, Souleymane Mboup, Phillis J. Kanki, and Mbowa R. Kalengayi, eds., *AIDS in Africa.* New York: Raven Press.

Allard, R. 1989. "Beliefs about AIDS as Determinants of Preventive Practices and of Support for Coercive Measures." *American Journal of Public Health* 79 (4): 448–52.

Allen, Susan, A. Serufilira, J. Bongaarts, P. Van de Perre, F. Nsengumuremyi, C. Lindan, M. Caraël, W. Wolf, T. Coates, and S. Hulley. 1992a. "Confidential HIV Testing and Condom Promotion in Africa." *Journal of the American Medical Association* 268 (23): 3338–44.

Allen, Susan, J. Tice, P. Van de Perre, and others. 1992b. "Effects of Serotesting with Counseling on Condom Use and Seroconversion among HIV Discordant Couples in Africa. *British Medical Journal* 304: 1605–09.

Allen, Susan, C. Lindan, A. Serufilira, P. Van de Perre, A. C. Rundle, F. Nsengumuremyi, M. Caraël, J. Schwalbe, and S. Hulley. 1991. "Human Immunodeficiency Virus Infection in Urban Rwanda: Demographic and Behavioral Correlates in a Representative Sample of Childbearing Women." *Journal of the American Medical Association* 266 (12): 1657–63.

Allen, Susan, A. Serufilira, V. Gruber, and others. 1993. "Pregnancy and Contraception Use among Urban Rwandan Women after HIV Testing and Counseling." *American Journal of Public Health* 83: 705–10.

Anderson, Roy M. 1996. "The Spread of HIV and Sexual Mixing Patterns." In Jonathan Mann and Daniel Tarantola, eds., *AIDS in the World II: Global Dimensions, Social Roots, and Responses.* The Global AIDS Policy Coalition. New York: Oxford University Press.

Anderson, Roy M., and R. M. May. 1988. "Epidemiological Parameters of HIV Transmission." *Nature* 333 (6173): 514–19.

Anderson, Roy M., S. Gupta, and W. Ng. 1990. "The Significance of Sexual Partner Contact Networks for the Transmission Dynamics of HIV." *Journal of Acquired Immune Deficiency Syndromes* 3 (4): 417–29.

Anderson, Roy M., B. Schwartländer, F. McCutchan, and D. Hu. 1996. "Implications of Genetic Variability in HIV for Epidemiology and Public Health." *Lancet* 347 (9018): 1778.

Anderson, Roy M., R. M. May, M. C. Boily, G. P. Garnett, J. T. Rowley. 1991. "The Spread of HIV-1 in Africa: Sexual Contact Patterns and the Predicted Demographic Impact of AIDS." *Nature* 352 (6336): 581–89.

Andriamahenina, Ramamonjisoa. 1995. "Enquête Epidemiologique sur l'Infection à VIH et la Syphilis dans les Villes d'Antananarivo, de Toliary et de Toamasina (Mai–Aôut 1995)." Laboratoire National de Réference MST/SIDS, Madagascar/Futures Group International, Madagascar.

Aral, S. O., and T. A. Peterman. 1996. "Measuring Outcomes of Behavioral Interventions for STD/HIV Prevention." *International Journal of STD and AIDS* 7 (suppl. 2): 30–38.

Archavanitkul, Kritaya, and P. Guest. 1994. "Migration and the Commercial Sex Sector in Thailand." *Health Transition Review,* 4 (suppl.): 273–95.

Arias, R., and Luis Servén. 1997. "Measuring Aid Flows: A New Approach." Working paper. World Bank, Policy Research Department, Washington, D.C.

Arrow, K. J. 1963. "Uncertainty and the Welfare Economics of Medical Care." *American Economic Review* 53 (5): 941–73.

Asiimwe-Okiror, Godwil, Alex A. Opio, Joshua Musinguzi, Elizabeth Madraa, George Tembo, and Michel Caraël. 1997. "Change in Sexual Behaviour and Decline in HIV Infection among Young Pregnant Women in Urban Uganda." Forthcoming in *AIDS.*

Asiimwe-Okiror, Godwil, J. Musinguzi, George Tembo, Alex Opio, Elizabeth Madraa, B. Biryahwaho, S. Okware, C. Byabamazima, and P. Turyguma. 1995. "Declining Trends in HIV Infection in Rural Areas in Uganda." Abstract WeC206. Ninth International Conference on AIDS and STDs in Africa, Kampala, Uganda. December 10–14.

Bagenda, D., F. Mmiro., F. Mirembe., C. Nakabito., D. Mugenyi, and L. Kukasa. 1995. "HIV-1 Prevalence Rates in Women Attending Antenatal Clinics in Kampala, Uganda." Abstract MoC016. Ninth International Conference on AIDS and STDs in Africa, Kampala, Uganda. December 10–14.

Baggaley, R., P. Godfrey-Faussett, R. Msiska, D. Chilangwa, E. Chitu, J. Porter, and M. Kelly. 1994. "Impact of HIV Infection on Zambian Businesses." *British Medical Journal* 309 (6968): 1549–50.

Bankole, Akinrinola, and Charles F. Westoff. 1995. "Childbearing Attitudes and Intentions." *DHS Comparative Studies* 17. Calverton, Md.: Macro International.

Barnett, Tony, and Piers Blaikie. 1992. *AIDS in Africa: Its Present and Future Impact*. New York: Guilford Press.

Barnum, Howard, and Joseph Kutzin. 1993. "Hospital Costs and Efficiency." In *Public Hospitals in Developing Countries: Resource Use, Cost, Financing*. Baltimore, Md.: Johns Hopkins University Press.

Barongo, L. R., M. W. Borgdorff, F. F. Mosha, A. Nicoll, H. Grosskurth, K. P. Senkoro, J. N. Newell, J. Changalucha, A. H. Klokke, J. Z. Killewo, J. P. Velema, R. J. Hayes, D. T. Dunn, L. A. S. Muller, and J. B. Rugemalila. 1992. "The Epidemiology of HIV-1 Infection in Urban Areas, Roadside Settlements: Villages in Mwanza Region, Tanzania." *AIDS* 6 (12): 1521–28.

Bastos, F. I. 1995. *Ruina e reconstrucção: AIDS e Drogas Injetaveis na Cena Contemporanea*. Brazil: Relume Dumara.

Beal, R. W., M. Bontinck, and L. Fransen, eds. 1992. "Safe Blood in Developing Countries: A Report of the EEC's Expert Meeting." EEC AIDS Task Force, Brussels.

Becker, Gary S. 1981. *A Treatise on the Family*. Cambridge, Mass.: Harvard University Press.

Beegle, Kathleen. 1996. "The Impact of Prime-Age Adult Mortality on Labor Supply." Michigan State University, East Lansing.

Behets, Frieda, Ramamonjisoa Andriamahenina, Jocelyne Andriamiadana, John May, and Andry Rasamindrakotroka. 1996. "High Syphilis and Low but Rising HIV Seroprevalence Rates in Madagascar." *Lancet* 347 (9004): 831.

Bernstein, Robert S., David C. Sokal, Steven T. Seitz, Bertran Auvert, John Stover, and Warren Naamara. 1997. "Simulating the Control of a Heterosexual HIV Epidemic in a Severely Affected East African City." Forthcoming in *Interfaces*.

Berstein, C. A., J. G. Rabkin, and H. Wolland. 1990. "Medical and Dental Students' Attitudes about the AIDS Epidemic." *Academic Medicine* 65 (7): 458–60.

Bertozzi, Stefano. 1995. "Global Challenge of AIDS: Ten Years of HIV/AIDS Research." In Y. Shiokawa and T. Kitamura, eds., *Proceedings of the Tenth International Conference on AIDS/International Conference on STD*. Yokohama. August 7–12, 1994. Tokyo: Kodansha.

Besley, Timothy, and Ravi Kanbur. 1988. "The Principles of Targeting." University of Warwick, Development Economic Research Centre, Department of Economics, Coventry, U.K.

Bhave G., C. P. Lindan, E. S. Hudes, and others. 1995. "Impact of an Intervention on HIV, Sexually Transmitted Diseases and Condom Use among Sex Workers in Bombay, India." *AIDS* 9 (suppl. 1): S21–30.

Biggar, R. J., P. G. Miotti, T. E. Taha, L. Mtimavalye, R. Broadhead, A. Justesen, F. Yellin, G. Liomba, W. Miley, D. Waters, J. D. Chiphangwi, and J. Goedert. 1996. "Perinatal Intervention Trial in Africa: Effect of a Birth Canal Cleansing Intervention To Prevent HIV Transmission." *Lancet* 347 (9016):1647–50.

Birdsall, Nancy. 1988. "Economic Approaches to Population Growth." In Hollis Chenery and T. N. Srinivasan, eds., *Handbook of Development Economics*. Vol. 1. New York: North-Holland.

Birungi, Harriet, and Susan Reynolds Whyte. 1993. "Injections, Control and Community Needs in Ugandan Health Care." In M. Bloem and I. Wolffers, eds., *The Impact of Injections on Daily Medical Practice*. Amsterdam: Free University Press.

Birungi, Harriet, Delius Asiimwe, and Susan Reynolds Whyte. 1994. "Injection Use and Practices in Uganda." WHO/DAP/94.18. World Health Organization, Action Programme on Essential Drugs, Geneva.

Blix, O., and L. Gronbladh, 1988. "AIDS and IV Heroin Addicts: The Preventive Effect of Methadone Maintenance in Sweden." Paper presented at the Fourth International Conference on AIDS, Stockholm, Sweden. June 12–16.

Bloom, David E., and Peter Godwin, eds. 1997. *The Economics of HIV and AIDS: The Case of South and South East Asia*. Delhi: Oxford University Press.

Bloom, David E., and Ajay S. Mahal. 1997. "The AIDS Epidemic and Economic Policy Analysis." In David E. Bloom and Peter Godwin, eds., *The Economics of HIV and AIDS: The Case of South and South East Asia*. Delhi: Oxford University Press.

Bloom, David E., Tim Brown, Paul Gertler, and Werasit Sittitrai. Forthcoming. "Employment and Earnings of Commercial Sex Workers."

Bobadilla, José Luis, Julio Frenk, Rafael Lozano, Tomas Frejka, and Claudio Stern. 1993. "The Epidemiologic Transition and Health Priorities." In Dean T. Jamison, W. Henry Mosley, Anthony R. Measham, and José Luis Bobadilla, eds., *Disease Control Priorities in Developing Countries*. New York: Oxford University Press.

Bolduc, Denis, Bernard Fortin, and Marc-André Fournier. 1996. "The Impact of Incentive Policies on the Practice Location of Doctors: A Multinomial Profit Analysis." *Journal of Labor Economics* 14 (October): 703–32.

Bongaarts, John. 1996. "Global Trends in AIDS Mortality." *Population and Development Review* 22 (1): 21–45.

Bongaarts, John, P. Reining, Peter Way, and Francis Conant. 1989. "The Relationship between Male Circumcision and HIV Infection in African Populations." *AIDS* 3 (6):373–77.

Bourdeaux, Richard. 1996. "HIV Ravages Drug-Torn Town in Belarus." *Los Angeles Times,* Sept. 29, 1996, p. A1.

Brandt, Allan M. 1987. *No Magic Bullet: A Social History of Venereal Disease in the United States since 1880.* New York: Oxford University Press.

Brown, David. 1997. "U.S. Advocates Triple Therapy To Fight AIDS." *Washington Post,* June 19, p. A1.

Brown, Tim, and Werasit Sittitrai. 1995. *The Impact of HIV on Children in Thailand.* Research Report 16, Program on AIDS. Bangkok: Red Cross Society.

Brown, Tim, and Peter Xenos. 1994. *AIDS in Asia: The Gathering Storm.* Asia Pacific Issues 16, Program on Population. Honolulu: East-West Center.

Brown, Tim, Werasit Sittitrai, Suphak Vanichseni, and Usa Thisgaborn. 1994. "The Recent Epidemiology of HIV and AIDS in Thailand." *AIDS* 8 (suppl. 2): S131–41.

Bulatao, Rodolfo A. 1991. "The Bulatao Approach: Projecting the Demographic Impact of the HIV Epidemic Using Standard Parameters." In United Nations and World Health Organization, *The AIDS Epidemic and Its Demographic Consequences.* New York: United Nations.

Bulterys, M., A. Chao, P. Habimana, A. Dushimimana, P. Nawrocki, and A. Saah. 1994. "Incident HIV-1 Infection in a Cohort of Young Women in Butare, Rwanda." *AIDS* 8 (11): 1585–91.

Burr, Chandler. 1997. "The AIDS Exception: Privacy vs. Public Health—The Case for Reinstating Traditional Rules for Fighting Epidemics." *Atlantic Monthly* (June).

Buvé, S. D. Foster, C. Mbwili, E. Mungo, N. Tollenare, and M. Zeko. 1994. "Mortality among Female Nurses in the Face of the AIDS Epidemic: A Pilot Study in Zambia." *AIDS* 8 (3): 396.

Buwembo, Joachim. 1995. "Uganda Government Media Move To Lift Ban on Condom Ads." *The East African*, March 6–12.

Caldwell, John C., and Pat Caldwell. 1993. "The Nature and Limits of the Sub-Saharan African AIDS Epidemic: Evidence from Geographic and Other Patterns." *Population and Development Review* 19 (4): 817–48.

———. 1996. "The African AIDS Epidemic." *Scientific American* 274 (3): 62–68.

Caldwell, John C., Pat Caldwell, and I. O. Orubuloye. 1989. "The Social Context of AIDS in Sub-Saharan Africa." *Population and Development Review* (2): 185–234.

Caraël, Michel. 1995. "Sexual Behaviour." Chapter 4 of John Cleland and Benoit Ferry, eds., *Sexual Behaviour and AIDS in the Developing World*. London: Taylor & Francis.

CARE (Cooperative for American Relief to Everywhere) and ODA (Overseas Development Administration). 1994. "Refugee Inflow into Ngara and Karagwe Districts, Kagera Region, Tanzania: Environmental Impact Assessment." Environmental Resources Management Report. ODA, London.

Carr, Jean K., Narongrid Sirisopana, Kalyanee Torugsa, Achara Jugsudee, Thippawan Supapongse, Cheodchai Chuenchitra, Sorachai Nitayaphan, Pricha Singharaj, and John G. McNeil. 1994. "Incidence of HIV-1 Infection among Young Men in Thailand." *Journal of Acquired Immune Deficiency Syndromes* 7: 1270–75.

Carrin, G., J. Perrot, and F. Sergent. 1994. "Influence of Financial Participation by the Population on the Demand for Health: An Analytical Tool for Countries in Greatest Need." WHO, Macroeconomics, Health and Development series. Geneva.

Carswell, J. W., and Lloyd G. Howells. 1989. "Prevalence of HIV-1 in East African Lorry Drivers." *AIDS* 3 (11): 759–61.

Chakraborty, A. K., S. Jana, A. Das, L. Khodakevich, M. S. Chakraborty, N. K. Pal. 1994. "Community-based Survey of STD/HIV Infection among Commercial Sex Workers in Calcutta (India). Part I. Some Social Features of Commercial Sex Workers." *Journal of Communicable Diseases* 26 (3): 161–67.

Chawla, Mukesh. 1993. "Physician Moonlighting in Egypt." Ph.D. diss. Boston University, Department of Economics, Boston, Mass.

———. 1997. "Dual-Job Holdings by Public Sector Physicians in India." Working paper. Data for Decision Making. Harvard School of Public Health, Department of Population and International Health, Boston, Mass.

Chela, C. M., Roland Msiska, Anna Martin, M. Sichone, B. Mwiinga, C. B. Yamba, S. Anderson, E. Van Praag. 1994. "Cost and Impact of Home-based Care for People Living with HIV/AIDS in Zambia." Unpublished report. World Health Organization, Global Programme on AIDS, Geneva.

Chen, Shaohua, Gaurav Datt, and Martin Ravallion. 1994. "Is Poverty Increasing in the Developing World?" *Review of Income and Wealth* series 40 (4, December): 359–76.

Chequar, Pedro. 1997. "Implications of Anti-retroviral Treatments—Brazil." Paper presented at the Informal Consultation on the Implications of Antiretroviral Treatments." World Health Organization, Geneva. April 29–30.

Ch'ien, J. 1994. "Street Surveys on Drug Users' Behaviour and Peer Counselling for Preventive Education and Crisis Intervention." Abstract PC0473. Paper presented at the Tenth International Conference on AIDS, Yokohama. August 14–17.

Chin, James. 1990. "Public Health Surveillance of AIDS and HIV Infections." *Bulletin of the World Health Organization* 68 (5): 529–36.

Choi, Kyung-Hee, and Thomas J. Coates. 1994. "Prevention of HIV Infection." *AIDS* 8 (10): 1371–89.

Choopanya, K., A. Vanichseni, D. C. DesJarlais, K. Plangsringarm, W. Sonchai, M. Carballo, P. Friedmann, and S. R. Friedman. 1991. "Risk Factors and HIV Seropositivity among Injecting Drug Users in Bangkok." *AIDS* 5 (12): 1509–13.

Cleland, John, and Benoit Ferry, eds. 1995. *Sexual Behaviour and AIDS in the Developing World.* London: Taylor & Francis.

Clemente, M., C. Ferreros, and M. E. L. Fernandes. 1996. "The Brazilian Condom Market: Positive Effects of Social Marketing." Abstract No. Tu.D.2887. Paper read at the Eleventh International Conference on AIDS, Vancouver, B.C, Canada. July 7–12.

Cohen, Myron S., Gail E. Henderson, Pat Aiello, and Heyi Zheng. 1996. "Successful Eradication of Sexually Transmitted Diseases in the People's Republic of China: Implications for the 21st Century." *Journal of Infectious Diseases* 174 (suppl. 2): S223–29.

Cohen, Myron S., Irving F. Hoffman, Rachel A. Royce, Peter Kazembe, John R. Dyer, Celine Costello Daly, Dick Zimba, Pietro L. Vernazza, Martin Maida, Susan A. Fiscus, Joseph J. Eron, Jr., and the AIDSCAP Malawi Research Group. 1997. "Reduction of Concentration of HIV-1 in Semen after Treatment of Urethritis: Implications for Prevention of Sexual Transmission of HIV-1." *Lancet* 349 (9069): 1868–73.

Coleman, Susan, Mary Lyn Gaffield, Frederic Robin, Goussou Koudou lazare, Rebecca Dillingham, and Carol Squire-Diomande. 1996. "PSI Research Report Summary: Côte d'Ivoire Condom KAP Survey, June. 1995." Population Services International, Washington, D.C.

Collins, D., J. Quick, S. Musau, and D. Kraushaar. 1996. "Health Financing Reform in Kenya: The Fall and Rise of Cost Sharing, 1989–94." *Stubbs Monograph Series* 1. Boston: Management Sciences for Health.

Conant, Francis. 1995. "Regional HIV Prevalence and Ritual Circumcision in Africa." *Health Transition Review* 5 (1): 108–12.

Connor, E. M., R. S. Sperling, R. Gelker, P. Kiselev, G. Scott, M. J. O'Sullivan, R. Van Dyke, M. Bey, W. Shearer, and R. L. Jacobson. 1994. "Reduction of Maternal-Infant Transmission of Human Immunodeficiency Virus Type 1 with Zidovudine Treatment." *New England Journal of Medicine* 331 (18): 1173–80.

Cuddington, John T. 1993. "Modeling the Macroeconomic Effects of AIDS with an Application to Tanzania." *World Bank Economic Review* 7 (2): 173–89.

Curtis, S. L., and Katherine Neitzel. 1996. "Contraceptive Knowledge, Use and Sources." *DHS Comparative Studies* 19. Calverton, Md.: Macro International.

Dabis, F., P. Msellati, P. Lepage, M. L. Newell, C. Peckham, and P. van de Perre. 1993. "Estimating the Rate of Mother-to-Child Transmission of HIV." A report of the Workshop on Methodological Issues, Ghent, Belgium. February 17–20. *AIDS* 7(8): 1139–48.

Dallabetta, Gina, Marie Laga, and Peter Lamptey, eds. 1996. *Control of Sexually Transmitted Diseases: A Handbook for the Design and Management of Programs.* Arlington, Va.: AIDSCAP/FHI.

Dallabetta, G. A., P. G. Miotti, J. D. Chiphangwi, A. J. Saah, G. Liomba, N. Odaka, F. Sungani, and D. R. Hoover. 1993. "High Socioeconomic Status Is a Risk Factor for Human Immunodeficiency Virus Type 1 (HIV-1) Infection but Not for Sexually Transmitted Diseases in Women in Malawi: Implications for HIV-1 Control." *Journal of Infectious Diseases* 167 (1): 36–42.

Dames and Moore. 1996. "Environmental Impact Assessment for Chad." Report. Washington, D.C.

De Cock, Kevin. 1993. "TB and HIV: An Overview." Excerpts from speech at the 1993 International AIDS Conference in Berlin, reproduced in "TB and HIV." *SIDALERTE* 27, October (suppl.).

De Cock, Kevin M., and Françoise Brun-Vezinet. 1996. "HIV-2 Infection: Current Knowledge and Uncertainties." Chapter 12 of Mann and Tarantola, eds., *AIDS in the World II: Global Dimensions, Social Roots, and Responses*. The Global AIDS Policy Coalition. New York: Oxford University Press.

De Gruttola, Victor, George R. Seage III, Kenneth H. Mayer, and C. Robert Horsburgh, Jr. 1989. "Infectiousness of HIV between Male Homosexual Partners." *Journal of Clinical Epidemiology* 42 (9): 849–56.

Deininger, Klaus, and Lyn Squire. 1996. "A New Data Set Measuring Income Inequality." *World Bank Economic Review* 10 (3): 565–91.

Demographic and Health Surveys, final reports, various countries and years.

De Muylder, X., and J. J. Amy. 1993. "Caesarean Section Rates in an African Country." *Paediatric Perinatal Epidemiology* 7 (3): 234–44.

Des Jarlais, D. C., S. Friedman, K. Choopanya, S. Vanichseni, and T. Ward. 1992. "International Epidemiology of HIV and AIDS among Injecting Drug Users." *AIDS* 6 (10): 1053–68.

Des Jarlais, D. C., H. Hagan, S. R. Friedmann, P. Friedmann, D. Goldberg, M. Frischer, S. Green, K. Tunving, B. Ljungberg, A. Wodak, M. Ross, D. Purchase, M. Millson, and T. Meyers. 1995. "Maintaining Low HIV Seroprevalence in Populations of Injecting Drug Users." *Journal of the American Medical Association* 274 (15): 1226–31.

Deschampes, M., J. W. Pape, A. Hafner, and others. 1996. "Heterosexual Transmission of HIV in Haiti. *Annals of Internal Medicine* 125: 324–30.

de Vincenzi, Isabelle, and T. Mertens. 1994. "Male Circumcision: A Role in HIV prevention?" *AIDS* 8 (2): 153–60.

DKT International. 1992. "1991 Contraceptive Social Marketing Statistics." Washington, D.C., June. Unpublished data.

———. 1993. "1992 Contraceptive Social Marketing Statistics." Washington, D.C., August. Unpublished data.

———. 1994. "1993 Contraceptive Social Marketing Statistics." Washington, D.C., July. Unpublished data.

———. 1995. "1994 Contraceptive Social Marketing Statistics." Washington, D.C., July. Unpublished data.

———. 1996. "1995 Contraceptive Social Marketing Statistics." Washington, D.C., August. Unpublished data.

———. 1997. *New Breakthroughs in Social Marketing: DKT International 1996–97 Progress Report.* Washington, D.C.

———. n.d. *How Social Marketing Changes Lives: The Stories Behind the Statistics.* Washington, D.C.

Dolin, P. J., M. C. Raviglione, and A. Kochi. 1993. "Estimates of Future Global Morbidity and Mortality." *Morbidity and Mortality Weekly Report* 42 (49). Summarized in *HIV and TB* 2, January–March 1994.

Dunn, D. T., M. L. Newell, A. E. Ades, and C. S. Peckham 1992. "Risk of Human Immunodeficiency Virus Type 1 Transmission through Breastfeeding." *Lancet* 340 (8819): 585–88.

Elias, C., and L. Heise. 1994. "Challenges for the Development of the Female-Controlled Vaginal Microbicides." *AIDS* 8: 1–9.

Ellis, Randall, Moneer Alam, and Indrani Gupta. 1997. "Health Insurance in India: Prognosis and Prospectus." Boston University, Boston, Mass.

Esparza, José, William L. Heyward, and Saladin Osmanov. 1996. "HIV Vaccine Development: From Basic Research to Human Trials." *AIDS* 10 (suppl. A): S123–32.

Ettiegne-Traoré, V., P. D. Ghys, M. O. Diallo, and others. 1996. "HIV Seroincidence and STD Prevalence during an Intervention Study among Female Sex Workers in Abidjan, Côte d'Ivoire: Preliminary Findings." Abstract Mo.C. 442. Eleventh International Conference on AIDS, Vancouver, B.C., Canada. July 7–12.

European Commission. 1995a. *Safe Blood in Developing Countries: The Lessons from Uganda.* Edited by Rex Winsbury. Development Studies and Research. Luxembourg: Office for Official Publications of the European Communities.

———. 1995b. "Safe Blood in Developing Countries: Principles and Organisation." Development Studies and Research. Written and edited by C. Gerard, D. Sondag-Thull, E. J. Watson-Williams, and L. Fransen. Luxembourg: Office for Official Publications of the European Communities.

———. 1997. *Considering HIV/AIDS in Development Assistance: A Toolkit.* Produced for the EC under contract B7.5046/94/06: Study of the Impact on the Social and Economic Development of Developing Countries. Brussels.

European Study Group on Heterosexual Transmission of HIV. 1992. "Comparison of Female to Male and Male to Female Transmission of HIV in 563 Stable Couples." *British Medical Journal* 304: 809–13.

Expert Group of the Joint United Nations Programme on HIV/AIDS. 1997. "Implications of HIV Variability for Transmission: Scientific and Policy Issues." *AIDS* 11 (4):1–15.

Farwell, Byron. 1989. *Armies of the Raj: From Mutiny to Independence,* 1858–1947. New York: Norton.

Feachem, Richard. 1995. *Valuing the Past, Investing in the Future. Evaluation of the National HIV/AIDS Strategy, 1993–94 to 1995–96.* AIDS/Communicable Diseases Branch, Commonwealth Department of Human Services and Health, Australia. Canberra: Australian Government Publishing Service.

Ferreira, Luisa, and L. Goodhart. 1995. "Incomes, Inequality and Poverty Alleviation in Tanzania, 1993." World Bank, Africa Technical Family, Human Development Division, Washington, D.C.

Ferry, Benoit. 1995. "Risk Factors Related to HIV Transmission: Sexually Transmitted Diseases, Alcohol Consumption and Medically-Related Injections." In John Cleland and Benoit Ferry, eds., *Sexual Behaviour and AIDS in the Developing World.* London: Taylor & Francis.

Fitzsimmons, David. 1996. "International AIDS Vaccine Initiative Launched." *International AIDS Vaccine Initiative Report* 1 (summer): 7.

Floyd, Katherine, and Charles F. Gilks. 1996. "Impact of, and Response to, the HIV Epidemic at Kenyata National Hospital, Nairobi." Report 1 (April). Liverpool School of Tropical Medicine.

———. 1997. "Cost and Financing Aspects of Providing Anti-retroviral Therapy: A Background Paper." Paper presented at the Informal Consultation on the Implications of Antiretroviral Treatments, World Health Organization, Geneva. April 29–30.

Ford, K., D. N. Wirawan, P. Fajans, P. Meliawan, K. MacDonald, and L. Thorpe. 1996. "Behavioral Interventions for Reduction of Sexually Transmitted Disease/HIV Transmission among Female Commercial Sex Workers and Clients in Bali, Indonesia." *AIDS* 10: 213–22.

Foster, Susan, Peter Godrey-Faussett, and John Porter. 1997. "Modelling the Economic Benefits of Tuberculosis Preventive Therapy for People with HIV: The Example of Zambia." *AIDS* 11: 919–25.

Fox L. J., P. E. Bailey, K. L. Clarke-Martinez, and others. 1988. "Prevention of Transmission of HIV in Africa: Effectiveness of Condom Promotion and Health Education among Prostitutes." *Lancet* 15: 887–90.

Futures Group International. 1995a. "SOMARC Peru's PIEL Condoms Soar to the Top of the Market." *SOMARC* [Social Marketing for Change] *Highlights* 3 (July): 1–2.

———. 1995b. "Working with Religious Opposition in Africa." *SOMARC* [Social Marketing for Change] *Highlights* 3 (July): 1–2.

Garbus, L. 1996. "The UN Response." In Jonathan Mann and Daniel Tarantola, eds., *AIDS in the World II: Global Dimensions, Social Roots, and Responses.* The Global AIDS Policy Coalition. New York: Oxford University Press.

Garnett, Geoff P., and Roy M. Anderson. 1995. "Strategies for Limiting the Spread of HIV in Developing Countries: Conclusions Based on Studies of the Transmission Dynamics of the Virus." *Journal of Acquired Immune Deficiency Syndromes and Human Retrovirology* 9 (5): 500–13.

Garrett, Laurie. 1994. *The Coming Plague: Newly Emerging Diseases in a World Out of Balance.* New York: Penguin.

Gerard, Christine, Danièle Sondag-Thull, Edward-John Watson-Williams, and Lieve Fransen. 1995. *Safe Blood in Developing Countries: Principles and Organisation.* Brussels: Office for Official Publications of the European Communities.

Gertler, Paul, and Jacques van der Gaag. 1990. *The Willingness To Pay for Medical Care: Evidence from Two Developing Countries.* Baltimore: Johns Hopkins University Press.

Gertler, Paul, and Jeffrey Hammer. 1997. "Strategies for Pricing Publicly Provided Health Care Services." World Bank, Policy Research Department, Washington, D.C.

Gilson, L., R. Mkanje, H. Grosskurth, J. Picard, P. Mayaud, J. Todd, D. Mabey, and R. Hayes. 1996. "Cost-Effectiveness of Improved STD Treatment Services as a Preventive Intervention against HIV in Mwanza Region, Tanzania." Abstract Mo.C. 444. Paper read at the Eleventh International Conference on AIDS, Vancouver, B.C., Canada. July 7–12. Johns Hopkins School of Public Health.

Giraud, Patrick. 1992. "Economic Impact of HIV/AIDS on the Transport Sector: Development of an Assessment Methodology." Consultation on Economic Implications of HIV/AIDS, U.N. Development Programme, Bangalore, India. May 25–28.

Gluck, Michael, and Eric Rosenthal. 1995. *The Effectiveness of AIDS Prevention Efforts.* Washington, D.C.: Office of Technology Assessment.

Gold, David. 1996. "Progresss in AIDS Vaccine Development." *International AIDS Vaccine Initiative Report* 1 (summer): 4–5.

Golz, J. 1993. "Current Experiences with Methadone in the Treatment of Injection Drug Users with HIV Infection in Germany." Abstract WS-D12-6. Paper presented at the Ninth International Conference on AIDS, Berlin. June 6–11.

Graham, Ron. 1990. "One African's Tragedy Focuses Attention on AIDS." *New York Times,* April 1, section 2, p. 35.

Greenberg, Jerome. 1972. "Venereal Disease in the Armed Forces. *Medical Clinics of North America* 56 (5): 1087–1100.

Grosskurth, H., F. Mosha., J. Todd., E. Mwijarubi, A. Klokke, K. Senkoro, P. Mayaud, J. Changalucha, A. Nicoll, G. Ka-Gina, J. Newell, K. Mugeye, D. Mabey, and R. Hayes. 1995a. "Impact of Improved Treatment of Sexually Transmitted Diseases on HIV Infection in Rural Tanzania: Randomised Controlled Trial." *Lancet* 346 (8974): 530–36.

Grosskurth, H., F. Mosha, J. Todd, K. Senkoro, J. Newell, A. Klokke, J. Changalucha, B. West, P. Mayaud, A. Gavyole, R. Gabone, D. Mabey, and R.

Hayes. 1995b. "A Community Trial of the Impact of Improved Sexually Transmitted Disease: The HIV Epidemic in Rural Tanzania—Baseline Survey Results." *AIDS* 9 (8): 927–34.

Gupta, Indrani, Phane Mujinja, and Mead Over. Forthcoming. "Distribution of Poverty and Mortality in the Presence of AIDS Epidemic in Kagera, Tanzania." Working paper. World Bank, Policy Research Department, Washington, D.C.

Hammer, Jeffrey. 1997. "Economic Analysis for Health Projects." *World Bank Research Observer* 12 (1): 47–71.

Hampton, Janie. 1991. *Meeting AIDS with Compassion: AIDS Care and Prevention in Agomanya, Ghana.* Strategies for Hope 4. London: ActionAid, in assoc. with AMREF and World in Need.

Hassig, Susan E., Joseph Perriëns, Ekungola Baende, Mbindule Kahotwa, K. Bishagara, and N. Kinkela. 1990. "An Analysis of the Economic Impact of HIV Infection among Patients at Mama Yemo Hospital, Kinshasa, Zaire." *AIDS* 4: 883–87.

Hatty, Suzanne E. 1993. "Australia." In Nanette J. Davis, ed., *Prostitution: An International Handbook on Trends, Problems, and Policies.* Westport, Conn.: Greenwood Press.

Haverkos, Harry W., and Robert J. Battjes. 1992. "Female-to-Male Transmission of HIV." *Journal of the American Medical Association* 268 (14): 1855–56.

Hersh, B. S., F. Popovici, R. C. Apetrei, L. Zolotusca, N. Beldescu, A. Calomfirescu, Z. Jezek, M. J. Oxtoby, A. Gromyko, and D. L. Heymann. 1991. "Acquired Immunodeficiency Syndrome in Romania." *Lancet* 338 (8768): 645–49.

Hethcote, Herbert W. 1976. "Qualitative Analyses of Communicable Disease Models." *Mathematical Biosciences* 28: 335–36.

Hethcote, Herbert W., and James A. Yorke. 1984. *Gonorrhea Transmission Dynamics and Control.* Lecture Notes in Biomathematics 56. New York: Springer-Verlag.

Hien, N. T. 1995. "Drug Use and HIV Infection in Viet Nam." Report of the WHO Drug Injecting Project Planning Meeting, Phase II. Bangkok, Thailand.

"Ho Chi Minh City Worries about Its Shooting Booths." 1996. *AIDS Analysis Asia* 2 (5): 1.

Holtgrave, David R., Noreen L. Qualls, James W. Curran, Ronald O. Valdiserri, Mary E. Guinan, and William C. Parra. 1995. "An Overview of the Effectiveness and Efficiency of HIV Prevention Programs." *Public Health Reports* 110 (2): 134–45.

Holtgrave, David R., Noreen L. Qualls, and John D. Graham. 1996. "Economic Evaluation of HIV Prevention Programs." *Annual Review of Public Health* 17: 467–88.

Hook, E. W. III, R. O. Cannon, A. J. Nahmias, F. F. Lee, C. H. Campbell, Jr., D. Glasser, and T. C. Quinn. 1992. "Herpes Simplex Virus Infection as a Risk Factor for Human Immunodeficiency Virus Infection in Heterosexuals." *Journal of Infectious Diseases* 165 (2): 251–55.

Human Rights Watch/Asia. 1995. *Rape for Profit: Trafficking of Nepali Girls and Women to India's Brothels.* New York: Human Rights Watch.

Hunter, Susan, and John Williamson. Forthcoming. "Developing Strategies and Policies for USAID Interventions for Support of HIV/AIDS Infected and Affected Children." U.S. Agency for International Development. Washington, D.C.

Hurley, Susan F., Damien J. Jolley, and John M. Kaldor. 1997. "Effectiveness of Needle-Exchange Programmes for Prevention of HIV Infection." *Lancet* 349 (9068): 1797–1800.

Ingham, Roger. 1995. "AIDS: Knowledge, Awareness and Attitudes." In John Cleland and Benoit Ferry, eds., *Sexual Behaviour and AIDS in the Developing World*. London: Taylor & Francis.

Institute of Medicine. 1985. *New Vaccine Development: Establishing Priorities*. Vol. 1. Washington, D.C.: National Press Academy.

IAVI (International AIDS Vaccine Initiative). 1996. "A View from Uganda: An Interview with Dr. Edward Mbidde." *International AIDS Vaccine Initiative Report* 1 (summer): 3.

IAVR (International AIDS Vaccine Research). 1997a. "IAVI Launches Scientific Program." *International AIDS Vaccine Research Report* 2 (1): 1–3.

————. 1997b. "Studying AIDS Vaccines in Thailand: An Interview with Natth Bhamarapravati." *International AIDS Vaccine Research Report* 2 (1): 4–5.

Ismail, Rokiah. 1996. "Changing Pattern of HIV Infection as Seen at the University of Malaya Medical Center." Eleventh International Conference on AIDS. Abstract Mo.C. 1502. Vancouver, B.C., Canada. July 7–12. University of Malaysia, Department of Medicine, Kuala Lumpur.

Izazola-Licea, José A. 1996. "AIDS: The State of the Art: A Review Based on the Eleventh International Conference on AIDS, Vancouver, B.C., Canada. July 7–12. Fundación Mexicana Para La Salud, A.C., Mexico.

Jackson, D., J. Rakwar, B. Richardson, and others. 1997. "Decreased Incidence of Sexually Transmitted Diseases among Trucking Company Workers in Kenya: Results of a Behaviour Risk-Reduction Programme." *AIDS* 11: 903–909.

Jacquez, J. A., J. S. Koopman, Carl P. Simon, and Ira M. Longini, Jr. 1994. "Role of the Primary Infection in Epidemics of HIV Infection in Gay Cohorts." *AIDS* 7 (11): 1169–84.

Jain, Manoj K., T. Jacob John, and Gerald T. Keusch. 1994. "Epidemiology of HIV and AIDS in India." *AIDS* 8 (suppl. 2): S61–75.

Jalal, Fasli, Hadi M. Abednego, Tonny Sadjimin, and Michael J. Linnan. 1994. "HIV and AIDS in Indonesia." *AIDS* 8 (suppl. 2): S91–94.

James, Estelle. 1982. "The Nonprofit Sector in International Perspective: The Case of Sri Lanka." *Journal of Economic Literature* 053 (123): 99–129.

Janz, Nancy K., Marc A. Zimmerman, Patricia A. Wren, Barbara A. Israel, Nicholas Freudenberg, and Rosalind J. Carter. 1996. "Evaluation of 37 AIDS Prevention Projects: Successful Approaches and Barriers to Program Effectiveness." *Health Education Quarterly* 23 (1): 80–97.

Johnston, Margaret. 1996. "Why an HIV Vaccine is Scientifically Possible." *International AIDS Vaccine Initiative Report* 1 (summer): 1.

Jones, Clair. 1997. "Current Costs of HIV/AIDS—Malawi: A Case Study." Business Exchange on AIDS and Development [BEAD], meeting notes. London, March 14.

Jones, Gavin W., Endang Sulistyaningsih, and Terence H. Hull. 1994. "Prostitution in Indonesia." ANU Working Paper 52. Department of Demography, Research School of Social Sciences, Canberra, Australia.

Kahn, James G. 1996. "The Cost-Effectiveness of HIV Prevention Targeting: How Much More Bang for the Buck?" *American Journal of Public Health* 86 (12): 1709–12.

Kamali, A., J. F. Kengeya-Kayondo, S. S. Malamba, A. J. Nunn, J. A. Seeley, H. U. Wagner, and D. W. Mulder. 1992. "Orphans and HIV-1 Infection in a Rural Population Cohort in S.W. Uganda." Poster presented at the Eighth International Conference on AIDS, Amsterdam, the Netherlands. July 19–24. *Poster Abstracts*, vol. 2, PoD5159.

Kambou, G., S. Devarajan, and M. Over. 1992. "The Economic Impact of AIDS in an African Country:

Simulations with a General Equilibrium Model of Cameroon." *Journal of African Economies* 1 (1): 109–30.

Kamenga, M., R. W. Ryder, M. Jingu, and others. 1991. "Evidence of Marked Sexual Behavior Change Associated with Low HIV-1 Seroconversion in 149 Married Couples with Discordant HIV Serostatus: Experience at an HIV Counseling Center." *AIDS* 5: 61–67.

Kang, Lai-Yi. 1995. "Economic Growth and Urbanisation in China Leads to an Increase in STDs." *AIDS Analysis Asia* 1 (5): 14–15.

Kaplan, J. E., K. K. Holmes, H. W. Jaffe, H. Masur, and K. M. De Cock. 1996. "Preventing Opportunistic Infections in Human Immunodeficiency Virus-infected Persons: Implications for the Developing World." *American Journal of Tropical Medicine and Hygiene* 55: 1–11.

Killewo, J., L. Dahlgren, and A. Sandstrom. 1994. "Socio-geographical Patterns of HIV-1 Transmission in Kagera Region, Tanzania." *Social Science and Medicine* 38 (1): 129–34.

Killewo, J., K. Nyamuryekunge, A. Sandstron, U. Bredberg-Raden, S. Wall, F. Mhalu, and G. Biberfeld. 1990. "Prevalence of HIV Infection in Kagera Region of Tanzania: Population-based Study." *AIDS* 4 (November): 1081–85

Kin, F. 1995. "Injecting Drug Use among Heroin Users in Malaysia: Summary of Research Findings." Report of the WHO Drug Injecting Project Planning Meeting, Phase II. Bangkok, Thailand.

Kirby, Douglas, Lynn Short, Janet Collins, Deborah Rugg, Lloyd Kolbe, Marion Howard, Brent Miller, Freya Sonenstein, and Laurie S. Zabin. 1994. "School-Based Programs To Reduce Sexual Risk Behaviors: A Review of Effectiveness." *Public Health Reports* 109 (3): 339.

Kitahata, M. M., T. D. Koepsell, R. A. Deyo, C. L. Maxwell, W. T. Doge, and E. H. Wagner. 1996. "Physicians' Experience with the Acquired Immunodeficiency Syndrome as a Factor in Patients' Survival." *New England Journal of Medicine* 334 (11): 701–706.

Kogan, Rick. 1990. "Final Crusade." *Chicago Tribune*, April 3, section 5, p. 7.

Konde-Lule, Joseph K., Maria J. Wawer, F. Nalugoda, Ronald H. Gray, and Rekha Menon. 1997. "HIV Infection in Rural Households, Rakai District, Uganda." Preliminary draft presented at the IUSSP [International Union for the Scientific Study of Population] Conference on the Socio-Demographic Impact of AIDS in Africa, Durban, South Africa. February 3–6.

Krueger, L. E., R. W. Wood, P. H. Diehr, and C. L. Maxwell. 1990. "Poverty and HIV: Seropositivity: The Poor Are More Likely To Be Infected." *AIDS* 4 (8): 811–14.

Kunanusont, Chaiyos. 1997. "National Anti-retroviral Program: Thailand Experience." Paper presented at the Informal Consultation on Implications of Anti-retroviral Treatments, World Health Organization, Geneva. April 29–30.

Laga, M., M. Alary, N. Nzila, A. T. Manoka, M. Tuliza, F. Behets, J. Goeman, M. St. Louis, and P. Piot. 1994. "Condom Promotion, Sexually Transmitted Diseases Treatment and Declining Incidence of HIV-1 Infection in Female Zairian Sex Workers." *Lancet* 344 (8917): 246–48.

Laga, M., A. Manoka, M. Kivuvu, B. Malele, M. Tuliza, N. Nzila, J. Groeman, F. Behets, V. Batter, and M. Alary. 1993. "Non-ulcerative Sexually Transmitted Diseases as Risk Factors for HIV-1 Transmission in Women: Results from a Cohort Study." *AIDS* 7 (1): 95–102.

Lamptey, Peter, and Peter Piot, eds. 1990. *The Handbook for AIDS Prevention in Africa*. Durham, N.C.: Family Health International.

Laumann, Edward O., Christopher M. Masi, and Ezra W. Zuckerman. 1997. "Circumcision in the United States: Prevalence, Prophylactic Effects, and Sexual Practice." *Journal of the American Medical Association* 277 (13): 1052–57.

Lavy, Victor, and John M. Quigley. 1993. "Willingness To Pay for the Quality and Intensity of Medical Care: Low-Income Households in Ghana." LSMS 94. World Bank, Policy Research Department, Poverty and Human Resources Division, Washington, D.C.

Laws, Margaret. 1996. "International Funding of Global AIDS Strategy: Official Development Assistance." In Jonathan Mann and Daniel Tarantola, eds., *AIDS in the World II: Global Dimensions, Social Roots, and Responses.* The Global Policy Coalition. New York: Oxford University Press.

Lazzarin, A., A. Saracco, M. Musicco, and A. Nicolosi. 1991. "Man-to-Woman Sexual Transmission of the Human Immunodeficiency Virus." *Archives of Internal Medicine* 151 (12): 2411–16.

Lee, S. S, W. W. Lim, and S. H. Lee. 1993. "Epidemiology of HIV Infection in Hong Kong: Analysis of the First 300 Cases." Abstract PO-C08-2780. Paper presented at the Ninth International Conference on AIDS, Berlin. June 6–11.

Leiner, Marvin. 1994. "AIDS: Cuba's Effort to Contain." In *Sexual Politics in Cuba: Machismo, Homosexuality and AIDS.* Boulder, Colo.: Westview Press.

Levine, W. C., G. Higueras, R. Revollo, and others. 1996. "Rapid Decline in Sexually Transmitted Disease Prevalence among Brothel-based Sex Workers in La Paz, Bolivia: The Experience of Proyecto Contra SIDA, 1992–1995." Abstract Mo.C. 441. Eleventh International Conference on AIDS, Vancouver, B.C., Canada. July 7–12.

Libonatti, O., E. Lima, A. Peruga, R. Gonzalez, F. Zacarias, and M. Weissenbacher. 1993. "Role of Drug Injection in the Spread of HIV in Argentina and Brazil." *International Journal of STD and AIDS* 4 (3): 135–41.

Linden, Eugene. 1996. "The Exploding Cities of the Developing World." *Foreign Affairs* 75 (1): 52–65.

Lockheed, Marlaine E., Adriaan M. Verspoor, Deborah Bloch, Pierre Englebert, Bruce Fuller, Elizabeth King, John Middleton, Vicente Paqueo, Alastair Rodd, Ralph Romain, and Michel Welmond. 1991. *Improving Primary Education in Developing Countries.* New York: Oxford University Press.

Low-Beer, Daniel, and Seth Berkley. 1996. "HIV." In Christopher J. L. Murray and Allan D. Lopez, eds., *Health Dimensions of Sex and Reproduction: The Global Burden of Sexually Transmitted Diseases, HIV, Maternal Conditions, Prenatal Disorders and Congenital Anomalies.* Cambridge, Mass.: Harvard University Press.

Lowenthal, Nancy, Angela Chimwaza, Saifuddin Ahmed, and Judith Timyan. 1995. "Malawi Contraceptive Consumer Knowledge, Attitudes, and Practices Survey, June–July, 1995." Population Services International, Washington, D.C.

Lurie, Peter, and Ernest Drucker. 1997. "An Opportunity Lost: HIV Infections Associated with Lack of a National Needle-Exchange Program in the U.S.A." *Lancet* 349 (9052): 604–608.

Lurie, Peter, M. Fernandes, V. Hughes, E. Arevalo, E. Hudes, A. Reingold, N. Hearst, and the Instituto Adolfo Lutz Study Group. 1995. "Socioeconomic Status and Risk of HIV-1, Syphilis and Hepatitis B Infection among Sex Workers in São Paulo State, Brazil." *AIDS* 9 (suppl. 1): S31–37.

Lurie, Peter, A. L. Reingold, B. Bowser, D. Chen, J. Foley, J. Guydish, J. G. Kahn, S. Lane, and J. Sorensen. 1993. *The Public Health Impact of Needle Exchange Programs in the United States and Abroad.* Vol. 1. San Francisco, Calif.: University of California.

Lwihula, George. 1994. "Variability in Saving and Assistance Behaviour in Households across Ethnic Groups in Kagera Region, Tanzania: A Focus Group Study." University of Dar es Salaam, Muhimbili University College of Health Sciences, Institute of Public Health, Department of Behavioural Sciences.

Maharjan, S., J. A. Peak, S. Rana, and N. Crofts. 1994. "Declining Risk for HIV among IDUs in Kathmandu: Impact of a Harm Reduction Program." Abstract 561C. Paper presented at the Tenth International Conference on AIDS, Yokohama, Japan. August 14–17. Lifesaving and Lifegiving Society, Kathmandu, Nepal.

Mann, Jonathan, and Daniel Tarantola, eds. 1996. *AIDS in the World II: Global Dimensions, Social Roots, and Responses.* Global AIDS Policy Coalition. New York: Oxford University Press.

Mann, Jonathan, D. Tarantola, and T. W. Netter. 1992. *AIDS in the World.* The Global AIDS Policy Coalition. Cambridge, Mass.: Harvard University Press.

Mann, Jonathan, H. Francis, T. C. Quinn, K. Bila, P. K. Asila, N. Bosenge, N. Nzilambi, L. Jansegers, P. Piot, K. Ruti, and J. W. Curran. 1986. "HIV Seroprevalence among Hospital Workers in Kinshasa, Zaire: Lack of Association with Occupational Exposure." *Journal of the American Medical Association* 256 (22): 3099–102.

Mansergh, G., A. Haddix, R. Steketee, P. Nieburg, D. Hu, R. J. Simonds, and M. Rogers. 1996. "Cost-Effectiveness of Short-Course Zidovudine To Prevent Perinatal HIV Type 1 Infection in Sub-Saharan African Developing Country Setting." *Journal of the American Medical Association* 276 (2): 139–45.

Martin, Anna L., Eric Van Praag, and Roland Msiska. 1996. "An African Model of Home-based Care: Zambia." In Jonathan Mann and Daniel Tarantola, eds. *AIDS in the World II: Global Dimensions, Social Roots, and Responses.* The Global AIDS Policy Coalition. New York: Oxford University Press.

Mastro, Timothy D., and Isabelle de Vincenzi. 1996. "Probabilities of Sexual HIV-1 Transmission." *AIDS* 10 (suppl. A): S72–82.

Mastro, Timothy D., Glen A. Satten, Taweesak Nopkesorn, Suebpong Sangkharomya, and Ira M. Longini, Jr. 1994. "Probability of Female-to-Male Transmission of HIV-1 in Thailand" *Lancet* 343 (8891): 204–07.

Mauskopf, J. A., J. E. Paul, D. S. Wichman, A. D. White, and H. H. Tilson. 1996. "Economic Impact of Treatment of HIV-Positive Pregnant Women and Their Newborns with Zidovudine: Implications for HIV Screening." *Journal of the American Medical Association* 276 (2): 132.

May, M. M., and R. M. Anderson. 1987. "Transmission Dynamics of HIV Infection." *Nature* 326: 137–42.

Mazzullo, J., S. Wroblewski, R. Rudd, and P. Fairchild. 1990. "Influencing Choices of Medical Students to HIV-related careers." Abstract no. S.D.910. Sixth International AIDS Conference, San Francisco. June 20–24.

McBrier, Page. 1995. "Children's-Health: Giving AIDS a Human Face." *Inter Press Service,* Unicef Feature 00153.UGA, September.

McCoy, Clyde B., Shenghan Lai, Lisa R. Metsch, Xueren Wang, Cong Li, Ming Yang, and Li Yulong. 1997. "No Pain No Gain: Establishing the Kunming, China, Drug Rehabilitation Center." *Journal of Drug Issues* 27 (1): 73–85.

Metzger, David. 1997. "Drug Abuse Treatment as AIDS Prevention." In NIH Consensus Development Conference, *Interventions to Prevent HIV Risk Behaviors: Program and Abstracts.* Bethesda, Md.: National Institutes of Health.

Miller, Norman, and Rodger Yeager. 1995. "By Virtue of Their Occupation, Soldiers and Sailors Are at Greater Risk." *AIDS Analysis Africa* 5 (6): 8–9.

Mills, Anne, Jonathan Broomberg, John Lavis, and Neil Soderlund. 1993. "The Costs of HIV/AIDS Prevention Strategies in Developing Countries." World Health Organization, Global Programme on AIDS, GPA/DIR/93.2, Geneva.

Minon, Jeffrey A., and Jeffrey Zwiebel. 1995. "The Economic Case against Drug Prohibition." *Journal of Economic Perspectives* 9 (2): 175–92.

Moffitt, Robert. 1991. "The Use of Selection Modeling To Evaluate AIDS Interventions with Observational Data." In National Research Council, *Evaluating HIV AIDS Prevention Programs*. Washington, D.C.: National Academy Press.

Moore, M., E. Tukwasiibwe, E. Marum, and others. 1993. "Impact of HIV Counseling and Testing in Uganda." Eleventh International Conference on AIDS. Abstract WS-C16-4. Vancouver, B.C., Canada. July 7–12.

Moore, Richard D., and John G. Bartlett. 1996. "Combination Antiretroviral Therapy in HIV Infection: An Economic Perspective." *PharmacoEconomics* 10 (2): 109–13.

Morris, Martina, Chai Podhisita, Maria J. Wawer, and Mark S. Handcock. 1996. "Bridge Populations in the Spread of HIV/AIDS." *AIDS* 10: 1265–71.

Morrow, Richard, Robert Colebunders, and James Chin. 1989. "Interactions of HIV infection with Endemic Tropical Diseases." *AIDS* 3 (suppl.): S79–87.

Moses, Stephen, Janet Bradley, Nico Nagelkerke, Allan Ronald, J. O. Ndinya-Achola, and Francis A. Plummer. 1990. "Geographical Patterns of Male Circumcision Practices in Africa: Association with HIV Seroprevalence." *International Journal of Epidemiology* 19 (3): 693–97.

Moses, Stephen, F. A. Plummer., J. E. Bradley., J. O. Ndinya-Achola., N. J. D. Nagelkerke, and A. R. Ronald. 1995. "Male Circumcision and the AIDS Epidemic in Africa. *Health Transition Review* 5 (1): 100–103.

Moses, Stephen, F. A. Plummer, E. N. Ngugi, N. J. Nagelkerke, Ao Anzala, and Jo Ndinya-Achola. 1991. "Controlling HIV in Africa: Effectiveness and Cost of an Intervention in a High Frequency STD Transmitter Core Group." *AIDS* 5 (4): 407–11.

Msamanga, G., E. Urassa, D. Spiegelman, E. Hertzmark, S. Kapiga, D. J. Hunter, and W. W. Fawzi. 1996. "Socioeconomic Status and Prevalence of HIV Infection among Pregnant Women in Dar Es Salaam, Tanzania." Abstract TuC2464. Eleventh International Conference on AIDS, Vancouver, B.C., Canada. July 7–12. Muhimbili University, College of Health Sciences, Dar-es-Salaam, Tanzania.

Mulder, Daan. 1996. "Disease Progression and Mortality Following HIV-1 Infection." In Jonathan Mann and Daniel Tarantola, eds., *AIDS in the World II: Global Dimensions, Social Roots, and Responses*. The Global AIDS Policy Coalition. New York: Oxford University Press.

Mulder, Daan, Andrew Nunn, Anatoli Kamali, and Jane Kengeya-Kayondo. 1995. "Decreasing HIV-1 Seroprevalence in Young Adults in a Rural Uganda Cohort." *British Medical Journal* 311 (30): 833–36.

Murray, Christopher J. L., and Allan D. Lopez. 1996. "The Global Burden of Disease." *Global Burden of Disease and Injury Series*, vol. 1. WHO, Harvard School of Public Health, World Bank. Cambridge, Mass.: Harvard University Press.

Mwabu, Germano M., Martha Ainsworth, and Andrew Nyamete. 1993. "Quality of Medical Care and Choice of Medical Treatment in Kenya: An Empirical Analysis." *Journal of Human Resources* 28 (4): 838–62.

Mwizarubi, B. K., C. L. Mwaijonga, and O. Laukamm-Josten. 1992. "HIV/AIDS Education and Condom Promotion for Truck Drivers, Their Assistants, and Sex Partners in Tanzania." Paper presented at the

Global Programme on AIDS, Effective Approaches to AIDS Prevention, Geneva. May 26–29. African Medical and Research Foundation, Dar-es-Salaam, Tanzania.

National Research Council. 1989. *AIDS: Sexual Behavior and Intravenous Drug Use.* Washington, D.C.: National Academy Press.

———. 1991. *Evaluating AIDS Prevention Programs.* Edited by Susan L. Coyle, Robert F. Boruch, and Charles F. Turner. Washington, D.C.: National Academy Press.

———. 1996. *Preventing and Mitigating AIDS in Sub-Saharan Africa.* Washington, D.C.: National Academy Press.

Nelson, Kenrad, David Celentano, Sakol Eiumtrakol, D. R. Hoover, C. Beyrer, S. Suprasert, S. Kuntolbutra, and C. Khamboonruang. 1996. "Changes in Sexual Behavior and Decline in HIV Infection among Young Men in Thailand." *New England Journal of Medicine* 335 (5): 297–303.

"NGOs Flout AIDS Control Policy." 1994. *The Telegraph* (India), Metropolitan section, September 8.

Ngugi, E. N., F. A. Plummer, J. N. Simonson, D. W. Cameron, M. Bosire, P. Waiyaki, A. R. Ronald, J. O. Ndinya-Achola. 1988. "Prevention of Transmission of Human Immunodeficiency Virus in Africa: Effectiveness of Condom Promotion and Health Education among Prostitutes." *Lancet* 2 (8616): 887–90.

Nicoll, Angus, D. Bennett, M. Catchpole, B. Evans, O.N. Gill, J. Mortimer, P. Mortimer, and K. Paine. 1996. "HIV, AIDS, and Sexually Transmitted Infections: Global Epidemiology, Impact, and Prevention." Health and Population Occasional Paper. Overseas Development Administration, London.

Nolan, Peter. 1993. "Economic Reform, Poverty and Migration in China." *Economic and Political Weekly* 28 (26): 1369–77.

Nold, A. 1978. "The Infectee Number at Equilibrium for a Communicable Disease." *Mathematical Biosciences* 46: 131–38.

Normand, J., D. Vlahov, and L. E. Moses, eds. 1995. *Preventing HIV Transmission: The Role of Sterile Needles and Bleach.* A Report of the Panel on Needle Exchange and Bleach Distribution Programs, Commission on Behavioral and Social Sciences and Education. National Research Council and Institute of Medicine. Washington D.C.: National Academy Press.

Oakley, A., D. Fullerton, and J. Holland. 1995. "Behavioural Interventions for HIV/AIDS Prevention." *AIDS* 9 (5): 479–86.

OECD (Organization for Economic Co-operation and Development). 1995. "Trends in International Migration: Continuous Reporting System on Migration." *Annual Report 1994.* Paris.

Ong, Jin Hui. 1993. "Singapore." In Nanette E. Davis, ed., *Prostitution: An International Handbook on Trends, Problems and Policies.* Westport, Conn.: Greenwood Press.

ONUSIDA (UNAIDS). 1997. *Diagnostico Epidemia VIH/SIDA Internacional y Nacional.* Santo Domingo, Dominican Republic.

Oppenheimer, Edna. 1995. "Drugs and HIV/AIDS in Myanmar: Land of the 'First Inhabitants of the World.'" *AIDS Analysis Asia* 1 (4): 5–6.

Osmanov, Saladin. 1996. "Implications of the Human Immunodeficiency Virus Variability for Transmission: Scientific and Policy Issues." Expert Group of the Joint United Nations Programme on HIV/AIDS. UNAIDS, Geneva.

Ou, C. Y., Y. Takebe, B. G. Weniger, C. C. Luo, M. L. Kalish, W. Auwanit, S. Yamazaki, H. D. Gayle, N. L. Young, and G. Schochetman. 1993. "Independent Introduction of Two Major HIV-1 Genotypes into

Distinct High-Risk Populations in Thailand." *Lancet* 341 (8854):1171–74; errata corrected in 342 (8865): 250.

Over, Mead. 1992. "The Macroeconomic Impact of AIDS in Sub-Saharan Africa." AFTPN Technical Working Paper 3. World Bank, Africa Technical Department, Population, Health, and Nutrition Division, Washington, D.C.

———. 1997. "The Public Interest in a Private Disease: Why Should the Government Play a Role in STD Control?" In K. K. Holmes, P. F. Sparling, P-A Mardh, S. M. Lemon, W. E. Stamm, and J. W. Wasserheit, eds., *Sexually Transmitted Diseases*. 3d ed. New York: McGraw-Hill.

Over, Mead, and Godlike Koda. Forthcoming. "Average Cost of Survivor Assistance Programs in Kagera, Tanzania." Working paper. World Bank, Policy Research Department, Washington, D.C.

Over, Mead, and Peter Piot. 1993. "HIV Infection and Sexually Transmitted Diseases." In Dean T. Jamison, W. Henry Mosley, Anthony R. Measham, and José Luis Bobadilla, eds., *Disease Control Priorities in Developing Countries*. New York: Oxford University Press.

———. 1996. "Human Immunodeficiency Virus Infection and Other Sexually Transmitted Diseases in Developing Countries: Public Health Importance and Priorities for Resource Allocation." *Journal of Infectious Diseases* 174 (suppl. 2): S162–75.

Over, Mead, Phare Mujinja, Daniel Dorsainvil, and Indrani Gupta. Forthcoming. "Impact of Adult Death on Household Expenditures in Kagera, Tanzania." Working paper. World Bank, Policy Research Department, Washington, D.C.

Over, Mead, M. Randall, P. Ellis, Joyce H. Huber, and Orville Solon. 1992. "The Consequences of Adult Ill-Health." In Richard A. Feachem, Tord Kjellstrom, Christopher J. L. Murray, Mead Over, and Margaret A. Phillips, eds., *The Health of Adults in the Developing World*. New York: Oxford University Press.

Padian, N. S., S. C. Shiboski, and N. P. Jewell. 1991. "Female-to-Male Transmission of Human Immunodeficiency Virus." *Journal of the American Medical Association* 266 (12):1664–67.

Padian, N. S., T. R. O'Brien, Y. Chang, S. Glass, and D. P. Francis. 1993. "Prevention of Heterosexual Transmission of Human Immunodeficiency Virus through Couple Counseling." *AIDS* 6 (9): 1043–48.

Pal, S. C., S. Sarkar, T. N. B. Naik, P. K. Singh, S. I. Tuchi, A. S. Lal, and S. P. Tripathy. 1990. "Explosive Epidemic of HIV Infection in North Eastern States of India, Manipur and Nagaland." *CARC* 3: 2–6.

Panos Institute. 1989. *AIDS and the Third World*. Philadelphia: New Society Publishers.

Parker, Richard G. 1996. "Historic Overview of Brazil's AIDS Programs and Review of the World Bank AIDS Project." Family Health International/AIDSCAP, Arlington, Va.

Pauw, J., J. Ferrie, and R. R. Villegas. 1996. "Controlled HIV/AIDS-related Health Education Programme in Managua, Nicaragua." *AIDS* 10: 537–544.

Peak, A. S., Rana, S. H. Maharjan, and N. Crofts. 1994. "An Indigenous Harm Reduction Programme for IDUs in a Developing Country, Nepal." Abstract PD0508. Paper presented at the Tenth International Conference on AIDS, Yokohama, Japan. August 14–17. Lifesaving and Lifegiving Society, Kathmandu, Nepal.

Peng, Xizhe. 1994. "Recent Trends in China's Population and Their Implications." CP 30. London School of Economics, Suntory-Toyota International Centre for Economics and Related Disciplines. London.

Perriëns, Joseph, Kenneth Hill, Nicholas Prescott, and Chaiyos Kunanusont. 1997. "Health Gains from Antiretroviral Therapy." Geneva: UNAIDS.

Philipson, Tomas J., and Richard A. Posner. 1993. *Private Choices and Public Health: The AIDS Epidemic in an Economic Perspective.* Cambridge, Mass.: Harvard University Press.

Phoolcharoen, Wiput, and Seri Phongphit. 1996. "HIV Prevention Works Report: Case Study Thailand." Paper presented at the Eleventh International Conference on AIDS, Vancouver, B.C., Canada. July 7–12.

Pickering, Helen, and H. A. Wilkins. 1993. "Do Unmarried Women in African Towns Have To Sell Sex, or Is It a Matter of Choice?" *Health Transition Review* 3 (suppl.): 17–27.

Pickering, Helen, M. Quigley, J. Pepin, and others. 1993. "The Effects of Post-Test Counseling on Condom Use among Prostitutes in The Gambia." *AIDS* 7: 271–73.

Pinkerton, Steven D., and Paul R. Abramson. 1996. "Implications of Increased Infectivity in Early-Stage HIV Infection." *Evaluation Review* 20 (5): 516–40.

———. 1997. "Effectiveness of Condoms in Preventing HIV Transmission." *Social Science and Medicine* 44 (9): 1303–12.

Piot, Peter. 1994. "Differences between African and Western Patterns of Heterosexual Transmission." In A. Nicolosi, ed., *HIV Epidemiology: Models and Methods.* New York: Rowen Press.

Pitayanon, Sumalee, Sukontha Kongsin, and Wattana Janjaroen. 1997. "The Economic Impact of HIV/AIDS Mortality on Households in Thailand." In David Bloom and Peter Godwin, eds., *The Economies of HIV and AIDS: The Case of South and South East Asia.* Delhi: Oxford University Press.

Plange, Nii-K. 1990. "Report on Prostitution in Fiji." University of the South Pacific, Department of Sociology, Suva, Fiji.

Plummer, Francis A., Stephen Moses, and Jackoniah O. Ndinya-Achola. 1991. "Factors Affecting Female-to-Male Transmission of HIV-1: Implications of Transmission Dynamics for Prevention." In Lincoln Chen,

Jaime Sepulveda Amor, and Sheldon Segal, eds., *AIDS and Women's Reproductive Health.* New York: Plenum Press.

Pokrovsky, V. V., I. V. Savchenko, N. N. Ladnaya, and others. 1996. "HIV Infection Surveillance in Russia in 1987–1995 (Statistics)." Russia AIDS Centre, Moscow.

Porapakkham, Yaowarat, Somjai Pramarnpol, Supatra Athibhoddhi, and Richard Bernhard. 1996. "The Evolution of HIV/AIDS Policy in Thailand: 1984–1994." *AIDSCAP* Policy Working Paper Series WP5. Family Health International, Washington, D.C.

Poshyachinda, V. 1993. "Drug Injecting and HIV Infection among the Population of Drug Users in Asia." *Bulletin on Narcotics* 45 (1): 877–90.

Prescott, Nicholas. 1997. "Setting Priorities for Government Involvement with Antiretrovirals." Paper presented at the Informal Consultation on The Implications of Antiretroviral Treatments, World Health Organization, Geneva. April 29–30. World Bank, East Asia and Pacific Department, Human Resources Operations Division.

Prescott, Nicholas, Chaiyos Kunanusont, W. Phoolcharoen, W. Rojanapitayakorn, Joseph Perriëns, and D. Boonyuen. 1996. "Formulating Rational Use of Antiretrovirals in Thailand." Abstract Mo.B. 533. Paper presented at the Eleventh International Conference on AIDS, Vancouver, B.C., Canada. July 7–12. Ministry of Public Health, Department of CDC, AIDS Division, Bangkok, Thailand.

Putnam, Robert. 1993. "The Prosperous Community—Social Capital and Public Life." *American Prospect* 13: 35–42.

Putnam, Robert, R. Leonardi, and R. Nanetti. 1993. *Making Democracy Work: Civic Traditions in Modern Italy.* Princeton: Princeton University Press.

Quinn, T. C., R. O. Gannon, D. Glasser, S. L. Groseclose, W. S. Brathwaite, A. S. Fauci, and E. W.

Hook 3d. 1990. "The Association of Syphilis with the Risk of HIV Infection in Patients Attending STD Clinics." *Archives of Internal Medicine* 150 (6): 1297–1302.

Quinn, T. C., A. Ruff, and N. Halsey. 1994. "Special Considerations for Developing Nations." In P. A. Pizzo and C. M. Wilfert, eds., *Pediatric AIDS: The Challenge of HIV Infection in Infants, Children and Adolescents*. Baltimore, Md.: Williams and Wilkins.

Radhakrishna, R., K. Subbarao, with S. Indrakant and C. Ravi. 1997. "India's Public Distribution System: A National and International Perspective." World Bank, South Asia Country Department 2 (Agriculture and Water Operations), Washington, D.C.

Reeler, Anne V. 1990. "Injections: A Fatal Attraction?" *Social Science and Medicine* 31 (10): 1119–25.

Refeno, Germain, Victor Rabeza, Gora Mboup, and Juan Schoemaker. 1994. *Madagascar: Enquête Nationale Demographique et Sanitaire 1992*. Antananarivo, Madagascar: Centre National de Recherches sur l'Environnement, Ministère de la Recherche Appliquée au Developpement, and Calverton, Md.: Macro International, Inc.

Reggy, Atieno, R. J. Simonds, and Martha Rogers. 1997. "Preventing Perinatal HIV Transmission." *AIDS* 11 (suppl. A): S61–67.

Rezza, G., C. Oliva, and H. Sasse. 1988. "Preventing AIDS among Italian Drug Addicts: Evaluation of Treatment Programs and Informative Strategies." Paper presented at the Fourth International Conference on AIDS, Stockholm, Sweden. June 12–16.

Robbins, A., and P. Freeman. 1988. "Obstacles to Developing Vaccines for the Third World." *Scientific American* 259 (5): 126–33.

Robinson, Noah Jamie, Daan W. Mulder, Bertran Auvert, and Richard J. Hayes. 1997. "Proportion of HIV Infections Attributable to Other Sexually Trans-mitted Diseases in a Rural Ugandan Population: Simulation Model Estimates." *International Journal of Epidemiology* 26 (1): 180–89.

Rojanapithayakorn, Wiwat, and Robert Hanenberg. 1996. "The 100% Condom Program in Thailand." *AIDS* 10 (1): 1–7.

Rowley, Janet, Roy Anderson, and T. W. Ng. 1990. "Reducing the Spread of HIV Infection in Sub-Saharan Africa: Some Demographic and Economic Implications." *AIDS* 4 (1): 47–56.

Ryder, Robert, V. Batter, M. Nsuami, and others. 1991. "Fertility Rates in 238 HIV-1 Seropositive Women in Zaire Followed for 3 Years Post-Partum." *AIDS* 5: 1521–27.

Ryder, Robert W., Mibandumba Ndilu, Susan E. Hassig, Munkolenkole Kamenga, D. Sequeira, M. Kashamuka, H. Francis, F. Behets, R. L. Cole-bunders, and A. Dopagne. 1990. "Heterosexual Transmission of HIV-1 among Employees and Their Spouses at Two Large Businesses in Zaire." *AIDS* 4 (8): 725–32.

Sarkar, Swarup, Anindya Chatterjee, Clyde B. McCoy, Abu S. Adbul-Quader, Lisa R. Metsch, and Robert S. Anwyl. 1996. "Drug Use and HIV among Youth in Manipur, India." In Clyde B. McCoy, Lisa R. Metsch, and James A. Inciardi, eds. *Intervening with Drug-Involved Youth*. Thousand Oaks, Calif.: Sage.

Sarkar, Swarup, N. Das, Panda, T. N. Naik, K. Sarkar, B. C. Singh, J. M. Ralte, S. M. Aier, and S. P. Tripathy. 1993. "Rapid Spread of HIV among Injecting Drug Users in North-Eastern States of India." *Bulletin on Narcotics* 45 (1): 91–105.

Sato, Paul. 1996. "Sentinel HIV Surveillance." In Jonathan Mann, and Daniel Tarantola, eds., *AIDS in the World II: Global Dimensions, Social Roots, and Responses*. The Global AIDS Policy Coalition. New York: Oxford University Press.

Sauerborn, R., P. Berman, and A. Nougtara. 1996. "Age Bias, but No Gender Bias, in the Intra-Household

Resource Allocation for Health Care in Rural Burkina Faso." *Health Transition Review: The Cultural, Social and Behavioural Determinants of Health* 6 (2): 131–45.

Scheper-Hughes, Nancy. 1993. "AIDS, Public Health, and Human Rights in Cuba." *Lancet* 342 (8877): 965–67.

Sepulveda, Jaime. 1992. "Prevention through Information and Education: Experience from Mexico." In Jaime Sepulveda, Harvey Fineberg, and Jonathan Mann, eds. *AIDS Prevention through Education: A World View.* New York: Oxford University Press.

Serwadda, David, Maria J. Wawer, Stanley D. Musgrave, Nelson K. Sewankambo, Jonathan E. Kaplan, and Ronald H. Gray. 1992. "HIV Risk Factors in Three Geographic Strata of Rural Rakai District, Uganda." *AIDS* 6 (9): 983–89.

Serwadda, David, M. Wawer, N. Sewankambo, R. H. Gray, C-J. Li, R. Kelly, and T. Lutalo. 1995. "Trends in HIV Incidence and Prevalence in Rakai District, Uganda." Abstract MoC085. Ninth International Conference on AIDS and STDs in Africa, Kampala, Uganda. December 10–14. Institute of Public Health, Makerere University, Kampala, Uganda.

Shaeffer, Sheldon. 1995. "Impact on Education." In *The Impact of HIV on Children in Thailand.* Program on AIDS, Research Report 16. Bangkok: Thai Red Cross Society.

Siegel, Joanna E., Milton C. Weinstein, and Harvey V. Fineberg. 1991. "Bleach Programs for Preventing AIDS among IV Drug Users: Modeling the Impact of HIV Prevalence." *American Journal of Public Health* 81 (10): 1273–79.

"Signs of Change as Taboo Subjects Get Air Time in Information Campaign. Country Profile: Pakistan." 1996. *AIDS Analysis Asia* 2 (3): 14–15.

Simonsen, J. N., D. Cameron, M. Gakinya, Jo Ndinya-Achola, L. J. D'Costa, P. Karasira, M. Cheang, A. R. Ronald, P. Piot, and F. A. Plummer. 1988. "Human Immunodeficiency Virus Infection among Men with Sexually Transmitted Diseases." *New England Journal of Medicine.* 319: 274–78.

Singh, Sujata. 1995. "Three Year Stint at Sonagachi." All India Institute of Hygiene and Public Health, Calcutta. August.

Sittitrai, Werasit. 1994. "Nongovernmental Organization and Community Responses to HIV/AIDS in Asia and the Pacific." *AIDS* 8 (suppl. 2): S199–206.

Smith, James, and Alan Whiteside. 1995. "The Socio-Economic Impact of HIV/AIDS on Zambian Businesses: Report for the BEAD [Business Exchange on AIDS and Development Group] and CDC." Commonwealth Development Corporation, London.

SOMARC (Social Marketing for Change), Office of Public Relations. 1996. "Crisis Communications Lessons Learned: Case Study of Niger and the Philippines." SOMARC Occasional Paper 20. Futures Group International, Washington, D.C.

Span, Paula. 1996. "Needle Exchanges Inject Controversy in AIDS Prevention." *Washington Post,* July 16, p. A1.

"Sri Lankan Condom Sales Higher but Not High Enough, Say AIDS Workers." 1996. *AIDS Analysis Asia* 2 (6): 15.

Ssengonzi, Robert, Martina Morris, Nelson Sewandambo, Maria Wawer, and David Serwadda. 1995. "Economic Status and Sexual Networks in Rakai District, Uganda." Abstract WeC259. Ninth International Conference on AIDS and STDs in Africa, Kampala, Uganda. December 10–14. Uganda Virus Research, Entebbe, Uganda.

Stanecki, Karen A., and Peter O. Way. 1997. "The Demographic Impacts of AIDS—Perspectives from the World Population Profile: 1996." IPC Staff Paper 86. International Programs Center, Population Division, U.S. Bureau of the Census. Washington, D.C.

Stewart, Graeme. 1997. "Compliance with Antiretroviral Therapy." Paper presented at the Informal Consultation on the Implications of Antiretroviral Treatments, World Health Organization, Geneva. April 20–30.

Stimson, Gerry V. 1993. "The Global Diffusion of Injecting Drug Use: Implications for Human Immunodeficiency Virus Infection." *Bulletin on Narcotics* 45 (1): 3–17.

———. 1994. "Reconstruction of Subregional Diffusion of HIV Infection among Injecting Drug Users in Southeast Asia: Implications for Early Intervention." *AIDS* 8 (11): 1630–32.

———. 1996. "Drug Injecting and the Spread of HIV Infection in South-East Asia." In L. Sherr, J. Catalan, and B. Hedge, eds. *The Impacts of AIDS: Psychological and Social Aspects of HIV Infection.* Reading, U.K.: Harwood Academic.

Stoneburner, Rand L., and Carballo, Manuel. 1997. "An Assessment of Emerging Patterns of HIV Incidence in Uganda and Other East African Countries." Final report of consultation. Family Health International, AIDS Control and Prevention Project. International Centre for Migration and Health, Geneva.

Stover, John, and Peter O. Way. 1995. "The Impact of Interventions on Reducing the Spread of HIV in Africa: Computer Simulation Applications." *African Journal of Medical Practice* 2 (4): 110–20.

Strauss, John, and Duncan Thomas. 1995. "Human Resources: Empirical Modeling of Household and Family." In Jere Behrman and T. N. Srinivasan, eds., *Handbook of Development Economics.* Vol. 3A. New York: North-Holland.

Subbarao, Kalanidhi, Aniruddha Bonnerjee, Jeanine Braithwaite, Soniya Carvalho, Rene Ezemenari, Carol Graham, and Alan Thompson. 1996. "Social Assistance and Poverty-targeted Programs: Lessons from Cross-Country Experience." World Bank, Poverty and Social Policy Department, Washington, D.C.

Taha, T. E., J. K. Canner, J. D. Chiphangwi, and others. 1996. "Reported Condom Use Is Not Associated with Incidence of Sexually Transmitted Diseases in Malawi." *AIDS* 10: 207–12.

Tan, Michael L., and Manuel M. Dayrit. 1994. "HIV/AIDS in the Philippines." *AIDS* 8 (suppl. 2): S125–30.

Tchupo, J. P., J. Timyan, C. Miken, J. G. Ouango, and S. Watts. 1996. "PSI Research Report Summary: Burkina Faso Condom KAP Survey, April. 1995." Population Services International, Washington, D.C.

Tembo, George, Hany Friesan, Godwil Asiimwe-Okiror, Rita Moser, Warren Naamara, Nathan Bakyaita, and Joshua Musinguzi. 1994. "Bed Occupancy Due to HIV/AIDS in an Urban Hospital Medical Ward in Uganda." *AIDS* 8: 1169–71.

Temmerman, M., S. Moses, D. Kiragu, and others. 1990. "Impact of Single Session Post-Partum Counseling of HIV Infected Women on Their Subsequent Reproductive Behavior." *AIDS Care* 2: 247–52.

Thomas, James C., and Myra J. Tucker. 1996. "The Development and Use of the Concept of a Sexually Transmitted Disease Core." *Journal of Infectious Diseases* 174 (suppl. 2): S134–43.

Thongthai, Varachi, and Philip Guest. 1995. "Thai Sexual Attitudes and Behaviour: Results from a Recent National Survey." Paper presented at the conference Gender and Sexuality in Modern Thailand. Australian National University, Canberra. July.

Titmuss, Richard. 1972. *The Gift Relationship: From Human Blood to Social Policy.* New York: Vintage Books.

Tokars, J. I., R. Markus, D. H. Culver, S. A. Schable, P. S. McKibben, and C. I. Bandea. 1993. "Surveillance of HIV Infection and Zidovudine Use among Health Care Workers after Occupational Exposure to HIV-infected Blood." *Annals of Internal Medicine* 118 (12): 913–19.

Townsend, R. 1994. "Risk and Insurance in Village India." *Econometrica* 62: 539–91.

UNAIDS. 1996a. "HIV and Infant Feeding: An Interim Statement." Geneva.

———. 1996b. "HIV/AIDS: The Global Epidemic." Fact Sheet. Geneva. December.

———. 1996c. "The HIV/AIDS Situation in Mid-1996: Global and Regional Highlights." Paper presented at the Eleventh International Conference on AIDS, Vancouver, B.C., Canada. July 7–12.

———. 1996d. "UNAIDS Fact Sheet, Mid-1996" Geneva.

———. 1997. *The Impact of HIV and Sexual Health Education on the Sexual Behaviour of Young People: A Review* (Updated). Joint United Nations Programme on AIDS. Geneva.

UNAIDS/Country Support. 1996. "Reported AIDS Cases as of 31 December 1995." World Health Organization/Global Programme on AIDS, February 1. Geneva.

United Nations. 1993. *Urban and Rural Areas by Sex and Age: The 1992 Revision.* New York: United Nations, Department for Economic and Social Information and Policy Analysis, Population Division.

———. 1995. "International Migration Policies, 1995." Poster. New York: United Nations Department for Economic and Social Information and Policy Analysis, Population Division.

U.S. Bureau of the Census. 1996. "World Population Profile 1996, with a Special Chapter Focusing on Adolescent Fertility in the Developing World." U.S. Department of Commerce. U.S. Government Printing Office, Washington, D.C.

———. 1997. "Recent HIV Seroprevalence Levels by Country: January, 1997." Research Note 23. Health Studies Branch, International Programs Center, Population Division. U.S. Bureau of the Census, Washington, D.C.

U.S. Bureau of the Census (database). 1997. "HIV/AIDS Surveillance Database." Version 1.1, Release 0. Population Division, International Programs Center, Washington, D.C.

U.S. GAO (General Accounting Office). 1993. *Needle Exchange Programs: Research Suggests Promise as an AIDS Prevention Strategy* (GAP/HRD-93-60). Washington, D.C.: U.S. Government Printing Office.

Valdiserri, Ronald O., David W. Lyter, Laura C. Leviton, Catherine M. Callahan, Lawrence A. Kingsley, and Charles R. Rinaldo. 1989. "AIDS Prevention in Homosexual and Bisexual Men: Results of a Randomized Trial Evaluating Two Risk Reduction Interventions." *AIDS* 3 (1): 21–26.

Van Dam, C. J., G. A. Dallabetta, and P. Piot. 1997. "Prevention and Control of Sexually Transmitted Diseases in Developing Countries." In K. K. Holmes, P. F. Sparling, W. E. Stamm, P. Piot, J. Wasserheit, and Per-Anders Mardh, eds., *Sexually Transmitted Diseases.* 3d ed. New York: McGraw-Hill.

van Praag, Eric, E. Katabira, S. Anderson, E. Ngugi, and D. Koy. 1996. "Can HIV/AIDS Care Initiatives Be Part of Integrated Care? Lessons from Developing Countries." Abstract Th.B.400. Paper presented at the Eleventh International Conference on AIDS, Vancouver, B.C., Canada. July 7–12. World Health Organization, Geneva.

Van der Gaag, J. 1995. *Private and Public Initiatives: Working Together for Health and Education.* Directions in Development series. Washington, D.C.: World Bank.

van de Walle, Dominique, and Kimberly Nead, eds. 1995. *Public Spending and the Poor: Theory and Evidence.* Baltimore: Johns Hopkins Press.

Van der Ploeg, Catharina P. B., Carina Van Vliet, Sake J. De Vlas, Jackoniah O. Ndinya-Achola, Lieve

Fransen, Gerrit Van Oortmarssen, and J. Dik F. Habbema. 1997. "STDSIM: A Microsimulation Model for Decision Support in STD Control." Erasmus University, Centre for Decision Sciences in Tropical Disease Control, Rotterdam, the Netherlands. Forthcoming in *Interfaces*.

Vlahov, David. 1997. "Role of Needle Exchange Programs in AIDS Prevention." In NIH Consensus Conference, *Interventions to Prevent HIV Risk Behaviors: Program and Abstracts*. Bethesda, Md.: National Institutes of Health.

Wawer, M. J., C. Podhisita, U. Kanugungsukkasem, A. Pramualratana, and R. McNamara. 1996a. "Origins and Working Conditions of Female Sex Workers in Urban Thailand: Consequences of Social Context for HIV Transmission." *Social Science and Medicine* 42 (3): 453–62.

Wawer, M., N. K. Sewankambo, R. H. Gray, and others. 1996b. "Community-based Trial of Mass STD Treatment for HIV Control, Rakai, Uganda: Preliminary Data on STD Declines." Abstract Mo.C.443. Eleventh International Conference on AIDS, Vancouver, Canada. July 7–12. Columbia University, Center for Population and Family Health, New York.

Weisbrod, Burton. 1977. *The Voluntary Non-Profit Sector*. Lexington, Mass.: Lexington Books.

Weniger, Bruce G., and Seth Berkley. 1996. "The Evolving HIV/AIDS Pandemic." In Jonathan Mann and Daniel Tarantola, eds., *AIDS in the World II: Global Dimensions, Social Roots, and Response*. The Global AIDS Policy Coalition. New York: Oxford University Press.

Weniger, Bruce G., K. Limpakarnjanarat, K. Ungchusak, S. Thanprasertsuk, K. Choopanya, S. Vanichseni, T. Uneklabh, P. Thongcharoen, and C. Wasi. 1991. "The Epidemiology of HIV Infection and AIDS in Thailand." *AIDS* 5 (suppl. 2): S71–85; erratum in *AIDS* 1993 7 (1): 147.

WHO (World Health Organization). 1996. *Investing in Health Research and Development, Report of the Ad Hoc Committee on Health Research Relating to Future Intervention Options*. Geneva.

WHO/EC Collaborating Centre on AIDS. 1996a. "HIV/AIDS Surveillance in Europe," First Quarterly Report 49 (March). Saint-Maurice, France.

———. 1996b. "HIV/AIDS Surveillance in Europe," Second Quarterly Report 50 (June), Saint-Maurice, France.

WHO/GPA. 1995. "Global Prevalence and Incidence of Selected Curable Sexually Transmitted Diseases: Overview and Estimates." WHO/GPA/STD/95.1. Geneva.

WHO, Program on Substance Abuse. 1994. "Multi-City Study on Drug Injecting and Risk of HIV Infection." A report prepared on behalf of the WHO International Collaborative Group. Geneva.

Wirawan, D. N., P. Fajans, and K. Ford. 1993. "AIDS and STDs: Risk Behaviour Patterns among Female Sex Workers in Bali, Indonesia." *AIDS Care* 5 (3): 289–303.

Wong, K. H., S. S. Lee, and W. L. Lim. 1993. "HIV Surveillance among Drug Users in Hong Kong." Paper presented at the Ninth International Conference on AIDS, Berlin. June 6–11.

World Bank. 1993a. "Staff Appraisal Report: Brazil AIDS and STD Control Project." Report 11734 BR. World Bank, Country Department I, Human Resources Division, Washington, D.C.

———. 1993b. *Tanzania AIDS Assessment and Planning Study*. World Bank Country Study. Washington, D.C.

———. 1993c. *World Development Report 1993: Investing in Health*. New York: Oxford University Press.

———. 1996a. "AIDS Prevention and Mitigation in Sub-Saharan Africa: A Strategy for Africa." Report

15569. World Bank, Africa Region, Technical Department, Human Resources and Poverty Division, Washington, D.C.

————. 1996b. "Tanzania—The Challenge of Reforms: Growth, Incomes, and Welfare." 3 vols. Vol. 1: Main Report. Africa Region, Eastern Africa Department, Country Operations Division. Washington, D.C.

————. 1997a. *World Development Report 1997.* New York: Oxford University Press.

————. 1997b. *World Development Indicators 1997.* Washington, D.C.

Wyatt, H. V. 1993. "Injections, Infections, and Sterility." In M. Bloem and I. Wolffers, eds., *The Impact of Injections on Daily Medical Practice.* Amsterdam: Free University Press.

Yeager, Rodger, and Craig Hendrix. 1997. "Global Survey of Military HIV/AIDS Policies and Programs." *Civil-Military Alliance Newsletter* 3 (1): 1–8.

Yu, Elena S. H., Qiyi Xie, Konglai Zhang, Ping Lu, and Lillian L. Chan. 1996. "HIV Infection and AIDS in China, 1985 through 1994." *American Journal of Public Health* 86 (8): 1116–22.

Zheng, Xiwan. 1996. "HIV/AIDS Epidemic in China." In *HIV/AIDS in China,* no. 4: 11–12. World Bank, Asia Technical Department, AIDS in Asia Unit, Washington, D.C.